The Economics of CHAOS

Eliot Janeway

The Economics of CHAOS

On Revitalizing
the American Economy

TT

TRUMAN TALLEY BOOKS
E. P. DUTTON
NEW YORK

Published in the United States by Truman Talley Books • E. P. Dutton,
a division of NAL Penguin Inc.,
2 Park Avenue, New York, N.Y. 10016.

Published simultaneously in Canada by
Fitzhenry and Whiteside, Limited, Toronto.

Library of Congress Cataloging-in-Publication Data

Janeway, Eliot.
    The economics of chaos : on revitalizing the American economy
    Eliot Janeway.—1st ed.
        p.   cm.
    "Truman Talley books."
    Includes index.
    ISBN 0-525-24711-4
    1. United States—Economic conditions—1945–   2. United States—
Economic policy.   3. Economic history—1945–   I. Title.
HC106.5.J36   1989
338.973—dc19   88-16063
                                                                                     CIP

DESIGNED BY EARL TIDWELL

10 9 8 7 6 5 4 3 2 1

First Edition

To Senator Sam Nunn, chairman of the Senate Committee on Armed Services and of the Senate Permanent Subcommittee on Investigations, with trust in his statesmanship, with admiration for his resourcefulness, and with reliance upon his leadership.

CONTENTS

PREFACE
Defusing
the Depression Danger

America's debts have grown more routinely bipartisan than the stress of emergency is likely to make her politics. Before Ronald Reagan removed the Democratic party stigma from deficit financing, stump speakers brushed off criticism of government borrowing with the rationalization that "we owe that much to ourselves anyway." America owes an increasingly disturbing portion of the debt burden Reagan left behind him to foreigners; not just her government debt, but the billions on top of the floating pyramid that her banks owe. It's a fact of American financial life that the government will pay all big debts that all big banks owe.

Historically, debt has come to play a role in the calculations of American economists as pivotal as lust once did in the calculations of otherworldly theologians. The clergy frowned on it, but their parishioners could not manage without it. The clergy had the good sense to turn their operating liability into a bargaining asset; they undertook to broker the terms for the saintly sanction of sinning for survival. America's borrowings have come to qualify as sanctioned sin. When St. Paul explained, "It is better to marry than to burn," he improvised the rationale justifying the indispensability of America's borrowings. In theory, the business of the world could be done more simply without debt. In practice, however, the world would stop doing business without it. In principle, less borrowings by America would have been desirable. In practice, more borrowings by America have become inescapable—until America activates policies calculated to reduce them. Vows of abstinence will not reduce America's borrowings.

Hence the urgency of a resolution to the crisis simmering to its climax with each new step-up in America's borrowings. To stop them altogether would hurl the world headlong into panic. To continue to indulge them would guarantee the same result on the installment plan. America has demonstrated an unruly instinct for drifting into dilemmas. But a keen preference for action again and again has brought her relief and room to grasp third alternatives to either horn.

Hence politics: beginning with what's wrong about democracy and what's salvageable. When Jefferson philosophized that the cure for democracy was still more democracy, he anticipated the cause of and the cure for the crisis brought to a head for the 1990s by America's borrowings. Granted that politics played weakly and wishfully among Americans has invited the crisis, nevertheless politics played resolutely and realistically with the world can resolve it.

The programmatic subtitle of this book—*On Revitalizing the American Economy*—claims prescriptive urgency over its analytical title— *The Economics of Chaos,* which is descriptive and, therefore, already history. In any race to head off a depression, ambitious proposals to change the system provoke arguments that distract from emergency pressures to reactivate the economy. Wrestling over principle opens the door to verbosity, but shuts the door to action. It wastes time when the effectiveness of any remedy hinges on saving time. Moreover, the urgency created by the debt-income squeeze dictates severe limitations on eligible policy solutions. No depression remedy proposal with a fighting chance to work in a precrisis atmosphere saturated with debt—private as well as public—can saddle the economy with still more taxes or the government with still more spending and, therefore, still more debt. Recessions are self-correcting, but depression is the danger when debts continue to rise while incomes start to fall, as they did beginning in 1986. The purpose of this book is to defuse the depression danger. My premise accepts the gravity of the problem, but asserts the readiness of solutions calculated to transform a creeping depression into a broad-based boom.

So, as with the Creator, the clergy, and the crowd of lore, the question comes down to terms on which life can be borne as tolerable and, beyond that, sighted as improvable and, one hopes, perfectible. To the credit of the race, theologians and economists notwithstanding, no society has ever managed to survive without harboring dreams of improvement to the point of perfectibility—not in their immediate surroundings, but on the contrary, when and as protected from pitfalls built into their environment. People being what they are, the dividing line between survival and utopia has always shifted as ambiguously as ideas of time and space. James Forrestal, the first American civilian to fly into Berlin after Germany surrendered in 1945, told me that the first sight he witnessed was a woman lying in the rubble alongside the runway giving birth: a Darwinian start toward Germany's reconquest of prosperity.

But the gates of Paradise do not open by themselves; only on terms negotiated as required, specified by the gatekeeper, and understood by entries. Economists have spread the pernicious illusion that the doors to markets are the worldly equivalents of the pearly gates, functioning automatically. America's borrowings, under the auspices of her economists, have shown her and the world that passports are needed, as well as market lubricants that market participants themselves cannot supply. Hence politics; hence governments, hence policies—but not in a vacuum: urgently, in time to keep the world moving and to launch its lurchings toward happy landings on terrain trusted to transform dissipated borrowings into productive investments.

Herbert Hoover's successor took up his responsibilities in the midst of the banking crisis of the 1930s unaware of the third alternative his New Deal would improvise to the dilemma of bankruptcy or profligacy. Nevertheless, confronted with the undeniable symptoms of financial paralysis, from his wheelchair FDR galvanized not only the government, but the entire country, into an electrifying one-hundred-day sprint, at once unprecedented and unforgettable, demonstrating the instant revitalization of the nation's arteries. On the eve of the two 1988 presidential conventions, the rising chorus calling on Reagan's successor to launch a repeat performance confirmed the formative influence of FDR's presidency and its New Deal program across the most triumphant, yet tumultuous and treacherous, half century of history ever recorded.

To the useful extent that America's history does provide comfort through parallels, it is, therefore, more than merely incidental intelligence that Ronald Reagan's successor took office just as unaware that a third alternative beckons as urgently to the dilemma of the 1990s: not merely whether America could dare to stop borrowing at her compulsive bankruptcy rate or would continue to try, but how she might renegotiate herself out of the political equivalent of debtors' prison. Mobilizing her formidable array of neglected domestic resources for an emergency effort at revitalizing her economy by an instant reversal of the terms of trade is the way for her to do it.

Free marketeers are allergic to talk of emergency government initiatives interrupting the continuity of markets performing on their own. If this divided world of ours were left to the mercy of markets, they would be right. But America is caught in a dilemma: how to adjust to market wisdom and, at the same time, how to deal with Soviet power. Soviet power is distracted by no such dilemma. It deals in markets, but no markets can deal with it. Only power can check power; only the power of the American government can deal with Soviet power. One way America can is to harness market power in her dealings with Soviet power, instead of allowing Soviet power to manipulate market forces against America.

Americans assume that competition is the rule of the road in markets and have become conditioned to accept their defeat with good grace as fair and square. America has learned to live with defeat at the hands of victors whom she herself shields and subsidizes. Consequently, the most authoritative of the commercialized instruction activities in America exploits the growth market hawking corporate gimmicks touted as copying Japan, whose decisive methods, however, are governmental. Forgotten is the uncontestable historical record of the humiliation the American economy visited upon Japan's militarism when America won the war of production after Pearl Harbor; forgotten also is how Japan planned its subsequent renewal by adopting America as its model of power rooted in productivity rather than conquest. Forgotten as well is Germany's record: two successive recoveries from defeat inflicted by America in wars of production, which were immeasurably more traumatic than America has tolerated in the spirit of fair play. True, Japan and Germany, at their lowest points, could mobilize discipline as an asset for rebuilding; though in each case, American assistance was more tangible. Discipline is not the American way; nor is generosity from foreign champions in the cards. But America has the cards to play. Only the incentives have been lacking, and incentives are the American way to get action.

America's success in orchestrating miracles for Japan and for Germany justifies confidence that she can manage revitalization for herself. Defeat at the market game has provoked many Americans—especially, but by no means only, those nursing pocketbook wounds—to get mad. America's challenge is not just to get even, it's to get ahead again, the American way—by the mobilization of incentives. Historically, each new crop of immigrants has struggled to get ahead individually. America's challenge in the 1990s is to use her government to forge the policies indispensable for America to get ahead again as a country: not merely as the market to be exploited, but once again as the power to be reckoned with. Either America will be the power again in the world, or chaos will dominate its politics as well as its economics during the nuclear age.

Presidents are elected to serve by learning that the Oval Office was designed as the world's most exacting classroom—and that its incumbent is a candidate for continuing education whose grades are on conspicuous display. Events are his library, experience alternately his guide and his trap, the voters his amateur teachers. Contrary to the teachings of economists, only one free market in the outside world remains open to him, or to them, to trust. It is not any market in which goods or services, let alone securities, are bought and sold. In the immortal words of Mr. Justice Oliver Wendell Holmes, as fresh for America's next century as for her last, this free world market is "the free market of ideas," especially now that Russia is being opened to it.

The word *vulgar* shares the same family tree as the Vulgate, the medieval vernacular version of the Bible, spoon-fed to the crowd. When the crowd learns from what it hears, urgency becomes the order of the day for its leaders. The idea that America cannot stop borrowing, but cannot continue to borrow as she has, is vulgar enough to have taken hold throughout the country. So has the companion idea that the time has come to trust her compulsive borrowings as necessities of life for sale to her foreign creditors. Both ideas are simple enough for democracy's examiners at large to transmit to their more or less responsive pupil in the Oval Office. The question is whether Congress—by definition, closer to some of the people than any president can be to all of them—will teach this lesson to him before events do. The members of Congress serve a dual purpose under the educational system provided by the Constitution: students themselves, with special licenses to conduct advanced studies for the benefit of each new arrival in the Oval Office.

These licenses stem from America's "peculiar institution" of congressional seniority. With the downgrading of history as a campus priority and as a popular pastime, Congress remains the one repository of institutional power in America guided by a living sense of history. Members of Congress share among them more knowledge of the problems inherited by any newcomer to the Oval Office (who may not have been a member) than he can possibly have absorbed along the way.

This book is meant to speed up America's educational process: both the teachings administered by the president's amateur mentors at large by political radar, and the professional seminars led by the members of Congress. It is dedicated to Chairman Sam Nunn of the Senate Armed Services Committee, whose seminar is the most pivotal on Capitol Hill, calling for the most urgent remedial action domestically and for the most reciprocal results internationally.

1

America's
Distinctive
Background

1

MARSHALL'S MAGIC
America's Managerial Challenge

Anthems rhapsodizing America the beautiful, America the brave, and America the bountiful have struck responsive chords, but have missed the theme song of America the distinctive and the refrain of America the pivotal. The omission has proved costly. The standard sentimental flag-waving appeals have been luxuries, moistening eyes, quickening pulses, and winning America a certain indulgence for her youthful naïveté. But this mushy approach has also revealed the immaturity of American leadership.

When the first flush of postwar ebullience cooled in the late 1950s, Washington was confronted with a twin challenge: to finance prosperity without inflation, and to thaw out the Cold War without appeasement. Midway through the 1980s, with static crackling in the air from old tunes gone stale, these challenges still confronted it. By that time, America was slumping into an unprecedented crossfire of debt inflation wasted on earnings deflation.

Previously, increases in debt had financed increases in earnings. Previously, too, arms buildups had gunned the economy. But the Cold War brought another reversal of form. Thanks to the nuclear buildup and related advances in weapons engineering, the escalation of costs, compounded with the acceleration of model changes, outweighed any scientific, technological, or economic benefits bought with it. The same benefits were buyable in commercial form at lower cost without the same risks.

Mathematical precision is at the opposite end of the analytical spectrum from the nuances of political and economic theory. Yet a basic geometric figure—a parallelogram—describes the responsibility of the modern American president: simultaneously to balance the interplay of political, economic, military, and social claims varying in importance from situation to situation. All of America's presidents since Franklin D. Roosevelt have failed to do so. All of them have been hampered by antiquated theories, failed experiments, or flawed policies. Most of them did not know the differences between experiments, theories, and policies. Harry S. Truman justified his decision to drop the bomb as an experiment. It set a policy unsupportable by any intelligible theory.

This book will recommend economic policies calculated to free America from unworkable economic theories. The purpose of these interrelated proposals will be fourfold: simultaneously to finance (1) expanded traffic with trading partners, (2) debt-free economic growth, (3) revitalized social programs, and (4) weapons systems to command fronts America cannot afford to lose—notably against terrorists. Only the government can activate them all.

Presidential practice outran the reach of established economic theory when Roosevelt entered the White House in 1933. Before him, the president was expected merely to govern; after him, to manage. The military resolution of the country's economic problem, followed by the nuclear resolution of its military problem, subjected the presidency to a new postwar learning process: determining how to finance continuous high-volume peacetime military spending on weaponry not destroyed in combat. The problem emerged as distinctively American; Russia, her only rival in the arms race, had no financing problem and carried no interest burden. Economists found themselves challenged to do the teaching without having grasped the rudiments of military finance or defense-related investment. None of the established economic theories, all imported from Europe, had anticipated the phenomenon of America in the Cold War. All of the imported theories expected recurrent bulges in military spending but dismissed them as interruptions of the business cycle.

America's accomplishments and expectations since Roosevelt's time have primed her leaders to meet two stern tests. The first, forced by the Depression, is the responsibility to keep pockets and stomachs filled, as well as bodies housed; the second, forced by Hiroshima, has been to put this economic feat to political use by defusing missiles at no risk to security. This record has also put her leaders on notice not to trust wars or preparations for wars to solve old economic problems, but to count on the cost of armament to create new economic problems. None of America's established economic theories anticipated her new postwar need to reconcile the high cost of expensive advanced military technology with her various methods of managing the peacetime business cycle.

Instead, America's distinctive economic strength has eroded steadily, rather than buying her a correspondingly distinctive stack of political bargaining chips. The roots of this failure to buy political power with economic strength go deeper than any policy stance taken by the succession of representative Americans who occupied the White House from the 1940s through the 1980s. In electing each of them, a shifting consensus of Americans put their money where their minds were: they voted their beliefs, necessarily jumbled as popular expectations and preferences always are, reaching for the best of both worlds without bracing for the worst. All these presidents, in turn, acted out the intellectual confusion over the purposes that guided the country as it wrestled to apply its European tutelage to its distinctive circumstances.

As the twentieth century progressed, America became increasingly disenchanted with the economic policies her government fashioned from the main economic theories in vogue, with no sensitivity to America's distinctiveness. These theories distracted America from realizing her potential as the pivot of power in the world. Americans have been ignorant of the distinctive American parallelogram of forces, as well as guilty about the world's need for their country to play a pivotal role. Long after the entire world acknowledged this uniqueness by turning to America as the donor of last resort, however, Americans retained their impractical intellectual dependence on Europe. The inability of European economic theories to translate into workable policies—even on their home turf!—did not deter their American devotees. The free-trade dogma advocated by Adam Smith is a discard; so is the permissiveness about debt invited by John Maynard Keynes. Neither theory was ever tried across the Pacific, where the challenge of the twenty-first century looms.

So long as the momentum of the great midcentury boom deferred America's problems, the intellectual baggage she carried with her imported economic theories was not a burden. But the worldwide economic deterioration of the 1980s confirmed suspicions of their irrelevance. Her failure to function as the pivot of power and, therefore, of prosperity brought not only her economic theories into question, but her sense of national self-interest. Desperation forced the world to demand that America make prosperity her main export. Nevertheless, complacency left Americans waiting for markets around the world to spread the prosperity that only their own government could activate.

The world is realistic in its practical perception; Americans are mush-headed in their intellectual naïveté. The country's recognition of its special characteristics is the precondition for identification and conservation of the distinctive national assets its successive administrations dissipated. To help the country revalue and mobilize these assets is the starting point of this book. These assets, enumerated in Section II, offer America the opportunity to organize prosperity and salvage security instead of waiting for events to proffer a gift of either or both on the smug assumption that mature powers retain the luck that has helped them to their ascendancy.

Between the Vietnam crisis of the 1960s and the oil war of the 1970s, America lost her reputation for managing events. The loss of her prosperity followed; paranoia over her military inadequacy spread. She failed to realize that the prosperity and, therefore, the peace of the postwar world hinged on America's resolve to be more opportunistic and less principled in harnessing her assets to further her own priorities, including her humanitarian priorities.

The way for America to switch her priorities is not by proclamation, but by stratagem, not by standing tall and posturing as Reagan did, but by scoring tangible marketplace breakthroughs and following through on them. These breakthroughs, by definition, would also serve her as bargaining tools with the rest of the world. Only measurable results will free the United States from the distracting influence of the economic theories evolved from the experience of European countries.

Professor Paul Kennedy, in his brilliant exposition of the rise and fall of empires from 1500 to 2000, has shown that modern Europe has its roots in a variety of decentralized, rival power bases. History has formed Europe's contemporary structural differences with the American system. Inescapably, its economic theories have also evolved in reflection of its decentralized structure and, therefore, express the vital dependence of each and every power center in Europe on foreign trade. Americans have been importing the theories on the classical but uncritical assumption that an economic theory that can work anywhere will work everywhere. They have shown no awareness that none of them has worked anywhere. Nor have Americans shown any surprise over the repeated revelation that none of these imported theories has worked to the satisfaction of a continuing American consensus.

Europe's importance in the world is magnified by its special relationship with America. Its cultural influence on America's thinking, its economic dependence on America's prosperity, and its political dependence on America's protection combine to suggest three simple new working premises for American policy-making. The first is that Europe's various economic theories, like its goods, enjoy premium export demand, especially in America. The second is that the United States has enjoyed her runs of prosperity in spite of her imported economic blueprints for ensuring it, not because of them. The third is that a declaration of intellectual independence for American economic policy-making offers the only hope of working a new economic miracle everywhere at once.

Europe is the world's most influential economic and financial opinion-making center. It is also the world's most vulnerable military target for Soviet firepower. Europe has had the sense to be petrified of incendiary talk of a retaliatory American economic offensive against the rest of the world. Its own history has taught it that economic aggression always ignites the military counterpart. Europe also has learned the hard way that it can't do well when America doesn't, and that it always does better when America does. Roosevelt reminded Europe of this truism at the worst of the Depression crunch of 1932–33; he set out to help America first, but he helped Europe simultaneously. With Truman, Europe gained the most from the Marshall Plan, but the overspill of American largesse benefited everybody, beginning with America, which launched it as a give-away, and including even Russia, despite Stalin's insane rejection of America's original offer to include Russia in the Marshall Plan.

When crisis in any form looms for Europe, its priority always forces the activation of new American priorities to resolve it, and therefore, to reaffirm the inescapability of America's leadership role in time of trouble. Its 1986–88 slump put Europe, like Japan, in need of a new American recovery. America had made money talk through the Marshall Plan, which overpowered Europe's suspicions about her motives. But a renewed Marshall Plan for America would speak for itself. No new government funds would be needed: only new government policies to reactivate the massive backup of capital in the private sector—a conspicuous contrast with the original Marshall Plan, which found Europe drained of money and Washington ready to pay for reconstruction. The entire world would greet a new American Marshall Plan as an opportunity and rush to share in its benefits. Every segment of American opinion would ask only one question: What took so long?

The answer is implicit in the difference between decisive military defeat, which is immediate, and its economic equivalent—depression—which always takes time to hit home. But once the depression alarm goes off, the impossibility of living with the resulting chaos hits home overnight, raising the cry for a new pragmatism designed to refinance, restructure, and redirect the shattered economy. By 1988, affluent Americans were sharing an unfamiliar insecurity with indigent Americans. The affluent were bracing for an onset of hard times, while the indigent were digging in to lower their standards of subsistence and to raise their tolerence of anxiety.

The resources left at America's disposal when the debate over depression formalized her difficulties in 1987 were ample to support a full-fledged recovery effort, subject to a recovery policy capable of mobilizing them. But the economic theories America tolerated as she went down to economic defeat and into social chaos provoked an angry political backlash. Only Reagan's charisma and his reputation for goodwill prolonged public patience with "market solutions" to market aggression in the form of dumping by foreign governments.

America was without any indigenous economic theory on which to build a political consensus during the long slide that began with her military defeat in Vietnam and culminated with her economic defeat at home. None of the theories tested during that decade—and all of them were—took her distinctive needs into account; none (especially Reaganomics) was affordable; none made sense to the non-jargon-slinging public; and none helped. But a Marshall Plan to do for America what America's first Marshall Plan did for Europe would turn the prevailing political attitudes in America from angry disillusionment to enthusiastic support.

No wave of economic expansion is ever generated by spontaneous combustion. The first spark is always lit in one country, and there it is lit by national self-interest. Any hopeful new start anywhere depends on the outward reach as well as the forward thrust of the economic energy it generates. But the United States remained the one country big enough to import distress originating simultaneously in all the countries dependent on her markets and to tolerate it up to a higher breaking point than any other industrial power has ever invited.

This distinctive capacity for absorbing distress explained her patient indulgence of it while the import deflation of the early 1980s was splintering the industrial base of her economy. Opportunism contributed to the economic Pearl Harbor America invited at her ports of entry; Americans rushed to buy import bargains, even while suspecting how much their savings as shoppers would cost them as income earners. Dogmatism rooted in the clichés of free trade hypnotized the victims into welcoming the losses as gains.

Optimism fed on the euphoric lure of America's presumed growth into a "service" economy free from the import threat—until depression struck the entire service industry, from restaurants to hospitals and even television networks. It jolted the country into learning a basic lesson her economists had never taught her: service industries are intertwined with manufacturing industries. Each relies on the other as a customer. Services cannot continue to enjoy expansion when the manufacturing industries, which produce income to be spent on services, suffer shrinkage. Realism made short shrift of the stubborn rationalization that America could import prosperity along with distress from any of the countries doing well at her expense. She was simply too big to lift herself on foreign bootstraps. By the same token, all the foreign bootstraps within reach of the American economy were too small.

The drag effect of distress in America—but especially of distress imported into America—is guaranteed to stifle prosperity everywhere. In the 1970s, the cost inflation America imported at OPEC's dictation invited Japan's saturation of the American economy, which by the mid-1980s boomeranged into the first failure suffered by the Japanese economy. That was when Japan discovered to its dismay the truism of economic geography that dooms any competitor of America's to lose any contest it may seem to win against America. While the media were weaving glamorous tales commemorating Japan's comeback from military surrender to economic victory in a mere forty years, the Japanese economy was swallowing its first bitter dosage of unemployment compounded with bankruptcies (including the unprecedented phenomenon of one suffered by a cabinet minister holding large government contracts in his private business).

Reagan swallowed the free-trade dogma, and the country choked. America was first dazzled by the import profiteering that always tops an inflationary boom, and then demoralized by the import dumping that, just as predictably, always leads a deflationary debacle. The import inflation that paced the U.S. sellers' markets of the 1970s collapsed into the import deflation that devastated them in the 1980s. Inescapably, however, the dumping countries suffered along with their target. Their economies started to contract despite the expanded outlet for their goods America was inviting them to buy with subsidized prices and credit. Though America was importing distress from every point on the map by the mid-1980s, her competitors underselling her in her own markets were scarcely exporting themselves into prosperity.

As the great midcentury boom flagged in the mid-1980s, America's lack of economic leadership revealed a still more disturbing lack of bipartisan policy direction. Cockney lore treasures a tale ridiculing the loser's logic that has misdirected America's economic stance. The perennial London cabbie suffers a bad case of aching feet and drags himself off to the doctor, who is quick to spot the trouble. "You've put your shoes on the wrong feet," he exclaims. "But they're the only feet I've got," his patient moans. The shoes are the foreign economic theories, and the feet are the moving parts of the American economy; like the hapless cockney, America has mistaken the mis-fitted theories she wears for the healthy limbs they have crippled. Carrying the analogy a step further to the age-old Chinese custom of binding the feet of young girls, America has been making her giant steps forward in shoes whose lasts were hand-made to fit the feet of smaller countries.

Yet once America learns to fit the right shoe to the right foot, and designs a last big enough to fit the distinctive size and contour of her feet, every other country will feel an immediate and irresistible urge to fall into step behind her. Why not? All of them are dependent on exports for prosperity and on imports for subsistence. The pragmatic test for any economic theory is the spreadability of its effectiveness from wherever it starts. Tragically, the history of economic theories demonstrates that the less they have worked at home, the more they have spread abroad. A practical, homegrown American economic policy would free national and regional economies everywhere from dependence on expectations and procedures outmoded in the mid-1980s by the failure of debt inflation to prevent earnings deflation—a phenomenon that confounded devotees of the various European brands of economic theories packaged for popular distribution on the American market. All European economic theories assume that debt inflation guns income inflation.

Workable policies evolve from theories that have run the gaunt-let of debate and experiments with their application. Economic theo-ries take hold as raw materials for arguments whose end products are actions: the beliefs beaten into people's heads by their campus teach-ers and professional advisers influence their adult behavior. The countries run by government direction—Japan, Russia (certainly pre-*perestroika*), China, the Third World dictatorships—do without theories. Marxism is an icon; capitalism, in its pure form, is an ideal, offering a standing invitation to converts, like Gorbachev, to discover the difference between policies based on it and experiments with variations of it. The countries that subject governments to popular veto thrive on theories—especially theories that neither work in the legal marketplace, where talk makes money move, nor click in the shadow markets, from innocent moonlighting to dope dealing, where money still talks.

Though British goods have lost their position in the booming American import market, the economic thinking that guides Britain has remained dominant as an export staple guaranteed to sell best in America. The deservedly high respect commanded by British culture has lulled Americans at both ends of the political spectrum into accepting British institutionalized economic thinking as the best, originating with Smith and culminating with Keynes. Britain's lamentable economic performance, unarguably the most disappointing in the world, throughout most of the twentieth century, has raised few suspicions about the relevance of the theories behind it; the writing was too good.

Britain's competitive performance was the best back when the pre-Revolutionary colonies adopted the economic thinking associated with "the nation of shopkeepers"; Napoleon's Grande Armée was no match for their success story. As Professor Joseph Dorfman, of Columbia, emphasized in his definitive history, *The Economic Mind in American Civilization,* the American economy from the colonial period until the Civil War depended not on industry, but on commerce, and commerce then meant foreign commerce. That was what English economic theorizing was all about, and that explains why America's acceptance of its trade-oriented premises came naturally to her during the formative decades of her intellectual conditioning. These theories copyrighted in Britain retained their ring for Americans long after America's industrial expansion inland stripped them of any relevance.

But economic theories have fed on their irrelevance to the problems perplexing their devotees. This built-in trail to their failure has endowed them with a provocative mystique of their own, perpetuating debate instead of resolving it by precipitating actions that speak louder than words can. By the same token, the occasional successes of economic theorists are forgotten as fast as they are embedded in the building blocks of history. Mark Antony's impassioned funeral oration at Caesar's grave tells us what we need to know in evaluating economic theories. "The evil that men do," he warns us, "lives after them. The good is oft interred with their bones." The moral Shakespeare pointed lives on as a realistic market letter for modern times.

The analogy continues with the posthumous fortunes of economic theorists who fail. Their books continue to sell.

Just as surely as all roads in the ancient world led to Rome, all trails in modern economic thought stem from Smith, even though he has been found wanting as a policy guide for modern America. All hands agree that *The Wealth of Nations* was one of the great books that formed the modern attitude toward people and money at work. But no text from *The Wealth of Nations* stands as his monument. The American Revolution does. The book itself turned out to be a masterpiece of advocacy for the colonists. He invested fifty years writing it and lucked out on the year he published it—just in time for the American Revolution to adopt as its bible for its revolt against excise taxes, and for the Constitutional Convention after that to enshrine as its monetary credo. In Chapter 2, I will unravel the tangled connection between Smith's unworkable fiscal axiom and the workable political tack the Founders took when they scrapped it.

As Professor James Oliver Robertson, of the University of Connecticut, has shown in his penetrating study, *America's Business,* when the rebellious colonists made Smith's credo their fighting slogan against King George, they made Smith a perennial best-seller. Never mind that they limited their principled propaganda against restrictive political practices to those imposed by remote control from across the Atlantic. They used the power they seized to impose restrictive practices of their own to protect their bustling new business baronies, as if they had never hailed Smith as their prophet. The first step authorized by their new government was a retreat to the very sales tax approach they had rejected from the Crown.

History has blurred the selective use the upstart American agitators made of Smith's theory. The revolutionaries sampled his antitax slogan, but the government they established did not buy it as a theory, let alone as a policy. Smith's latter-day disciples, intoxicated with his rhapsodies to free trade running freely, have advertised their ignorance of history as well as of politics. To sharpen the irony, they emerged as ardent advocates of an aggressive U.S. sales tax policy as an alternative to U.S. income tax increases.

Even when government was tiny, Smith's premise failed as a blueprint for economic policy. The growth of government is reason enough why it always will. Every increase in the reach of government results from the failure of a market function, he argued, and is the direct cause of new interference. Smith's alibi for market failures was that governments lack the discipline and, therefore, the patience to wait for market solutions. But Smith is a hanging judge when the value of government intervention into market functions is on trial. The debate over his ideas has raged on despite their impracticality in the simple world revolutionized in his time by the steam engine, let alone in the complicated world revolutionized in our time by the space engine. His alibi has held an audience, but it has not held water.

Of the trails that have led from Smith, the one blazed by Keynes led to the left. Keynes was at least as persuasive a polemicist as Smith. He was also much funnier, and he was effective as an improviser of practical policy expedients, as Smith never tried to be. Yet his theories have thrived on the same overexposure in debate and failure in practice. The authority he has exerted over American opinion, directly and through his disciples, stems from the perception of him as standing in the great tradition of British economic thought and dominating the debate over how to finance alternatives to depression and war. By an eerie coincidence, however, the challenge responsible for his institutionalization in America duplicated the one that catapulted Smith into political fame: financing war for America. Their respective ruminations on the workings of the economy had nothing to do with the successive adoption of each by American war administrations. In both cases, isolated recommendations were pulled out of context when the normal workings of the economy were suspended. Their prestige followed.

Keynes's recipe for the Depression was deficit financing to feed the spending stream and, more specifically, to pay for public works. He was too early and too rational. In America, the scale of government borrowing and spending needed to overcome the inertia of peacetime "slumpflation" boggled mentalities conditioned to tolerate its drag. His "borrow-spend" solution was impractical in underestimating the magnitude of the injection needed. The tragedy of Keynes's recommendation for overcoming the Depression and avoiding the war was that the only source of money on the scale needed to fill his recipe for peace was war itself.

When, finally, the war forced deficit financing in the form of emergency public works on the scale needed to justify Keynes's claim, the beneficiaries of the new wartime prosperity had lost their interest in giving any economic theory credit for ending the Depression. The fact of the war, not the lure of the theory, superseded the old evil of depression with a new evil of aggression that demanded confrontation. In any case, Roosevelt won political credit for Keynes's two imaginative and effective programs: managing the war without inflation and ending it without deflation.

Ironically, the peacetime application of the massive depression cure Keynes had proposed as a liberal was deferred until the onset of the conservative inflation crisis of the 1980s. By then, however, circumstances had come full circle from the 1930s, when the debt load was washed out by the Depression. In the 1980s, despite borrowings more bloated than any Keynesian would have dared suggest, much less condone, the deficit financing Keynes had advocated as a cure for depression boomeranged into a cause of it.

In addition to Smith and Keynes, two other sources of economic theorizing have won an influential hearing in America. Like Keynes, Friedrich von Hayek and Karl Marx fall within the shadow of Smith's family tree, but represent a more extreme study in contrasts than Smith and Keynes. Hayek out-Smithed Smith; Marx made Smith grist for his mill. Hayek added a new dimension to Smith's idea that markets not only work but work in the same way everywhere and always, regardless of distinctive national geographical and institutional characteristics. He set up a balance sheet: financial morality, defined as pay-as-you-go and tax (sparingly) before you spend (austerely), versus financial immorality, defined as borrowing money to spend that hasn't been collected in taxes. Hayek guaranteed financial collapse as retribution for prosperity by political artifice. His moralizing won him a surprising reputation for practicality. America's weakness after Vietnam, accentuated with each overload to her debt burden, was taken as confirmation.

Marx thundered that the very efficiency with which markets can be trusted to work explains the inequities they proliferate. With more righteousness than the Devil could be imagined mustering to quote Holy Scripture, Marx went on to agree with his acknowledged mentor, Smith, that capitalist governments tempted or pressed to repair damaged markets will destroy them instead. No Reaganomist could have asked for anything more.

Marx merely generalized about how communist governments would work. But he did reserve the same power to them that Smith claimed for capitalist markets. Marx believed that communism could work everywhere, just as Smith believed capitalism could. Both ideologues shared the premise that their respective systems would work best in advanced countries, though Marx won his hearing in a backward country. His late twentieth-century followers claimed confirmation for his theory on the grounds that power pays; Hayek, echoing Smith, claimed confirmation for his theory on the grounds that profligacy doesn't.

None of these conceptual European relics inherited by America retained any relevance, much less applicability, to the distinctive problem the depression looming for the late 1980s posed for engineering a recovery in America big enough to export profitably. Yet all of these anachronisms retained a hypnotic hold over American opinion. The stubborn vogue commanded by catchphrases of European origin recalled the standard joke about the acceptance enjoyed by the Austrian economists in pre-Hitler Berlin: "Why is their prestige so high in Germany?" "Because no one ever asks their advice." The contemporary American twist on the answer is more revealing still and more frustrating: "Because everyone in America takes the advice of economists, and none of it works, but things work out anyway."

Things have indeed worked out for America. Certainly, the country muscled through the century following the Civil War with good reason to believe it was doing something right that came naturally. Credit was not due Americans for being wiser than people who lived elsewhere; they were just lucky enough to be born with more chips, not necessarily financial, which freed them to be dumber. In fact, that's how America got ahead of herself. When she was still generating momentum never demonstrated before by any upstart power, her unpreparedness for the responsibilities that come with power had provoked Otto von Bismarck to grumble: "Providence has a special regard for children, drunkards, and the United States of America." With the arrival of economic maturity and the continuation of political immaturity in the last third of the twentieth century, she grew comfortable allowing familiar imported economic cure-alls to enjoy their innings of favor. So long as she could muster a consensus of confidence that she was doing well, none of these foreign theories was in danger of attracting blame for getting her into trouble, while all of them were bound to share credit for keeping her out of trouble.

America's long, slow drift into trouble suggested the appropriateness of a hard look, not inward but backward, to her one original economic thinker of major stature, Thorstein Veblen, by the 1980s long forgotten. The impact of his insights on the naïveté of his colleagues cast him as a satirist. "Conspicuous consumption" stands as one of his striking contributions to the language. It applies not just to the flaunting of idle wealth but to the spectacle of America's economists on display—the more of them the better; the angrier their disputes, the less likely the risk of their counsel's being adopted. Chapters 3, 4, and 5 focus on his revolutionary contributions to more than just the language. His approach to America's distinctiveness, updated, remains the one body of thought offering guidelines for an American economic theory translatable into workable policy.

Books galore have been written on America's distinctive political system, with its dual sovereignty: legislative as well as executive, congressional as well as presidential. But no books have been written on the distinctive American economic system that evolved with the assumption of superpower status in the nuclear age. The Cold War forced the dual sovereignty to negotiate annual defense obligations requiring future disbursements too large to justify when contracted or to accept when made.

Foreign economic theories have ignored the independent constitutional power of Congress, and Smith of course had no way of anticipating it. The prominence enjoyed by America's president has determined her status abroad. But even the most charismatic American presidents function under continual pressure to demonstrate the ability to lead by compromising with their congressional partner in sovereignty. Recurrent losers in this running negotiation have been political suicides. Taft fell when he "lost his congressional base": the very expression Nixon used to punctuate his own departure from the White House. In between, Wilson lost his presidency when he went to war with Congress; Harding lost his when Congress repudiated him. Hoover lost Congress before the country lost its prosperity. Reagan prolonged his myth longest, but lost it fastest stumping for a Senate he could control, like Wilson before him. Of the losers in the presidential contest with Congress, Reagan was the one stuck with an explicit bet on his ability to manage the American economy on an upward roll, and without apparent concern for congressional opinion.

Armed with foreign economic theories, America stands alone as the superpower guided by no national policy and endowed with a supereconomy guided by no relevant theory; each reinforces and limits the other. This unique dual capability has positioned America to make marketplace moves her rival political superpower in Russia cannot counter. But America cannot neutralize Russia's aggressive drives, or get her money's worth from providing protection for the rest of the world, unless her government develops economic policies tailored to her distinctive needs. Her economic malaise of the 1980s, triggering unmanageable social complications and political defeatism as it deepened, advertised her failure to exploit her distinctive leverage as the superpower armed with the supereconomy. She also failed to persuade Russia to talk business at the political negotiating table (and did not even try to extract reciprocity from Kremlin protectorates whose economies were joining the pack doing her in). Each failure guaranteed the other.

As America's debt inflation accelerated the deflation of the worldwide price structure, America's first reaction to the sense of crisis saturating the atmosphere was to clutch at reworkings of the various imported economic theories that had had their day. But by the 1980s, America was floundering between the challenge to do more expected of a political superpower and the marketplace setbacks likely to undermine an economic superpower. The resulting disillusionment forced a once-confident country to look past its entrenched beliefs and search its experience in increasingly desperate efforts to improvise expedients, however unwise or unnecessary, with a chance of working at home or abroad (for instance, oil import taxes). Economic and financial leverage displaces the temptation to flex political or military muscle, and, vice versa, economic drift and financial dependence kindle the temptation to assert political and military muscle. Witness Central America.

Bewilderment and frustration went with America's failure to sift rival theories through experimentation into practical policies. By the mid-1980s, proof was spreading that government was doing something wrong and that business wasn't doing very much right. Entire sectors of the economy were shaken by debt pains and income shortfalls, omens of the debt wipeouts and income stoppages that accompany full-fledged collapse.

But revived memories of the 1929 trauma, though suggestive, did not apply. Reaganomics reconstructed the 1929 crash as an American export resulting from high U.S. tariffs. This was an optical illusion. The seeds of the crash had been written into the foredoomed Versailles peace treaty a decade before. America had been buying time by financing the terms of the treaty, but without realizing that she was throwing good money after bad. When the collapse of the treaty broke Europe's banking structure, the debris of debt repudiation shattered the market in New York.

True, Wall Street's speculative excesses had already jeopardized it. But Wall Street's call money brokers fueled the boom by reborrowing from Europe expensive margin money, which New York banks had lent Europe in the first place. When the New York banks stopped lending cheap money to Europe, Europe stopped relending it as expensive money to the New York brokers. Despite all that has happened since 1929 to make the world of the 1980s different, Reaganomists were still using their distorted domestic theory of that collapse as proof that their unworkable free-trade policy would avoid a repeat performance.

No special historical insight was needed to grasp the vital difference between the 1929 collapse and the even more massive breakdown that loomed in the late 1980s. A second world depression of the century was the only product left in the battered markets of the world with MADE IN AMERICA stamped on it. Only an antidote bearing the same slogan had any chance of restoring vigor to the economy America had let slide and momentum to the world economy dependent on it. The world had no choice but to wait as America stumbled.

2

HAMILTON'S HERITAGE
Managing the Debt

Janeway's law for American policymakers defines how to get and stay ahead of events: wherever possible, update deals that have worked for America in the past. This pragmatic approach silences objections based solely on theory. The use of experience beats reinventing the wheel. Unfortunately, thanks to the abandonment of history—especially economic history—in what is laughingly called American education, America's past policy achievements are by no means common knowledge.

Alexander Hamilton devised the first deals needed to get the country going: managing the huge debt the federal government inherited; inducing holdouts from the new sovereignty to buy into the system; and relying on bank credit instead of income taxes to lubricate the new mechanism. With the world watching and America stumbling, all three interrelated domestic and foreign urgencies were haunting America again in the 1980s, inviting the application of Janeway's law to Hamilton's solution. So read this chapter as a guide to American policy-making in the late twentieth century, not as a mere history lesson. The formative experience of the Continental

Congress was a preview of Reaganomics: borrowing without taxing to pay for it.

America was born carrying the tainted seed of her first financial crisis. The original Articles of Confederation of the thirteen revolutionary states guaranteed that instant market disaster would follow eventual military victory. The Continental Congress has faded into a blurred memory of passing references, mainly deprecatory, in high school textbooks. But it served the indispensable purpose of issuing the IOUs that paid for the war. Of course, everyone knew that the Continental Congress had no way of making them good except by issuing even more as old ones fell due. It managed to renew just enough to carry the colonists across the finish line. The taint followed the kited paper, which the Continental Congress could not back up, but which was needed literally to paper over the gap left by the absence of a government wielding sovereign power.

No doubt any revolutionary regime would have faced its post-war creditors with a zero credit rating. The thirteen individual states were bankrupt when they organized the Continental Congress into their instrument for wielding the revolutionary powers they had seized. They qualified the Continental Congress as their designated scapegoat by authorizing it to borrow the money they needed, but withholding from it the power to tax. The states proceeded on the pragmatic premise that putting up the troops to win the war was enough of a contribution for them to make without putting up the money as well.

The states had rushed into their war of independence frightened that they might lose it, but without stopping to think how they would pay for it if they won. They allowed their Continental Congress to sink into fiscal irresponsibility during the six years of haggling in chaos that started with the victory in 1783. The states replaced it at the Constitutional Convention with the federal government, which they endowed with the trappings of fiscal responsibility in order to hand over the prerogative of paying for their war. But under cover of the umbrella they designed the federal government to hold over them, the states kept the powers to collect taxes and charter banks authorized to issue folding money: a milestone event obscured by the powers the federal government has asserted over these sovereign functions in modern times.

Accordingly, the constitutional union the states negotiated took the form of an innovative partnership that put their old debt into the custody of the new federal agent they designated to dispose of it. The establishment of the federal government proved a brilliant fiscal success in the country's start-up phase. It solved the immediate problem by giving the new government the borrowing power needed to finance itself. Once the government set up in business, the economy asserted the same need and turned to the same solution: borrowing. But the failure of the new partnership to make any supervisory provision for the borrowing needs that grew with the economy left the country open to the monetary crises that punctuated its growth.

The troubled heritage of the budding sovereignty stemmed from its search for the power to define. Elizabeth Janeway describes this power in her *Improper Behavior* as the most fundamental, the most far-reaching, and yet the most elusive of all forms of power. The Founding Fathers used the crispest, most straightforward, and unambiguous language imaginable in asserting their claim to the power to create and regulate money. Their definition made it seem simple.

Constitutions begin with definitions, but whether for show or for use is the secret of their draftsmen. The strategists in control of the Constitutional Convention wrote into the Constitution the same definition that Adam Smith wrote into *The Wealth of Nations*: the power to create money is the power to mint coin, and vice versa (ARTICLE I, SECTION VIII, CLAUSE 5 of the Constitution empowers Congress "To coin money, regulate the value thereof, and of foreign coin . . ."). This exercise in simplicity screened a masterpiece of ambiguity. The secret of the draftsmen was that they knew better than to believe that minting money was the only way, or even *a* way, to finance the government, let alone the economy. But they also knew that these words—which became known as the Gold Clause—would placate the more reluctant signatories of the Constitution as well as the angry creditors they all represented. The power to define is also the power to deceive. This stipulation of the Founding Fathers seemed a model of incisiveness: at the outset of the era of paper money, none could have been more misleading—by intent.

From the beginning, therefore, the Constitution sanctioned a strategic exemption from the supposedly airtight power to mint money that it reserved for the federal government. Although Smith had been aware that businesses and banks issued bills of exchange negotiable at discounts on merchandise in transit, he never anticipated that "bank money" would take over from minted coin, establishing paper money as real money, not tied to tangible merchandise moving in front of it or to hard value sitting behind it. Consequently, the language of the Constitution avoided the institutionalized reality of paper money its draftsmen had a vested interest in exploiting.

Looking backward, the Founders felt righteous in guaranteeing their constituents security against a repeat performance of unrestrained borrowing with its continuing dilemma: inflation, if not supported by taxes, and deflation if it is. Looking forward, they felt free to trust their restriction of federal money to the minted variety as insurance against either danger (inflation permitting the flow of too much paper money; deflation, too little). For the present, however, they felt justified in authorizing one big public borrowing to launch the new government to make good the bad debts of the Continental Congress.

So far so good, the Fathers of the Republic told themselves and their audience, and they had reason to take pride in their fiscal engineering. But while congratulating themselves on barring the front door to the dilemma of inflation or taxation, they kept the back door open to continued floods of paper originating under the jurisdiction of the states. They opened a loophole for the states to accommodate an enormous expansion in private borrowing and the credit abuses bound to come with it. The loophole opened the channel through which the system rushed on its boom-and-bust course. The corollary axiom of twentieth-century experience—that private borrowing requires government credit backed by the taxing power in order to keep the banks solvent—has yet to be learned by their heirs.

The pioneering "dig" into the purposes of the Founding Fathers was made by Charles A. Beard in his provocative classic, *An Economic Interpretation of the Constitution of the United States,* published in 1913. He described their main and most compelling objective as the prohibition of paper money, supported, as he put it, "with the requirement that the gold and silver coin of the United States shall be the legal tender." But as late as the book's thirteenth edition in 1941, even Beard, for all his investigative shrewdness, was not sensitized to spot the convenient loophole left open to the states. One by one, they slipped through by chartering banks to function as financial speakeasies. The new crop of state banks breached the ban by creating paper currency: they demonstrated, as the Continental Congress had not, that the power to take deposits and make loans was the power to create money. The state legislatures, supposedly subdued by the supposed transfer of power to the federal government, showed their approval by creating more banks. The states begat the banks, and the banks begat the paper needed to finance the future.

The history textbooks have blacked out the original respectability and revolutionary justification of the paper-money issue. Pedagogic neglect reflects the caliber of the wisdom spoon-fed to the young and is responsible for the folklore that blends simplistic assumptions about economics with righteous myths about politics. Thanks to the latter-day preoccupation of economics with computer model building, the study of the interlocking history of economic thought and political crusading has suffered. Neither historians nor economists nowadays are equipped to reconstruct the ploys of plotters or the premises of politicians woven into our history.

Despite their larger size and greater complexity, America's modern problems share a common substance with those Hamilton faced—pinpointed by the bankruptcy crisis that spread across America's regional economies in the 1980s and whipsawed her troubled banking system into a fracas between the federal government and the states. A study of Hamilton's solution to the earlier imbroglio will shed light on a way to manage its modern offshoot. Hamilton finished the operation launched decades earlier by America's most prominent deal broker, Ben Franklin.

The American Revolution was suffered and celebrated as a patriotic ordeal, but it unfolded as a business proposition. The part Franklin played, first in slowing down the break with the Crown and then in speeding it up, has been on display in the chronicles of the pre-Revolutionary years to all who knew where to pry. His role was tied to the two-sided issue on which the break came: the right of the individual colonies to issue paper of their own for businesses to use in financing private commercial operations, and the power of the colonies together to persuade Parliament to redeem it at par in silver or in sterling. (*Silver* was money throughout Europe, and *sterling* was paper—more convenient for transactions, but backed by silver at a Royal Mint.)

The catch was that Parliament repeatedly disqualified the "loan bills" the colonies issued as money; the sales pitch for colonial paper turned on the discounts that deepened with the bills flooding the market and the mixed credit ratings of their endorsers; and the play climaxed with the redemption of the bills at par in London. An important side issue hinged on the Crown's claim to monopolize the trade of the colonies, and on their counterclaim to trade wherever they pleased, but this was subsidiary to the terms of financing. Trade could not be regulated until its flows had been financed.

Paper money was given a bad name midway through the eighteenth century, which stuck, but the habitual uses of paper money were written into the history of the colonies. In 1712, Massachusetts issued the first trade bills and briefly authorized their holders to pass them off as legal tender. During the following half century, the need grew for the colonies to issue paper money faster than they could earn silver from exports or from the redemption of loan bills in London. As early as 1729, Franklin made his mark explaining the need of the colonies for legalized paper money. By 1751, nevertheless, the London establishment decided that the colonial interlopers were making too much of a good thing out of the racket of accumulating bills at discounts at home and collecting on them in good money in London. Parliament cracked down, forbidding new ones from being issued, requiring old ones to be called in, and subjecting new bills, issued as nonlegal tender, to taxes levied to liquidate them.

In 1757, Franklin, already welcome in Whitehall and installed as assistant postmaster general in the colonies, was dispatched to London by his fellow merchants to lobby for his progressive thinking, and theirs, on the money issue. As was usually the case with Franklin, it was a double-think: to win his adopted Commonwealth of Pennsylvania the right to collect the taxes Parliament wanted and needed from "the Proprietor" (the Penn family), and to win all the colonies the right to issue the paper needed to do business. The economic insensitivity of the chroniclers is reflected in *The Columbia Encyclopedia*'s omission of any reference to this second purpose or to how much of Franklin's eighteen-year mission it claimed.

Dorfman notes that Franklin "dined and conversed with Adam

Smith and his intimate, the philosopher-historian David Hume," and "even convinced them of the relative soundness of the Pennsylvania variety of paper money." But the emergent monetary practices of the colonists made no impression on the theories of England's foremost classical economist. Dorfman also ignored the lesson of the American experiment: that paper money was indispensable in an expanding democratic society; Smith and Hume missed it too. Franklin, wary about treading on delicate ground after his revelation, raised no question about the need to mint money from bullion (a more realistic position than the Constitution was to take a third of a century later). His shrewdness prompted him to pick Hume as the more penetrating of the two men; Dorfman reports that Franklin cut him in on one of his better deals.

But whatever progress he made for his thinking and the colonists' cause was reversed in 1764, when Parliament again forbade any colony to issue bills as legal tender. Despite this rebuff, as late as 1768, Franklin, who was, Dorfman reminds us, "without peer in moving with the course of events, still felt practical incentives to describe George III as 'the best king any nation was ever blessed with.' " As kings went, the question about George III was how much better than the worst he was. The accepted view, supported by formidable contemporary evidence, was that he went insane. The more charitable modern scholarly interpretation is that his bizarre medical problems, subsequently diagnosed as a cause of mental confusion, drove his doctors to drive him to distraction. (He pissed purple, not as a salute to his royal blood, but because he suffered from the obscure hereditary disorder known as *porphyria.*) At any rate, he performed less well than the mythical best of kings, and Franklin was not above stooping to flattery.

In 1764, Franklin packaged the incentives he took with him into an offer to the British government. He evidently thought it too good to be refused because he resubmitted it several times as the colonial confrontation sharpened. Dorfman summarizes it as a three-way balancing act among the need of the Crown for revenue, the need of the colonies for paper money, and the pressure from conservative colonial opinion for restraint in issuing paper. Franklin proposed to resolve this clash of interests by organizing a bank managed from London, but issuing bills as legal tender in ten-year loans, at 5 percent, collateralized by land worth twice as much, and redeemed with bullion the colonies earned and paid to the government: the rudiments of a modern monetary system subject to responsible government regulation—still needed (especially the land-lending feature)!

This aborted vision of Franklin's revealed the distinctive departure that land ownership in America was to take from the pattern dictated for capitalist land use in older feudal societies (not just in Europe, but in Japan and throughout Latin America as well). Feudal owners had routinely rented user rights to upstart entrepreneurs. Marx's description of this standard practice as "ground rent" has baffled his American readers, despite the brilliance of his style and the unrivaled mastery of the evidence he marshaled. Yet Marx was merely describing the routine process by which capitalism had sprung from its feudal roots. Never mind that American capitalism in its maturity adopted this same feudal practice.

But Franklin's prophetic concept of banking land as collateral in order to finance its development seized on the distinctive innovation needed to promote the opening of America's frontier. The older societies that evolved into capitalism through the conveyance of feudal landholdings already in use avoided the problem of development, literally, from the ground up. But groundbreaking in advance of any hope of earned income was just as literally where America began. Therefore, the start-up problem fated to plague her breakthroughs and haunt her breakdowns in her westward expansion was exactly what Franklin's proposal said it would be: how to bank land, in the absence of community-furnished infrastructures, for the period of time needed to support debt with earned cash flows. Because Franklin was a prophet, and because the endless honors bestowed on him provided comfortable distractions from this most fundamental and most practical of all his insights, America was left to work and worry her way through the history of flimsy finance and banking booms and busts awaiting her.

One of Franklin's collaborators anticipated the complaint of the monetarists (voiced again in the twentieth century by Professor Milton Friedman) that government borrowing is a hidden tax. "It will operate as a general tax on the colonies, and yet not actually be one," he wrote, explaining that the borrowers would make money on the turnover and that the richest would rightly pay the most. Thus

Franklin's team anticipated the case Hamilton later won for a permanent federal debt.

Franklin for once miscalculated. He overestimated the shrewdness of his friend Lord North, whose government opted for the excise tax alternative instead of welcoming Franklin's practical compromise that might have prevented the war that cost England so dear. His visionary practicality in suggesting that usable land be collateralized into a productive credit base was also brushed aside and forgotten. By the time Franklin was able to report that the government was ready to reconsider, the rhetoric stirred up in the colonies by the Stamp Act had taken hold. The paper-money issue that had brought the breach to a head was out of sight as a revolutionary slogan, but by no means out of mind as a postrevolutionary aim.

When the states demonstrated their power to revolt, they reasserted their right to borrow, but subject to none of the prudent prewar conditions Franklin had offered. Their wartime abuse of this power institutionalized their bankruptcy. The popular sneer about the currency they printed to carry their debts—"Not worth a Continental"—echoed their embarrassment.

The precepts of the revered Smith were lost in the political shuffle, but the conservative element continued to pay lip service to his libertarian view of the interplay between markets and governments. Intellectual consistency, coin rarely traded in political economics, would have called for American devotees of Smith to back the correspondingly libertarian politics of Thomas Jefferson, who abhorred the idea of a permanent debt. Instead, the so-called mercantile interest put its political chips on Hamilton, who picked up where Franklin left off. He operated behind the façade of President Washington's quasi-regal aloofness as the equivalent of a prime minister. His eloquence sold American opinion the British institution of a permanent government bond float, passing muster as sound because paying interest and free from the test of repaying principal. Franklin pronounced his blessing on the Hamiltonian solution when he told the Constitutional Convention that he perceived the "sun rising," not setting. Two hundred years later, Reagan, in his defensive 1987 State of the Union address, singled out Franklin's expression, ignorant that it implied acceptance of a permanent government debt.

The infant republic endowed its first crop of Treasury-subsidized millionaires without exposing itself to the risk of policy innovation or them to the risk of speculation. (In 1778, Franklin reckoned his Continentals worth 15 cents on the dollar, after he had gone "several years" without his 6 percent interest.) The bill jobbers bought up the Continental paper for pennies on the educated assumption that their government would build character for itself by making the Continentals good, dollar for dollar, plus interest; and, once the government was established, it promptly vindicated their confidence.

Hamilton guaranteed a buy-in of the entire float of dubious Continental paper and bought the loyalty of the traffickers with the profits. He made no bones about arguing that the way to win the loyalty of money is to buy it. Hamilton's first report as Treasury secretary boasted or admitted, depending on the point of view, that the new federal Treasury had paid out the stupendous sum of over $13 million in back interest, just short of half the principal it assumed at par. The same opportunity for a buy-in, though on a vastly larger scale, beckoned the managers of the U.S. Treasury in the 1980s. They reckoned its mountainous deficits in a political vacuum, without counting the rivers of revenue sloshing around in the underground economy. Chapter 15 offers an updated version of America's classic solution: sudden solvency for the U.S. Treasury and a safe-conduct pass to profits for the people emancipating the Treasury from dependence on foreign creditor governments.

At the time of Hamilton's original buy-in, riskless speculation in Continentals generated the dynamism that propelled the country's pioneers toward the frontier. Influential historians, notably Frederick Jackson Turner, have written persuasively about the frontier as America's gateway to growth. But with all deference to Turner's vision, America's launching pads from which she financed her future growth were the established coastal financial centers the trailblazers left behind them. In Boston, Philadelphia, and, especially, New York, the conservative claque of new creditors greeted the new Treasury by raiding it—with its blessing. All through the nineteenth century, their clients in Europe backed and joined them. The growth of the country continued to catch Europe's imagination and attract its money; money from Europe always bought America out of her bankruptcy crises, and the bankers with reputations for raising it in

Europe, particularly J. P. Morgan, called the tune in New York and Washington.

The former colonies made history in more than the obvious sense. They not only won their war, but used their victory to reinstate the paper-money issue they had fought for, and to keep it won, a distinctive historical achievement never recognized by the British economic theorists whose names the colonists made household words. History is full of shrewd but failed efforts to turn irrepressible conflicts into profitable compromises. The one Franklin failed to sell in London would have given the colonies what everyone wants, the best of both worlds: a paper currency, backed not by bullion but by land, the colonies' proprietary asset, circulating at discounts in the colonies (equal to devaluation), but redeemable at par in London.

The frontier republic stumbled onto the compromise that resulted from the haggling over the division of federal and state powers in the Constitution. The states were freed to permit the banks they chartered to issue unlimited amounts of paper money in fact, while joining the federal government in denying the practice in principle. They could point to the solemn commitment of the Constitution to the gold standard as proof of the country's fiscal integrity. History has viewed this calculated transition from coin to paper as a natural evolution, and it is certainly true that the familiar lag of encrusted economic theory, which defined money as minted, was overtaken by the novel economic practice, which printed it.

Hamilton and his coterie had dealt themselves a winning hand and they played every card in it. The minted money provision gave them sound credentials that they used to legitimize the paper money that they knew was indispensable for daily use, as opposed to the showcase in which gold sits. But the delegates to the Continental Congress and the states that sent them knew better than to repeat the error of the Continental Congress by issuing currency directly. Though the Constitution reserved the minting power to the federal government, its silence on the subject of the engraving power was golden. That eloquent silence covered the tracks of the states when they let the banks they chartered do their dirty work for them and for the new federal Treasury as well; they avoided blame for the disastrous banking panics that followed. Moreover, they succeeded so well in establishing their control over banking that pressure for federal regulation did not appear until 1907, when, true to form, the pressure of panic forced it.

This persistent prophetic divergence between the needs of the future and the premises of the past haunted the country as it grew. Two hundred years later, when the country approached the bicentennial of the Constitution, its initial abdication of authority over bank regulation was still weakening the domestic structure of finance, just when international overextension of the banks demanded the strengthening of the national banking system. All through the 1980s, the war between the states over the distribution of responsibility for bank regulation between Washington and the fifty state capitals was still very much on. No battlefield staked out by the lobbying fraternity was more lucrative: the federal franchises on interstate banking ripened among the richest plums to be plucked in the Washington orchard. The challenge to policy-making was ignored.

In Hamilton's time, progress ran far enough ahead of dogma to make creative policy experimentation unnecessary. The mother country that America had just repudiated was already providing the model to be copied, and Hamilton copied it. England's Treasury floated a permanent debt to tie the country's monied interests to it. Hamilton's Treasury did exactly the same with exactly the same results.

Where Smith had rated the financial practices of governments by the values of responsible individuals and proprietorships, Hamilton recognized the higher political morality of observing a double standard between private and public financial conduct. Private borrowers are always well advised to pay their debts (or at least to beguile their creditors with plans of repayment). Public borrowers are always better advised never to pay more than the interest—borrowing it, if necessary, by compounding the principal. There's no doubt that Hamilton, like Franklin, was a thorough student of Smith's theories, and there's even less doubt that, like Franklin too, he was a practical observer of their limitations.

Hamilton, citing the debtor's prison, politically speaking, in which the Continental Congress had entrapped itself, operated on the simple premise that a government able to peddle its paper to buyers motivated to make money from it has a longer life expectancy than one unable to do so. He anointed himself the patron saint of American financial conservatism by making the states an offer calculated to convert disgruntled creditors into enthusiastic bankers ready to play again—this time for their private accounts, not just as dealers in dubious state government paper. In form, Hamilton's offer was a one-way trade: federal assumption of the states' back debts, plus forgotten interest. It made them a gift of their lost credit and credibility, but without putting up a cent of the hard coin stipulated by the Constitution. In substance, however, Hamilton kept immeasurably more than he gave away for the simple reason that he created the currency for the purpose. His coup in persuading the past creditors of the states to buy it enabled him to sell it to his own future creditors.

Under cover of the hard line Hamilton talked toward the states, the soft treatment he gave them healed their financial wounds and massaged their political muscles; the states were labeled as fiscally impotent, though recognized as politically omnipotent. Two centuries after his time, this assertion of debtor power demonstrated the enduring realism of the Hamiltonian rule that the bigger a debt is, the surer a debtor is of a bailout, and the shrewder a creditor is for offering buy-ins. He established the principle that anticipated the power play the Third World's bankrupt borrowers put on in their dealings with their presumably solvent creditors among the financial powers during the 1980s. Washington found itself in the same position vis-à-vis its own busted banks.

Hamilton's realism was limited to the realm of fiscal politics. Though his achievement stands as a landmark in the evolution of centralized financial institutions as national economic and political power centers, it left the new federal government a supplicant in the federal-state bargaining process. He paid the states—more precisely, their creditors—in money-good paper as political tribute for the lip service the states had paid to the U.S. Treasury. This scarcely qualified as an assertion of national sovereignty. On the contrary, Hamilton's peace offering to the states acknowledged their undisturbed sovereignty; the responsibility the federal government took over for paying their back bills consolidated it. Claimants are normally expected to collect on their claims. But the federal government emerged as a novel claimant: instead of asserting claims on which it could collect for its own account, it insisted on paying claims on which its hapless predecessors, the states, newly elevated into its sovereigns, had gone bad.

Hamilton could not have encroached on the prerogatives the states kept if he had tried, and he knew better. The fact is that he did not try. His sophisticated double dealing showed that he knew what he was doing in buying in the Tories and using their money to wipe the slate clean for the states, putting them back in business as borrowers. He himself was one of the founders of the Bank of New York, the most venerable institution in New York City, a prime beneficiary of the power the states retained to charter banks. The widely publicized fights that raged in Washington in the early 1800s over franchising a U.S. national bank were sideshows. No pretentious aristocrat from Philadelphia bearing a name associated with inherited wealth, like that of Nicholas Biddle, had a chance of breaking into the charmed political circle of jobbers in state bank charters. On the contrary, Biddle's claim galvanized the local operators of every political stripe into united action against him. The experienced political operators on whom he relied, notably Henry Clay and Daniel Webster, were practical in accepting the patronage he lavished on them and philosophical in observing the frustrations he suffered.

Oncoming generations were thrown on their own to discover the unnerving lesson that life in America would turn into a race against a succession of debilitating banking breakdowns with violent consequences. Even in the best of times, rumblings of the next round of bank runs were always heard around the corner, but how closely or how seriously, no one could ever tell in advance. The business of America may have been business, as Calvin Coolidge pontificated after he removed himself from the line of fire in 1928 and before the banks were burned by it after 1929. But, as his successors soon discovered, the business of government was certainly not to save the banks from the consequences of their own follies.

Though Hamilton's shrewd maneuvering with the states was limited to their financial ties to the federal government, its consequences surfaced on a much broader political terrain. In the name of national sovereignty, he accepted the primacy of states' rights over banking. Specifically, his institutionalization of states' rights as the controlling factor in the banking system established the principle as the force responsible for the Civil War. As fast as the country prospered, property values rose, and the prices of slaves rose with them. Soon, the right all the states claimed—to charter banks empowered to print money—was sharpened into the right some states demanded to legalize slavery. The mercantile North bought freedom for its markets by legitimizing the slave markets of the South. As Emerson philosophized, cotton threaded the North and South together into one regime. The federal government started out as the middleman between the sovereign states.

The political side of states' rights, from 1815, when disgruntled right-wing remnants of the Federalist party held its rump convention in Hartford, Connecticut, until the Civil War, obscured the financial side. The damage from footloose currency engraving by the state banks hit the country more continuously and over a longer time, surviving even the assertion of federal authority over the states during Reconstruction. The collapse of the country's banking system in 1932, state by state, stemmed directly from Hamilton's sleight of hand. Moreover, the collapse of the banks in considerable part reflected their inability to hold government securities: inconceivable though it may seem, the public debt was too low, relative to the growth of bank deposits and investments. But Hamilton foresaw the need for growth in the banking system, stabilized by holdings of government paper. He identified the need of the banks for an ample float of government debt as the opposite side of the coin from the need of the Treasury for an ample network of banks able to absorb it. The history of America's financial crises has fluctuated between an imbalance on one side or another.

America's formative financial crisis had left her with a makeshift banking system. She was still looking for banking stability in the spring of 1933, and the financial turbulence of the late 1980s found it still lacking. In 1933, the flimsy foundations originally improvised for the banking system after the Constitutional Convention were about to collapse. The states were powerless to cope with this national emergency. Once the lame duck administration of Herbert Hoover was engulfed in the wreckage strewn around by the broken banking system of the early 1930s, the New Deal surfaced full-blown as the modern lifesaver the country was waiting for.

That was the good news FDR brought to Washington with him. The bad news greeting him was that, by the spring of 1933, postmor-

tems to ascertain whether America had exported her banking collapse to Europe, or whether Europe had hit America with the backlash from its debt repudiation, were academic. Both power centers
were locked in paralysis together. Germany, normally the strongest
country in Europe, had been left the weakest. Hitler took over in
Berlin exactly when Roosevelt did in Washington, and on the same
promise to declare war against depression and to restore national
vigor.

Both men found that fulfilling the second promise was easier
than the first. Before either could declare victory against the Depression, each was vowing war against the other. The backlash from the
most destructive banking collapse of all time precipitated a corresponding military revival, proof that the reach of America's first
financial crisis of public debt management stretched long across time
and far across space. The simultaneous collapse of the American
banking system and the European debt structure in 1932 brought on
the eventual climax of their joint tragedy on the battlefields of the
1940s. The gravestones are a reminder that the fundamental cycle to
which the economy is subject is not economic at all, but political and
military: once economies slip their business-cycle moorings and
plunge into depression, governments go to war.

The genetic defects in America's banking system grew into
permanent structural hazards to her prosperity. Nor was that all. As
her distinctive impact has expanded, these weaknesses have jeopardized the prosperity of the world as well. Fallout from the overseas
overextension of the multinational American banks has splintered
major segments of America's domestic credit base—notably agriculture, oil, lumber, real estate, manufacturing, mining, and the financial markets: a long sick list! But the domestic base of America's
internationally involved banking system remained structured along
state lines. The states' prerogatives were "grandfathered" by default
in the deal Hamilton gave them. The banking crisis of the 1980s
shaped up as a replay of the historic argument over whether to
permit the states to continue to regulate the banks or to allow a few
nationwide chains with international subsidiaries to take over.

History has blurred the ongoing impact of America's birth pains
on her growing pains. The country's banking system outgrew the
regulatory reach of the individual states long before the country
plunged headlong into its financial convulsions of the 1930s, to be

repeated in the 1980s. The structural anachronism that flawed its heritage empowered the states to claim supervisory rights over parts of the banking Leviathan. In the 1980s, one such Wall Street–based multinational giant evaded federal regulation by scooping up a synthetic local base as remote from the orbit of international banking as South Dakota.

This confusion over the basic need for federal bank regulation, and the hectic nationwide lobbying resistance it incited well into the 1980s, was the direct result of the Founding Fathers' calculated refusal to cope with the evolving definition of money from the minted to the printed variety. The Constitution has been researched and saluted for the wisdom of the system it crafted. But its most costly heritage, by no means merely financial, has resulted from ceding to the states banking powers they were not and are not equipped to handle. The intentions of the Founding Fathers were complicated and obscured by their political wheeling and dealing, in the shadow of the then-muted issue of slavery.

Everyone is aware of the success story of their legacy: the fact is that the country has survived, that we are here, and that, in good times and bad, we are rich. On the other side of the ledger, America has failed to manage the three-directional tug on her: to finance the Treasury, to meet personal banking needs throughout the regionalized economies, and to provide the concentrated financing required to sustain continuing overall growth. Recurrent opportunities for the government to finance buy-ins at discounts testify to recurrent breakdowns in the banking system. By the 1970s, the needs of the Treasury had outrun the capacity of the banks to invest in its paper despite the enormous inflation of the money supply. In Section V, devoted to solutions, I propose methods of relieving the banking system in the 1980s from the pressure on it to absorb an endless float of government paper. I also propose a method for the Treasury to repeat Hamilton's achievement of refinancing the American system by making buy-in offers that holdouts at home and dependents abroad could not refuse.

VEBLEN'S VISION
The Technological Revolution

America floundered into the twentieth century fueling her new continental-scale industrial economy with the banking system that had failed to sustain even her original eastern mercantile system. Every run of commercial prosperity ran the banks out of money and broke their customers. This anomaly was accentuated by the technological breakthroughs that forced the freezing of larger sums of capital for longer periods of time than the average-size businesses of the day could raise or even the largest banks could accommodate. The crying need of the expanding society was its insatiable appetite for industrial capital to keep up with the continuous revolution in its laboratories and toolsheds.

For lack of such long-term capital, the old-time wolves of Wall Street greeted each commercial credit crisis as an opportunity to impersonate Little Red Riding Hood's grandmother as they gobbled up orphans of the storm. The ongoing controversy over the country's anachronistic, panic-prone banking system touched off a new argument over the new trusts, built during the first wave of "merger mania" at the turn of the century. The banks did not have enough

37

money to keep pace with the demand for takeovers, but they had the next best thing: investment banking affiliates, ready to sell lush welcome mats in the stock market. Credit panics paved the road the stock market built toward the sky in 1929.

The industrial revolution mobilized capital and labor on a scale that was puny by modern standards. Toward the end of the nineteenth century, it spawned the technological revolution, an altogether different animal because it was automated, impersonalized, and endowed with powers of limitless growth. For the first three decades of the twentieth century, technology and finance ran a high-speed marathon against one another. Nothing in any formal economic text prepared business, economic, or political opinion to recognize that technology, by lengthening the investment cycle, had changed the workings of the business cycle and threatened exhaustion of the credit reservoir whenever the economy went full steam ahead. The academic establishment was more unprepared to envision the technological revolution than its graduates were to grapple with it.

Before the technological revolution, the quest that dominated the researches of economists was the narrow one for equilibrium between supply and demand, within the marketplace as a whole as well as inside the markets for each and every item. The technological revolution superseded the practicality of gauging market equilibrium by subjecting the resources of the economy to the tug of conflicting political, economic, military, and social claims, all proliferated by the leaps in productivity. The economists continued to bemuse themselves with the metaphysics of equilibrium, but government found itself challenged to manage this parallelogram of forces fluctuating in relation to one another.

Thorstein Veblen foresaw that the technological revolution would develop enough momentum to dispel the mirage of equilibrium in American economic thought. He explained how the need of technology for capital activated the four competing claims within the parallelogram of forces juggled by government.

He emerged from the obscurity of provincial academia in 1899 with the publication of *The Theory of the Leisure Class: An Economic Study in the Evolution of Institutions,* his first iconoclastic best-seller. He died in 1929, just two months before the crash he predicted, in the face of proclamations from Professor Irving Fisher, of Yale, and lesser luminaries of the academic economic establishment, that a new era of unbroken and unbreakable prosperity had arrived.

In 1933, the New Deal arrived instead, chanting the catchphrases Veblen coined as its fighting words in behalf of "the common man." This striking phrase of his had the ring of a slogan, more so than either "workers" or "proletariat," the terms used interchangeably by Marx, and it invoked a broader appeal; Americans resent feeling downgraded into a lower class. Veblen minted this new political coin in a single laconic sentence summing up the boom-bust consequences of the war-peace cycle in the 1919 shutdown of the economy, when he wrote in *The Engineers and the Price System*: "The common man won the war and lost his livelihood."

The New Deal institutionalized Veblen's distinctive thinking and his spicy vocabulary when it managed its memorable hundred-day flying start by adopting his bold strategy of reform as its recipe for recovery. Veblen was merciless in lampooning "the kept classes" at the apex of the establishment. They shared the same target in his sights as their hired hands on display as "the captains of industry" whose "trained ignorance" of technical processes and progress inside their own businesses guaranteed mismanagement. He quipped that any workingman would be sent to jail for attempting the same damage to the wheels of industry with a monkey wrench. He enriched the language of politics with a new verve as a legacy to the Depression decade: "planned scarcity" strangling "the potential for plenty," "rigged prices" protecting aged plant, "conspicuous waste" indulging the insupportable tastes of "the leisure class," alternatively satirized as "absentee owners." The mouthpieces for both conservative and liberal lobbies have continued to sloganize Veblen's common-purpose phrases with comfort and effectiveness—without ever having read him, or perhaps even heard of him.

When the 1933 banking crisis gave the New Deal its mandate, it vindicated Veblen, a prophet acclaimed but unheeded in his lifetime. His restless ghost haunted the drama in which FDR played the leading part. His thinking, in its breadth, anticipated the interconnections between FDR's New Deal and war administrations. These interrelated pressures forced Roosevelt's operatives to rethink America's approach to her political economy, as they restructured her economic society and as they renegotiated her political role in the world. Veblen completed the script they were to follow before they started to read it.

Veblen was the first American social thinker whose observations and theories changed the course of history without an assist from a burning grievance, such as slavery or unemployment. He was the only American economic thinker to stir up revolutionary controversy in his own right and to dominate it. In the first third of the twentieth century, he exerted an unparalleled influence over the attitudes of opinion-making Americans toward their economy while it was still bursting its seams.

In the interplay between economic theories and social institutions during the first two decades of the century, the conflicting theories of Smith and Marx shared a common influence, but a common irrelevance. The theories of Keynes and Hayek, unveiled in the third decade, were guaranteed an influential audience by the mere fact of their opposition to the first two at a time when the turbulence of events unsettled long-settled habits of thought. By contrast with all four of these philosophies, however, Veblen's novel ideas and novel ways of expressing them won such powerful sponsorship and wide acceptance in his lifetime that the system absorbed his thinking, leaving no subsequent trace of his intellectual influence within a mere five years after his death.

During the decade between the 1914–18 war and the 1929 crash, abuses by "the malefactors of great wealth," as President Theodore Roosevelt branded them, made the country receptive to Veblen's merciless satire and his functional realism. The rising generation of restless, disgruntled, and opinionated academics, armed with Veblen's radical empiricism, enlisted first as his disciples and then as Roosevelt reformers. Through them, his ideas guided the New Deal to build defenses against depression and to renew the country's energies for the great leap forward awaiting it at midcentury and beyond.

Veblen was as indigenously American in the modern social sense as the symbolic portrait by Grant Wood of a corn belt farmer with his wife on one side and his pitchfork on the other. He was a highly individualistic product of the melting pot, raised not in an urban slum, as stereotypes of the melting pot suggest, but by a reasonably affluent family in a Norwegian-speaking compound in rural Wisconsin, with the Viking sagas in his past and his pungent

English prose style in his future. Throughout his controversial career, he presented a Janus-like double image: an oddball ungainly country bumpkin, consorting with fellow outcasts in Bohemian squalor, and an urbane savant, comfortable in a long list of languages and cultures, universally respected by his peers, except on sensitive state occasions in academia when tenured sinecures were being parceled out in prestigious universities. Appropriately, this flamboyant son of the emergent culture of "hyphenated Americans" sounded the first call for a distinctive American analytical and policy stance toward America's distinctive social problems and economic opportunities.

Veblen's familiarity with the evolution of social institutions in Europe made him a rarity among American economists. It helped him distinguish those European practices destined for adoption by America—notably, the arms business—from the distinctively American innovations needed instead. It also helped to explain the "loner" status to which he was relegated by departments of economics all his life. The purveyors of the then prevailing economic wisdom in the United States, while preaching the gospel that markets work the same everywhere, sneered at quasi-socialistic practices of capitalism everywhere in Europe, such as subsidizing transportation and communications, paying Social Security, and accepting unionization.

At the same time, their dedication to the dogmas of laissez faire led them to resist Veblen's advocacy of new Washington initiatives calculated to speed the maximization of America's distinctive potential. Prophetically, he called for federal sponsorship of technical development, federal regulation of new securities offerings, minimization of risk to bank deposits by the separation of commercial and investment banking functions, and vigorous enforcement of the antitrust laws—all far out at the time. The New Deal responded to every one of these calls with major programs.

Despite his itinerant academic status, Veblen always had the knack of focusing the spotlight on his ideas. Perhaps because of his flair for publicity, his reputation as a radical of the nuttier variety traveled ahead of him, belying the awe he inspired among serious students for the range of his erudition as well as the practicality of his uniquely magnetic powers of persuasion. More unsettling still, irony was his favorite métier, and his fanciful thrusts—sometimes for serious debate, sometimes for satiric effect—were forever being taken literally by the conformist pack baying at his heels.

America was clearly taking over the leadership role in the world in the first three decades of the twentieth century. Yet, just as clearly, she did not know what to do with it. The national debate over policy direction pitted the forward-lookers, who were pressing for America to take initiatives abroad, against the standpatters, who were insisting that she mind her own increasingly considerable business at home. But the internationalists did not see that the rise of America, though achieved under the protective cover of England, nevertheless portended the fall of the British Empire and the devaluation of Britain's leadership. America's isolationists, though often identified on the progressive side of domestic debate, had eyes only for immediate problems at home and refused to recognize that her solutions lay in facing the challenges that confronted her overseas.

Universal usage refers to the 1914–18 war as World War I. But only the 1939–45 war had a global reach. The outbreak of the 1914–18 war in Europe found Veblen for once in step with flag-waving opinion. He argued that victory over Germany called not only for her defeat in battle but for the dismantling of the internal hierarchic structure of German militarism. Traditionally, militarism had mobilized manpower. Veblen saw that imperial Germany (along with Japan!) had industrialized feudalism and developed transoceanic striking power, challenging America to harness the technological revolution to military priorities. He was the one independent radical unreservedly against imperial Germany, and for U.S. entry into the war on the side of France and England, despite the popular revulsion in America over their alliance with the czar.

Veblen had put President Woodrow Wilson on public notice against the disaster that a German victory would bring before Wilson himself recognized the only alternative: to take the United States into the war. Wilson proclaimed it a war to make the world safe for democracy. Veblen analyzed it as a war necessary to uproot the industrialized warmaking power that imperial Germany had already demonstrated, and that imperial Japan was building, as he warned. American industry and finance promptly enabled England and France to hold their shaky lines against the superior force Germany fielded, then won the war for them, revealing Veblen as a prophet.

At the outbreak of the 1914–18 war, Japan was still acting the part of the quaint and docile paper ally of the Western powers, unready to challenge any of them, much less all, yet quick in 1915 to take advantage of their preoccupation in Europe to slice off strategic colonial enclaves in China as bases for its future marauding. When the peacemakers at Versailles disclosed the deal they had made to pay off Japan for doing nothing to help win the war, it was Veblen, not Wilson, who was ready with the verbal ice pick to describe it: "vulpine secrecy."

In *The Nature of Peace,* published in 1919, Veblen accused Washington of conniving with its allies in schemes to tolerate and, worse still, employ the vestigial remains of Germany's imperial militarist system, while continuing to sacrifice lives wholesale in the crusade to dethrone the kaiser. Any such prospect seemed unbelievable when Veblen scouted it, while the superpatriots were riding high against "the Huns." But the Western Allies did exactly that when their victory won them the option of enlisting the services of the hard core of Prussian military professionalism as their shock troops against Bolshevism in Russia. More unbelievable still, yet more predictable, according to Veblen, they dragged America in with them even though she was then confirming her isolation with her angry rejection of the League of Nations. The victors demonstrated that history bears at least one resemblance to truth: it is stranger than fiction. They acted out the bizarre scenario Veblen had laid out: first, stripping the spoils of war from German militarism, then actually subsidizing its campaign to recoup its lost loot at their expense. Within a mere two decades, Hitler gave strident reconfirmation to Veblen's thesis.

In one of the most perspicacious X rays of Realpolitik ever revealed, Veblen published an in-depth review in an obscure economic journal of Keynes's *The Economic Consequences of the Peace* (1920), at the height of its succès de scandale. Veblen, a reject from the corridors of power, saw deeper into the workings of power than the rising star of the British establishment. Keynes, with the practiced economist's eye for national competitiveness, had taunted his political employers at Versailles for crippling the export earning power Germany needed to pay its reparations bills.

Veblen chided Keynes for taking "political documents at their face value." He grasped the functional reality that the treaty, far from freezing the terms of victory, started as a sop to public opinion in England and France but evolved into a mechanism for continual negotiation. First, it secured the position of caste and property under the revolutionary German socialist regime. Then, it arranged a military alliance between the victors and Germany to contain Bolshevism. Finally, it opened the way for repeated, massive infusions of capital by the Allies to offset the German trade deficits and payments Keynes had foreseen. This critique put Keynes in an ivory tower and showed Veblen worthy of a voice in the White House situation room.

At the same time, Veblen was an inveterate spoofer. One of his playful Socratic tricks was to suggest unconditional surrender to the kaiser: a farcical alternative to the Draconian peace he was advocating. His prank whipped up a storm among the literal minded. Veblen survived it in better shape than Socrates had managed himself: he went scot-free.

That was the good news for the budding audience of those able to read him and laugh with him. But the great news came from the attacks on him from those who missed his jokes. Their red-baiting helped to establish him as the folk hero he became during the interlude of false prosperity between Wilson's dispirited departure and Roosevelt's triumphant arrival. Veblen's vogue was a reminder that a disturbingly representative cross section of the country was enjoying this false prosperity without believing in it.

Veblen could also be serious in dismantling the solemn façade of institutionalized ideology, though he and bureaucratic citadels of inertia were not meant for one another. Dorfman wrote Veblen's definitive biography as a prelude to his *Economic Mind in American Civilization.* He recounts an episode revealing how impressive Veblen could be in action. It occurred in 1918, during Veblen's brief tour of government service with the Food Administration. The ineptitude of Wilson's war administration reached a new high when it managed to run the country into a food shortage, unleashing a predictable inflation of food prices, which sent wartime wages through the roof and accentuated the cost spiral that culminated in the calamitous 1919 bust. The difficulty centered on shortages of farm labor in an economy distorted by overregulation.

The Justice Department, in charge of protecting the flag, was harassing farm and forest workers believed to belong to various radical groups, notably the IWW. The "Wobblies," as they were called, responded by scattering. The AF of L, which had been awarded protective custody over the Labor Department by the White House, was encouraging the witch hunt, even though its own members were up in arms against the runaway in food prices spurred by this government agitation against the departed agitators. Veblen's projected solution cut through the cross fire of partisanship. He offered a suggestion, backed by a letter he was careful to secure from a friendly IWW official, that the government open direct negotiations with the farm workers it was harassing; invite them to organize on a majority-rule basis, and make the arrangements the direct responsibility of the secretary of war. No official in the war administration could have been more remote from conflicts of interest arising from the jurisdictional and ideological labor disputes crippling the war effort.

Veblen, in the course of pushing his recommendation, found his way into the upper reaches of the Justice Department, and, once there, into the office of John Lord O'Brian, subsequently one of the most respected leaders of the American bar, then in charge of enforcement of the Espionage Act. Dorfman records this profile of Veblen's briefing O'Brian on the exorbitant wartime cost of allowing flag-waving to sabotage plowing. Its source was an assistant attorney general, certifiably no "red" and, as his account of the meeting showed, representative of the literate and responsible professionals in all walks of life reached by Veblen's work.

My first impression of the physical appearance of the stranger was not particularly favourable, by which I mean that it did not give the expectation that I would hear anything particularly interesting or important. As he spoke, I became alertly conscious of the fact that, whoever he was, here was a man of keenest wit and intelligence, clear and picturesque vocabulary, somebody exceptional, and I became eager to know who he was and managed to find out by a side glance at the visiting card which he had placed on Mr. O'Brian's desk, and lo and behold, he was none other than Thorstein Veblen! Of course, I immediately became doubly alert and interested, and think I said to Mr. O'Brian after the interview, that there was the keenest-minded person, who had come into the Department while I was there.

Two by-products of this vivid episode left their traces on history. The first was personal: Veblen's assistant on this brief project in search of the scattered agricultural and forest labor force of the radicalized western states, who accompanied him on his fieldwork, was Isador Lubin, who emerged in public view as a charter member of the New Deal when FDR appointed him commissioner of labor statistics. The second was political: Wilson suffered the crowning humiliation in his fight for the lost cause of the League of Nations in Everett, Washington, the center of IWW forest worker backlash against his administration's harassment. When he brought his campaign to sell the League to this regional economic center, the Wobblies lined the main street as he passed and stunned him with "the silent treatment." Veblen was no politician, but neither was Wilson. If Wilson's political advisers had listened to Veblen's wise economic counsel, they might have spared their chief the shock of personal repudiation that brought on his crippling paralytic stroke. Sometimes the best political tactic is the simplest economics: the kind that works.

One reason Veblen's economic advice on this critical occasion was workable is that it was rooted in America's distinctive sociology and called for correspondingly distinctive American solutions. In no other country blessed with uncountable food and forest reserves could a helpful civilian underling have been allowed to initiate an out-of-channel, interdepartmental proposal for direct collective bargaining between an anarchistic labor organization and a wartime government. In no other country would the bureaucratic underling, a well-known dissenter himself, have been allowed to act out the irony of nominating the secretary of war to the loyalty enforcement officials as the guarantor of due process and civil rights in the midst of an outburst of war hysteria of which the dissenting initiator himself was a target. Even if the helpful initiative had come from a spoof played by Charlie Chaplin, its occurrence in real life, in, of all places, the somber chambers of the Justice Department, the power center of the future Orwellian state, suggested that despite outbreaks of bigotry and despotism, America still offered more room for dissent and maneuver than its older European counterparts could imagine, let alone tolerate. This was a flexibility of which Veblen was thoroughly aware and which he was shrewd in exploiting.

Veblen's most far-reaching discovery was the revolutionary role of the engineers in creating technology. In the process, they enriched the workaday world with the visionary concept of Francis Bacon, who, at the outset of the modern era in the early seventeenth century, foresaw old sciences and crafts spawning new ones. From the time of Smith over a century before Veblen to the time of Keynes and Hayek a generation after, economists coming to grips with capitalism have found themselves wrestling with the impact of technology on the labor market. Nevertheless, they have continued to focus on the consequences for the labor market, not on the technological causes of those changes.

Marx, always alert to seize upon contradictions in capitalism, entrapped himself in a tricky one of his own in describing how it worked. He foresaw the unemployment problem ahead for capitalism, but he also realized that only the avant garde of the employed work force could make the revolution, and that it might not. Keynes, crusading to save capitalism from the political backlash of its own malfunctions, focused on the failure of forced cuts in wages to activate the labor market as the spur to recovery promised by classical theory. His theoretical insights inspired active monetary and fiscal responses to the social problems created by unemployment. By contrast, his critical peers welcomed unemployment as guaranteeing economic stability, with no sensitivity to the social instability it also guaranteed in the emerging media society financed on credit, vulnerable to strains in the work force, and dependent on stable operations to absorb increasingly high overhead costs.

The key to any view of the economy is in its theory of value. Before Veblen's time, and even during the cultural lag afterward, this meant the labor theory of value. Smith had shown labor squeezing value out of materials cut, formed, and packaged into products for market consumption. Marx raged that labor would reclaim the value stolen from its productivity by capital. Hayek worried about how labor would save enough to finance the future investments of employers in keeping it employed. Keynes devised methods of supplementing its spending power. At bottom, through all the jokes and sneers about the endemic disagreements among economists, all of them agreed on the indispensable utility of a labor theory of value, or at least a theory of the income-savings-investment process encompassing the labor market.

Except Veblen: he gave it the decent burial that the rise of the engineers made overdue. He showed how the engineers controlled the increasingly complex interconnections that automated end products to make new end products. He superseded the labor theory of value with the realistic modern insight that technology is the explosive ingredient in business activity that creates value—not only for services, which are mainly labor-intensive, but also for goods, which are more capital-intensive. It is axiomatic that new technology eats capital faster than capital invested in old technology can earn its keep. Capitalism has proved more vulnerable to the shortened life of investment than to the falling rate of profit that Marx emphasized. Veblen foresaw that technology would guzzle capital with the voraciousness of a wolf and build muscle with the speed of a tiger.

The last decade of Veblen's life was one of ebullient investment confidence dominated by a record boom in stock market speculation, climaxing in the promotion of new issues promising the moon. Managements and their bankers did not yet foresee the problem of financing technological innovation faster than its rising cost and shortened life expectancy could earn its replacement, let alone pay its way. The main operating preoccupation of management was still to devise methods of speeding up labor in the production of goods. Managers had no sensitivity to the danger that depression would collapse the demand side before *scientific methods*—meaning systematic labor speedups—could deflate costs on the supply side. The country's engineering schools were gaited to fill demands for time-study managers, and this cultural imbalance prodded them to turn out "efficiency engineers" in droves.

Even Henry Ford, the tinkerers' model of the successful engineer and the self-proclaimed benefactor of the workingman, ended by falling back on speedups in his drive to cut costs at whatever cost to safety, provoking the *New Republic* to publish a scorching attack on Detroit as "the city of three-fingered men." The frenzy of the pressure for speedups on the factory floor inspired an early ethnic joke, its protagonist a woman assembly-line worker from "Little Warsaw" in Detroit, brighter than the "college-boy" efficiency experts pressing her to make simultaneous new timesaving motions with both hands and feet. The next time she sees them coming, she shouts: "Stick a broom up my ass, and I'll sweep the floor, too." I heard this joke first from Walter Reuther, when he was still fighting his way up the ladder of the still directionless United Auto Workers.

But while management remained blind to the technological opportunities ahead of it, while capital was demanding more work for less pay but offering more products at richer markups, and while engineers were echoing calls for crackdowns on labor instead of explaining their options on technology, Veblen anticipated the pivotal role awaiting the engineers in the technological revolution. He also anticipated the political cleavage it would deepen within the work force.

Ironically, Marx had recognized the same division of purpose among the proletariat that Veblen predicted among the engineers and found filtering down into the work force. Though Marx inflated the labor theory of value into a political credo, he had been careful in drawing the lines of class war between capitalists and workers to leave room for militant consciousness as an essential ingredient of class confrontation. Despite his emphasis on the materialistic interpretation of history, he conceded that the collapse of capitalism he expected would not precipitate the revolution he predicted until the workers first understood, then resolved (if they ever did!) to make the revolution in their minds before they made it in the streets.

Veblen noted the same ambiguity in social attitude among the engineers: they made their livings on the payrolls of capitalists, but they made their calculations by putting themselves in the shoes of their employers, thinking rich while living decidedly less so. They routinely bound over their birthrights, in the form of their breakthroughs, to their corporate benefactors without even checking whether their superiors had a glimmering of the potential in those breakthroughs.

Veblen attributed the continual acceleration in the technological revolution to the permanent stall in the proletarian revolution. He concluded that only the engineers could change the capitalist system, and he showed that they were—but by making it bigger while trying to keep it the same. Their intent was merely to do their jobs: neither to take charge of a reformed system nor to take responsibility for changes in the system precipitated by the jobs they did. All technicians at bottom invite entrapment in "the Eichmann syndrome": "I just work here." The syndrome is by no means restricted to the Nazi terror: it was first identified by Plato in his description of "auxiliaries" as instruments of sovereigns. The nuclear physicists who worked on America's first atomic bomb confessed succumbing to it too.

Veblen explained that the engineers were catapulted by the dynamics of technology into functioning as carriers of the technological revolution just by doing their jobs, echoing the opinions of their employers while they did, as Horatio Alger had before them. The overhaul they performed on capitalism outfitted the old money system with a new power switch. It sent the changed system jittering into an economic St. Vitus' Dance when capital flows that transformed blueprints into products were turned on, and into a paralytic stall when they were turned off.

Anatole France had satirized the rigid caste structure of Europe in the nineteenth century with his sardonic grant to the millionaire of the same right to sleep under a bridge in the rain as a tramp. Early in the twentieth, the Ford Motor Company cranked up the imagination of America with its prophetic claim that its radical $5-a-day wage scale would give the workingman not only the same right but the same ability to own a car as his employer. Overnight, the cult of the engineer won converts in search of a new hope, and Veblen was hailed as its prophet.

No doubt the startling breakthroughs scored in the 1914–18 war—on wheels, on wings and with fins, all backed by blueprints specifying standardized parts—revealed an unforgettable illustration of Veblen's insight into the role of the engineers as an instrument of the revolution Americans could feel changing their daily lives, without intruding on their economic motivations or political preferences. So did the breathtaking diversification of new luxuries, automated and sanitized for mass marketing as conveniences or necessities to the consuming public. As Veblen predicted, the spread of "conspicuous consumption" proved a more influential economic force than the fundamentals of economic development.

In the space of a single generation, America's working population had been whirled from Mark Hanna's promise of "a full dinner pail" in the 1890s to the bizarre luxury of "The Cow in Your Kitchen" at the onset of Prohibition: from bare-bones subsistence to the synthetic convenience of evaporated milk substituting for the routine of daily milk delivery or the even earlier drudgery and slop of cows in the yard. Overnight, marketing gimmickry found popular new uses for tool-room gimmickry; a new version of Renaissance man known as the huckster turned up, making news, markets, and money. Credit for creating the sales pitch for evaporated milk went to Albert Lasker, the advertising tycoon lampooned in the muckraking best-selling novel, *The Hucksters,* which enriched the language as his method of operating enriched him.

Lasker exemplified the new high-powered, high-cost consultant, moving in on corporate managements and inflating the unproductive charges of corporate operations: a trend that Veblen foresaw would cost management its flexibility on the marketing side and lose economies, won on the plant floor, in the executive suites. Lasker himself, who netted even more by the month from commissions than from his very considerable flow of dividends, was larger than life and loquaciously cynical. Like Hoover, he had established a public wartime presence with the Democrats; Wilson made him the head of the Shipping Board. Like Hoover, too, he then did well with the Republicans, though, as he once ruefully remarked to me, "I got more for my money picking vice presidents than presidents." Always practical, he sold his last wife on the idea of marrying him by telling her that he "would leave her the richest widow in the country." When he did, she emerged as a prominent contributor to humanitarian health causes pushed by Democrats: proof that progress is cyclical.

Veblen's cult of the engineer, commercialized by the huckster, won acceptance from all ideological sides. His radical thinking set the tone for the political mainstream when Herbert Hoover, then Calvin Coolidge's secretary of commerce, decided to base his bid for the Republican presidential succession in 1928 on his credentials as the Great Engineer: the alter ego of Economic Man, doubling in brass as the Great Humanitarian. During and after the 1914–18 war, he made his name in public service by feeding the victims of the famine Veblen had foreseen and tried to avert. Among the dissenting intelligentsia, avant-garde members of economics faculties began to pay more attention to their engineering colleagues than to their traditional mentors in history and philosophy. Margaret Mead recalled her father ruminating on the porch, porkpie hat perched on his head, absorbed in Veblen when not on the romantic prowl. He was a far-out professor of economics at the University of Pennsylvania's new Wharton School, and she regarded his work as alien to hers as a pioneering anthropologist, unaware that Veblen's unique combination of originality and erudition had prompted him to forge formative links between her father's field and her own.

Not only Ford, but his fellow-travelers in the focus of the media—Thomas Edison, Harvey Firestone, and a host of lesser tinkerers hitting the jackpot from their woodshed workshops—emerged as real people acting out the success stories of the engineers. They created a distinctively new American folk hero tailored to fit the new sociology into traditional legends, and they riveted media attention on their long-since-forgotten eccentricities of opinion. Among the progressive professionals, Alice Hamilton, one of the first women to qualify as a physician and "the mother of industrial medicine," declared that she would want any son of hers to become an engineer, but only after spending a couple of years among the Hull House social workers in Chicago learning how the other half lived "before he started managing men." She personified the blend of do-goodism and functionalism that characterized the positive side of Veblenism.

The negative side of Veblenism advertised his cynicism and irritated the insecurity of the establishment's reigning wit. Scoffers provoke mockers, and H. L. Mencken, the abrasive arbiter of the emergent counterculture of the era of Prohibition and prudery, took Veblen's merciless satire of the status quo during its treacherous glow of prosperity as a threat to his franchise. Mencken presented a baffling profile of radicalism in morals, know-nothingism in politics, Bourbonism in economics, enlightenment in literature (repeatedly enlivened by his battling in the streets and courts of Boston against the city's custom of banning books deemed dangerous to public morality), and erudition in journalism. He was an editor of the enterprising *Baltimore Sun,* as well as the proprietor of the high-brow *American Mercury,* and more than incidentally, the Henry Ford of the pornography publishing business of the future.

With malicious perversity, he singled out Veblen as his bête noir, in callous disregard of the striking parallels in their work. Both served as Paul Reveres of the brewing crisis, Veblen scrutinizing its economic causes and Mencken its cultural consequences, and each dramatized the distinctive evolution of American society from its polyglot antecedents. Mencken's scholarly work, *The American Language,* stands as a monumental contribution to the comparative sociology of philology, and his fierce defense of the distinctively American writing of Sinclair Lewis, Theodore Dreiser, and Eugene O'Neill emancipated American readers from the ingrained assumption, which American economists have continued to accept, that everything European is better. Mencken's incessant, irritable knocking gave Veblen an additional boost with the influential audience they shared. It also helped to channel Veblen's fateful posthumous influence on the New Deal.

An accident of geography explained how. Mencken had mounted a second-front war against the newspapers as they reported Veblen's "every wink and whisper," and against campus intellectuals as they chanted "rah, rah, rah for Professor Dr. Thorstein Veblen." But the Veblen fans Mencken singled out for his special sting were "the humble Columbia instructors" addicted, he sneered, to spending their petty cash salary increases on Veblen's books and to showering their dimes on round-trip subway rides from Morningside Heights to Greenwich Village for evening talks at the Veblen Clubs springing up among the intelligentsia. These same humble Columbia instructors—notably, Raymond Moley, Rexford Tugwell, and Adolf Berle—emerged as the "Brains Trust" responsible for the structural changes the New Deal engineered in 1933.

Roosevelt turned to them for his economic blueprints when he was still the attractive and ambitious incumbent of the governor's mansion in Albany, New York. His predecessor, Alfred E. Smith, had put the political power vested there in FDR's safekeeping when he sallied forth to make his foredoomed run against Hoover in 1928, and to stake out his claim to a return engagement in 1932. Smith did not discover that he had given the fox the keys to the chicken coop until Roosevelt stole the option on the White House from him. By then, Veblen was dead, but the country was on notice that the Roosevelt town house in New York City was halfway between the Columbia campus on Morningside Heights and the avant-garde ferment in Greenwich Village. When the entourage decamped to Washington, the blueprints were unveiled. By then, all popular traces of Veblen's copyright on them were blacked out. Veblen's work was done; that of his protégés was just beginning.

4

SCHUMPETER'S SHORTSIGHTEDNESS
The Technological Imperative

Of Veblen's rich assortment of prophetic insights into the processes of institutional modernization, I have already mentioned the take-over of the pivotal role in the economy by the engineers, the techno-logical theory of value, and the evolution of the beneficiaries of entrepreneurs into the "kept classes," who were paid interest and dividends to stay away from corporate headquarters. High-priced hired hands, paying high-priced hucksters to promote them as "cap-tains of industry," had taken over from the legendary tooth-and-claw capitalists who knew their own businesses and dealt with their own money. This new breed of management, trained to pass itself off as professional, qualified to join the world's oldest profession. As Veblen jeered, "trained ignorance" was the hallmark of these rootless front men. They lacked the incentive and the stomach to resist cost-inflating pressures, as well as the expertise to evaluate compet-ing claims for investment in new technology.

Two changes in function within the corporate establishment explained the technological theory of value: the delegation of respon-sibility for corporate decision making to managements without the knowledge to make informed decisions and the delegation of power to engineers to force investment decisions without responsibility for financing them. The technological theory of value, in turn, explained the insatiable appetite of the new corporate mechanism for infusions of capital on which to run when it did run, and then, to gallop.

With these insights, Veblen, though exiled by the academic establishment, put his finger on a momentous reversal in the basic investment process: from "save first–invest afterward," to "borrow first–invest immediately–earn afterward"—with luck. Nevertheless, a century after he debunked the prime puritanical tenet of the folklore of capitalism, saving was thought to remain the limit on investment, even in most professional perceptions. As Veblen saw, back when public markets were still tiny, the pressure on corporate financing to catch up with the technological revolution required managements to inflate the market valuations of continual offerings of new stock issues, with unforgettable consequences in 1929. This same scramble to invest before earning, let alone saving, contrary to classic teaching and contemporary belief as late as 1988, resulted in corporations inflating their debt burden a generation before the sequence of depression and war forced the government to follow suit with a vengeance in the 1940s and with abandon in the 1980s. Complaints from the corporate establishment against overborrowing by government invite the retort that corporations set government the example. Bipartisan insistence by the economic establishment that the rate of personal savings must set the limit on the rate of corporate investment diverted the 1988 presidential economic debate into a campaign to cut consumption in order to finance investment without inflation; the Harvard professor who was economic spokesman for the Democratic candidate concurred.

What was distinctive about Veblen's approach to the business cycle was, first, his recognition of how to quantify it; then, his judgment that it started and stopped with capital functions—borrowing and investing for inventorying and expansion—rather than with income functions—earnings, disbursements, and savings—or with market functions—primed by the demand of hypothetically liquid consumers fortified by high savings. By contrast, the conventional approach to the business cycle views it as a fixed mechanism operating at varying speeds. Veblen saw that the mechanism changes along with its speeds. The conventional approach treated investment as just another statistical item in the macroeconomic mix. But Veblen saw how investment, by changing the behavior of the business cycle, changed the institutional structure of the economic mechanism.

The pressure to finance continuous investment—by overborrowing and/or stock-watering—reflected the transfer of corporate initiatives from the boardroom to the laboratory. This shift left managements increasingly unsure of the fundamental franchise that everyone regarded as the inalienable right of corporations to control: the rate at which they save and/or invest their own capital. Historically, the prerogative of investing did remain private so long as the money businesses invested was their own. But as the technological revolution put them under competitive pressure to raise new money in projects exhilarated by their promotional efforts, managements were compelled to throw new money after old to perform on their promises. The technological revolution was precipitated by political decisions that seemed economic in character because not merely tolerating corporate bigness but encouraging the early trusts to gouge. The managerial process juggernauted out of the control of corporations originally habituated and motivated to restrict reinvestment until *after* the odds on profitability increased. Corporations turned into compulsive investors regardless of their profitability; the price of staying in business became continuous investment.

Marx, in one of his seminal perceptions as an economist, had anticipated a falling rate of profit for capital as management invested more. Typically, though, this practical judgment was flawed by his impractical political expectations, prompted by his tunnel vision of the labor theory of value, that capital would pass along the squeeze to labor and provoke the outraged masses until they took to the streets. Instead, labor did better and stayed home cultivating the suburban gardens it found itself accumulating as capital placed more winning bets on technology: the opposite of Marx's surplus-value theory and the opposite of labor's own resistance to automation.

Veblen's formulation of the technological theory of value explained how capitalists lost control of their corporate franchises under the routine, superficial supervision of the high-cost bankers, economists, lawyers, accountants, lobbyists, and publicists they themselves hired to share their "trained ignorance." Until the technological revolution engineered this transfer of power after the turn of the century from the frock coats to the lab coats, the rate of return, measured by cash intake from customers, remained the criterion of competitive survival. After the transfer, the rate of investment, measured by cash outgo for technology, took over. In the era of the Fortune 500, accepted accounting practice achieved the financial equivalent of transubstantiation: it grafted investment cost overruns onto bulging fixed asset accounts, instead of charging them against operating income. The pressure on corporate finance to accommodate the hazards of continuous investment turned cash losses into paper markups.

But the various modern versions of classical economic theory—including the radically modified versions of it formulated, respectively, by Marx and Keynes—continued to assume that capital retained the same decision-making power over investment after the technological revolution that it had exercised before. Economists were still regarding capitalists as free to ask how much their businesses could do for them, long after the lowly engineers in their anonymity were forcing managements to say how much they would do for their corporations—with other people's money. The labor theory of value conceded the decision-making power to capitalists. Veblen invited the Galileo treatment from the economic fraternity when he superseded the labor theory with the technological theory of value. He described the change in the competitive test as it occurred: from how much cash individual capitalists could pocket from their businesses to how much credit giant corporations could load onto their projects.

The Rockefellers and the du Ponts knew how to buy or beat hostile politicians, break strikes, fix prices, and neutralize critics. But they were bewildered and felt threatened by the challenge of technology to force the tempo of investment—especially because it originated with subordinates commanding special knowledge they themselves could not fathom, suggesting the potential of independence and, worse still, the threat of competition. But the new breed of hired executives, deprived of options over how, when, and even whether to tap treasuries to invest capital, did not even know enough to feel threatened. When Veblen blasted synthetic magnates as "captains of sabotage" conspiring to limit production and inflate prices, he endowed them with a sinister mystique commanding awe and inspiring fear. But when he ridiculed them as passing for "captains of industry" while serving as paymasters to the "kept classes," he reduced them to ripe targets for the budding reformers-to-be of the New Deal.

In the plain and simple world described by Adam Smith's commonsensical sermonizing on behalf of the old virtues, debt was always looked down upon as inescapable for life's losers and never tolerated as preferable by its winners. Depressions were accepted as the wages of commercial sin. For the government to opt for debt as a policy to spread relief and bring recovery was unthinkable. Traditionally, the inflation of a company's debt had guaranteed the automatic deflation of the equity subordinated to it. But the unfamiliar complications cascading with the technological revolution forced corporate boards and their bankers to fall back on debt and dilution, the standard techniques of distress, as the new method for financing growth. Amidst the frenzy that lured Wall Street to collapse in 1929, stock offerings were inflated in order to justify the inflation of borrowings, and the inflation of borrowings was covered by inflating stock offerings. The pyramiding that resulted struck amateur market minds as a combination of intoxicant and sedative. They fantasized that this stock watering would broaden the equity base under the debt load and steady the top-heavy apex of debt, just when both were being swamped by the flood of new securities outrunning the flow of earning power.

Accordingly, all through the 1920s, establishment economists, profiting from the market inflation but insensitive to its roots and risks in the technological revolution, continued to pay lip service to Smith's puritanical principles and to violate them. Veblen warned that any time the financial markets stopped their inflationary spiral, the system would collapse. When they did, it did collapse, but without stopping the technological revolution. On the contrary, the scope of technological breakthroughs continued to broaden and their speed to quicken, despite the long freeze on investment activity forced by the Depression.

While the boom of the 1920s was on, however, the economic debate focused on counting marketplace numbers and ignored technological developments, especially the stirrings of the most momentous chapter of the technological revolution: the nuclear revolution. It was the first in history to carry the seed of universal destruction, as gunpowder or the submarine or even firebombing did not. In the intertwined history of economics and technology, the nuclear frontier opened for technology in the 1930s, just when all openings to growth seemed sealed off for economics. Yet, if only economists had enjoyed the benefit of technological orientation back then, their anxieties would have centered on shortages of capacity, not on failures of demand. The nuclear revolution dramatized the startling divergence between the depression-proof momentum of technology and the depression-prone top-heaviness of finance. Its progress in the laboratory accelerated even after activity in the marketplace all but stopped in the 1930s.

When Veblen's disciples vaulted into the saddle with the New Deal, they caught up with his alert to the financial consequences of the economic changes chain-reacting from the technological revolution. Politically, however, the Brains Trusters were recoiling with embarrassment from FDR's exercise in campaign trail overkill, and so was he; he had lambasted Hoover for "spending the country into the Depression." Now he himself was proposing to spend the country out of it. Berle had just published *The Modern Corporation and Private Property* (in collaboration with Gardiner Means, another prominent New Deal recruit from academia). Their researches were hailed as monumental, as they were, because they detailed the usurpation of corporate power by bureaucratic management that Veblen had publicized in principle; and they documented the emergent need for government protection of the passive investing public from "insider" abuse. They blueprinted the concept formulated by Veblen just as this need was festering into a burning depression issue. The proliferation of misery from the crash had set up Wall Street as a ripe political target for Roosevelt, but the stir made by the Berle and Means book gave him the respectable policy guide he sought as a rationale for governing.

FDR's various groups of intellectual advisers—principally, the Columbia faction (notably Berle) advocating bigness in government and the Harvard faction attacking bigness in business—agreed in persuading him that the answer to his twin problem of how to run a political circus and an effective administration was to build the New Deal around the new Securities and Exchange Commission (SEC). Under cover of the New Deal's political preoccupation with guaranteeing opportunity and protection for labor, farmers, small businesses, and minorities, the SEC centered its policy probes on past malpractices explaining the Depression-time unemployment of capital and on present problems blocking its future reemployment. Consistent with the history of capitalist analysis, conservatives focus on the willingness of labor to take less in starting recoveries, and liberals on the willingness of capital to give more.

While the SEC was still wrestling with this policy mission, its most important chairman, Justice William O. Douglas (as he was soon to be), told me that once the next bull market received the institutional momentum it would need and get, the mission of the SEC would "shrink into the financial equivalent of a garbage collection agency." How it did, and how, when it did, the new generation of institutional abuses responsible for the depression of the 1980s eluded its scrutiny, is part of another story. But in the critical chapter of history that resulted in the transfer of power over the American economy from Wall Street to Washington, the SEC functioned as the New Deal's original think tank, bridging the gap between Veblen's intellectual legacy and Roosevelt's political legacy and carrying out the program Veblen had conceived of recovery by reform.

The public utility industry, dominated by half a dozen holding companies, emerged as the main source of publicly financed capital investment in the 1920s. It accounted for the biggest borrowings, the most insupportable capital structures, and the largest accumulation of backed-up investment needs. Inescapably, it produced the most spectacular insolvencies of the 1930s. All the holding companies landed under the supervision of the SEC, which was empowered to break them up. The SEC ordered them to redistribute their controlling shares in the various utility operating companies to their own investors. By the time the SEC executed this "death sentence," the holding companies were floundering under vast accumulations of unpaid preferred stock dividends. No equity was left for their common stocks, nor any borrowing power, let alone cash to support new investment. In fact, their subsequent breakup opened the way for a crescendo of nationwide catchup capital investment by the country's local and regional operating utility companies. That climaxed with the tremendous wave of heavy power installations during the Korean wartime buildup of 1950–53. They reinvested as fast as they were recapitalized. Momentous political consequences followed from this fight, which acted out Veblen's analysis of boom-time corporate finance dictated by the technological revolution.

Of Veblen's various explorations enjoyed by posterity, none proved as influential for the future of the economics profession as the corollary he developed from the technological theory of value; namely, the need for a model of transactions and activities (especially inventory accumulation and/or liquidation), stripped from their familiar dollar tags, and measured in physical terms, for purposes of analysis and forecasting, by units in bulk. The product of this insight has long since been standardized into the *inflation deflator,* adjusting the value of financial transactions in any given period for the physical volume of work done, and computing an "input-output analysis" of it. More than half a century after Veblen's death, Professor Wassily Leontief, of New York University, the Nobel Prize–winning innovator and virtuoso practitioner of this analysis, was at pains to explain to the professional audience of *Challenge* that the key to modern input-output analysis is in the concept of the technological theory of value. He complained that "most input-output tables are still done in dollar terms," but promised that "more and more we are introducing straight physical units of information, and this is where . . . engineering data become so helpful . . . as a source of observations to determine the input-output coefficients which then define the technologies and the effects of technological change."

Before Veblen developed the long-overdue concept for this basic procedure, economics had been tied together by morals and arithmetic. It had relied on metaphysics to provide the stitching, in the form of the nebulous notion of equilibrium between forces kept, by definition, in continuous disequilibrium by every price change in the real-life marketplace. Veblen freed economics from metaphysics, scoffed at morals, and redirected economics toward engineering, which is applied mathematics. Decades before engineers first designed computers that could accelerate implementation of his input-output concept, he found engineering in need of ideas as badly as he lampooned economics for needing modern tools.

Once computers substituted electronic brains for human brains in standardizing the control of business activities and financial transactions, the effort to quantify this idea of the inflation deflator took over as the main preoccupation of the economics profession among the academics as well as their progeny in the corporate establishment. If any of the "pinstripers" vaulting on their MBAs into affluent careers of "trained ignorance" had heard of Professor Paul Douglas, of the University of Chicago (later U.S. senator from Illinois), it would have been as a liberal politician, damned in their eyes on both counts, and not as a father of econometrics to be thanked for helping to provide the source of their daily diet of cake. None of them would have recognized Veblen as their professional grandfather; only a few would even have heard of him.

In 1923, Douglas launched a campaign to nominate Veblen to the new honorary presidency of the American Economic Association, an effort which culminated in a fiasco of academic politics. The offer did materialize, but subject to two conditions: that Veblen join the association and deliver a presidential address to it. In this way, the economics establishment left him free to express the heretical rudeness that was his trademark—immeasurably more generous treatment than the Inquisition gave Galileo, indeed a mark of progress.

Irony was implicit in the one-man circus Veblen staged against the association. The progressives in the economics profession had founded it in 1885, for the explicit purpose of freeing economists from the exhortations of preachers and moralists. They resolved to dedicate themselves to scientific standards and to base their findings on processes discernible and measurable in society. Veblen's work not only met their standards, but set goals for their students. His reward in his prime was their derision as a "sociologist"; the association's recognition of him in his last decade came as a token. Veblen was abrupt in his refusal to accept even this nominally conditional surrender.

In 1926, Douglas took up the cudgels for another lost cause when he lobbied the association to finance a memorial to Veblen, whose response was to snap: "I am not dead yet." Veblen proceeded to flaunt the decision of the economics profession to blackball him as a sociologist by demonstrating his virtuosity as an anthropologist and a philologist as well. He produced a translation of a Viking saga which he used as a "dig" to point up a pet theory of his own: how the emergent law of contracts institutionalized the legalized piracy of the Vikings, as well as the slave trade and other future blessings of civilization. These pagan beginnings evolved into the feudal foundations of capitalism, which he found to have secularized the twin principles of sin and servility promulgated by the Church.

Professor Joseph Schumpeter, of Harvard, in his encyclopedic *History of Economic Analysis,* published posthumously in 1954, credited Douglas with "one of the boldest ventures in econometrics ever undertaken." But Douglas's exploration of the inflation deflator in *The Theory of Wages,* published in 1934, during the last decade before the development of the computer, resulted from the pursuit of trails opened for him in Veblen's classroom at the University of Chicago. With uncharacteristic injudiciousness, however, Schumpeter, a grandee of the American Economic Association, ignored the uncompromising initiatives Douglas repeatedly undertook in behalf of Veblen with the hostile majority of their establishment colleagues.

Inconsistency was the least of the criticisms Schumpeter invited by his cavalier attitude toward Veblen. He himself credited the development of capitalism to the processes of "creative destruction," by which he meant the inventions of the prototypical individual entrepreneur. This was a simplistic version of Veblen's description of the institutionalized technological process of capital accumulation and liquidation. Moreover, Schumpeter was at pains to begin his monumental history with elaborate explanations of the connections between economic theories and sociological observations, and to deplore the harm done to both callings by "mutual vituperation." He went so far as to point out that "many of the best sociologists preferred to call themselves something else . . . in order to stress . . . professional competence in the face of the indictment of dilettantism": plain reference to the invidious label of sociologist put on Veblen.

The failure of consensus econometric analysis to sense the portents of speculative collapse in 1987 recorded the success of the economic fraternity in insulating sociology from economics. Arbitrarily, this fatal success outlawed all economic behavior not deemed in line with "rational expectations" programmed into computers. In the intellectual tradition of the Inquisition, the fraternity turned a blind eye to change. Where Galileo merely murmured, "Nevertheless, the earth moves," to induce recognition of astronomical change, Veblen forced recognition of social change by scoffing. The consensus of academic luminaries missed the evidence of the crash in the making during the spring and summer of 1987 because its members operated with their eyes glued to the statistics and with their perceptions numbed.

Schumpeter's influential one-way feud with Veblen was uncharacteristic and unprofessional, as well as inconsistent with his advocacy of recognition for sociologists. His strictures dampened professional interest in Veblen's work when its relevance was on the rise, along with its political impact on the New Deal. The search for Schumpeter's motive has assumed a new importance with the rise of his influence (with a timely assist from lectures I arranged for at Princeton's Woodrow Wilson School dedicated to his broad conception of economics as "historical sociology"). Schumpeter, in his role as the resident elder of the classic European tradition of economic inquiry, may have resented the challenge of the grass roots, home-grown approach to "historical sociology" that Veblen personified. Another more subtle angle is that Veblen represented a personal threat to him. He did, but not only to Schumpeter, and not only on the score of their respective claims to erudition or their rival Continental roots; Schumpeter emphasized the importance of inventions to the process of "creative destruction" to which capitalism owed its growth, but he neglected to credit Veblen with having pioneered the point.

Schumpeter, notwithstanding his scholarly eminence, was given to strutting, and a colorful brag from his youth in Vienna followed him to the summit of the academic hierarchy: a determination to win recognition as the greatest economist, the greatest horseman, and the greatest lover in all Austria. Veblen, for his part, was a natural writer overflowing with inspiration and insight; he was as learned as Schumpeter. But he had no interest in horses, and no need to play the Romeo. Women threw themselves at him as if in heat. He personified the satyr pursued and, once snared, was devoid of any powers of resistance. More troublesome still, taken in passion, he was innocent of any wiles aimed at disengagement from stud duties. He habitually traveled with pockets stuffed full of love letters, inviting jealous peeks from irritated rivals for his affections.

In 1937, well after the dust had settled on Veblen's vendettas, I asked Charles Beard to explain why Veblen had been blackballed for professional status by one faculty after another despite his distinction. Beard had demonstrated in his own career that a professor who won a public for his writings could manage without a campus job. I regarded him as the friend most likely to have a plausible answer, not merely because he had resigned from Columbia in protest over the firing of a militant antiwar socialist professor of economics, but also because Veblen's influence had saturated the Columbia faculty and, not least, because Beard's own contacts with New Dealers of influence were so close.

His reply was that academic freedom had nothing to do with Veblen's troubles over tenure. "The union of faculty husbands and fathers ran out of locks for their wives and daughters," he snorted.

Schumpeter, conservative that he was, had no difficulty accepting and evaluating socialist and even communist theories of economics. He had no hesitation in entrusting assistants who were declared Marxists, or more individualistic types of radicals, with his manuscripts, of which he kept no copies. He was courageous in his praise of Marx's professional virtuosity. He knew enough about Veblen's thinking, as Veblen's zany first wife did not, to know that Veblen was not a socialist (though she herself was, in her idealistic fashion; she was reduced to asking Veblen's students whether he was). On the contrary, Veblen's individualistic radicalism, which defied classification, sent Schumpeter up the wall. He rejected Veblen's effort to develop a distinctive American approach to economic "Sozial-politik": a critical test of Veblen's originality and Schumpeter's mortgage to pedantry. I believe that Veblen represented triple jeopardy to Schumpeter's amour propre—masculine and professional as well as cultural. I deplore the result. It was unworthy of Schumpeter as well as unfair to Veblen.

The "union ban" on Veblen endowed him with a popular mystique. This boomerang effect embarrassed even the most authoritative deans of the economics profession. Perhaps the sociological coincidence of young women entering the economy had something to do with the growth of his audience. The new generation of young women with editorial jobs lionized him. Elizabeth Janeway, a perceptive critic, surmised that women's interest in him was kindled by his willingness to listen to them and to correspond with them, though many of them dispensed with this formality. His need for a chair commensurate with his standing was offset by continual offers of beds to share.

During the generation before the computerization of econometrics took over the dominance of economic analysis academically and commercially, the acknowledged pathbreaker who standardized the official recording of business-cycle fluctuations in America was Professor Wesley C. Mitchell, of Columbia. (His protégé and collaborator was Professor Arthur Burns, the very personification of conventional academic distinction, who graduated into public service under the sponsorship of President Dwight Eisenhower and established himself as a major figure in the staid world of international finance during his long tenure as chairman of the Federal Reserve Board.) Like Douglas, Mitchell was one of Veblen's students at the University of Chicago. Schumpeter acknowledged Mitchell's contributions, but remarked, in another abrasive departure from judiciousness, that Mitchell did not make his mark until he had outgrown "his early associations with Veblenite tendencies . . . in the eyes of the profession."

Mitchell never did. In 1925, when he received an honorary degree from his alma mater, he did not hesitate to beard the faculty lions in their den by paying tribute to "the disturbing genius of Thorstein Veblen—that visitor from another world. . . . No other such emancipator of the mind from the subtle tyranny of circumstance has been known in social science, and no such enlarger of the realm of inquiry." All through his career, Veblen invited words of praise from established figures more readily than offers of employment with status from their colleagues. The Chicago faculty listened with respect to Mitchell. But it started no stampede to draft Veblen to fill a chair.

Schumpeter ignored what Mitchell said about Veblen. He also disregarded the hard fact of Dorfman's remarkable dedication of his history to "the Pioneering Spirit of Thorstein Veblen and the First Born of His Intellectual Heirs, Wesley C. Mitchell." Schumpeter was the unique master of the entire literature. He knew that Dorfman was also Veblen's biographer, but blithely quoted Dorfman's history with blatant insensitivity to Dorfman's wholehearted devotion to Veblen. (Schumpeter's widow, who edited his history with care reflecting naïveté, disclosed Schumpeter's familiarity with the volume of Dorfman's history covering Veblen's work and his intention of citing it, too, if he had lived to complete his text.) But Schumpeter carried his vendetta against Veblen to the point of lumping him with some deservedly forgotten German economist, and he deplored "glaring defects . . . natural and acquired" in both men, clearly a blow aimed below the belt.

Sociologically speaking, Veblen's conception of the business cycle enjoyed a remarkable cyclical progress of its own. Mitchell provided it in the direct line of succession from Veblen. Burns administered it in the direct line of succession from Mitchell. From Veblen's shadowy status beyond the fringe of academic acceptance, his concept evolved into the embodiment of academic authority en route to becoming the official instrument of government power, although Veblen's authorship was buried under the achievement embodied by his disciples who institutionalized it. The profile of American "historical sociology" in the monument he left behind suggests a parallel at the summit of the economics profession with Proust's epic tale recounting the blossoming of courtesans into duchesses— except that Veblen was denied the honors he had earned after his ideas were institutionalized. But his ghost had the last laugh on the academic establishment which had refused to take him seriously during his lifetime. The politicians put his thinking to work even though the academics refused to put him to work.

When Veblen died, a particularly perceptive obituary was written in "Wallace's Farmer" by Roosevelt's future secretary of agriculture and vice president, Henry A. Wallace, who later injected Veblen's concept of the common man into national political debate. In 1933, Wallace represented a great catch for a Democratic administration and, by the same token, a fatal defection for the Republicans, not least because his distinctions included his considerable accomplishments as an agronomist and a plant geneticist. Iowa likes its evangelism to ring to the echo of the cash register, and Wallace had participated in the development of hybrid corn.

Memories of Henry Wallace's contributions have been tarnished by the tawdry and bizarre fiasco in which his career collapsed. In 1948, he let himself be used as the presidential candidate of a "spite," one-time Communist-front coalition ticket against Truman, who nevertheless routed Governor Thomas E. Dewey, of New York. In 1929, however, Wallace commanded enormous political and intellectual prestige in Iowa and throughout the farm belt. He was a third-generation Republican populist publisher-editor whose father had died in 1924 while serving alongside Hoover as secretary of agriculture in the odor of Republican sanctity.

Subsequent Democratic party domination of national politics erased memories of FDR's sensitivity to its handicap as the minority party in the farm belt, which it had been since the Civil War despite the roots of William Jennings Bryan in the area. Where Wilson, like Grover Cleveland before him, had been content to operate subject to this limitation, FDR was at once realistic and imaginative enough to build the Democratic party into the vehicle for a nationwide coalition of a majority more representative than its parochial past suggested.

Wallace was catapulted into national prominence when Roosevelt took over as the mainstream farm belt representative of this new Democratic coalition. His prior enlistment in Veblen's constituency during the 1920s had unfolded as a study in contrast between the Babbittry that kept the surface of American life bland and the Babel that threw its intellectual life into ferment. Accordingly, his subsequent enlistment in the New Deal added enormous credibility, not only to Roosevelt's coalition, but to the intellectual endowment Veblen left for the New Deal to use.

In the obituary, Wallace described Veblen as "one of the first men in the United States to demonstrate that there was such a thing as a business cycle, and many of the men who are now working with the business cycle received their original impetus either directly or indirectly from Veblen." In 1929, the idea of the business cycle was still a relatively bold new hypothesis. Notwithstanding Hoover's eight years on display as the master statistician, reducing the ambiguities of economics to the precision of engineering, no official statistics supported the idea. (The closest approximation came from a widely quoted provincial bank economist by the name of Ayres, who explained the fluctuations of the stock market by correlating them with the recurrent shutdowns and reopenings of steel mills. But Ayres used this technique to anticipate and measure the behavior of the stock market, which everyone knew was erratic. No private substitute for an official gauge of cyclical fluctuations in the economy was available either.)

At the time Wallace paid this tribute to Veblen, American agriculture had collapsed into a slump more severe than any mere business-cycle recession. Veblen attributed the distress of American agriculture, in its fifth year by 1929, to the technological revolution as it spread "poverty in the midst of plenty" (still another Veblenism incorporated into the language). The application of machine, chemical, and especially electric power to the soil exploded agricultural productivity, while its market reach remained static and imports undercut domestic farm prices. Wallace's immersion in Veblen's work led him to see the need for the far-reaching institutional reforms the New Deal engineered in American agriculture in 1933.

Wallace used his cabinet position to implement Veblen's thinking. He picked an econometric adviser at the Department of Agriculture, Mordecai Ezekiel: identified by Professor Lawrence Klein, of the Wharton School at the University of Pennsylvania, the Nobel Laureate of econometric technique, as one of the founders of the method of business-cycle analysis formulated by Veblen. (He also found the first successful political polltaker in Louis Bean, the only pollster to call Truman's landslide victory over Dewey.) I have memories of frustrated meetings with Ezekiel, Bean, Wallace, and their respective counterparts throughout the government at the worst of the 1938 recession. A consistent complaint of these discussions centered around the absence of usable national statistics on physical inventories. Early in the century, Veblen warned that the lack of them would expose the system to recurrent financial and economic mishaps. As this book went to press in 1988, they were in more urgent need than ever, but still missing. Still no one in high office noticed their absence, at the climax of the biggest inventory bubble in history, from the statistical clutter. Such is the routine of progress.

It was a far cry from the 1920s to the 1980s, and it's always a far cry from economic theorizing to the practicalities of how the system works as it changes. In the half century after Veblen's death, the inner workings of the system underwent unrecognizable functional changes; he anticipated the big ones. One of Veblen's analytical firsts evolved as a refinement of one of Marx's shrewder insights. Early in the evolution of the modern system, Marx distinguished between mere recessions in what we would call cyclical activity and full-fledged financial crises, which he attributed to banking breakdowns.

At their root, he found borrowing bubbles, which he identified as bust-prone. He blamed them on excesses of greed by newer breeds of capitalists inflamed by prospects of overnight killings in speculative dealings more exciting than the more reliable recurrent margins of profit squeezed from sweating workers. Veblen, enjoying the advantage of hindsight from the system's later growth, dug deeper and found a compulsion behind its once-in-a-generation borrowing sprees on "abnormal credit" from external sources. Veblen foresaw these compulsive sprees flaring up into climactic catastrophes. The excessive enthusiasm of 1929 not only stripped the gears of the system but also burned out its bearings and its brakes. The protective devices built into the system between the 1930s and the 1980s proved flimsy in the face of the incomparably larger inflation of "abnormal credit" in the 1980s.

The ferment in the laboratory set in motion irreversible shifts within the structure of corporate power. In a logical world, engineering innovations would strengthen the control of production management. But in the topsy-turvy, overindebted world of insupportable overhead costs that Veblen described, the breakthroughs in the laboratory swung the power of decision making within the corporate hierarchy to financial management, and financial management placed its reliance on marketing management to mark up prices and bring in cash to cover borrowings. The effort to inflate profit margins ended, as Veblen showed, in a wasteful inflation of marketing costs and a ruinous deflation of profit margins. Corporations responded to irresistible competitive pressure to finance investment in new plant faster than they could earn back investment committed to old plant.

Breathtaking increases in the cash cost of bringing new products to market shrunk the reach of even liberalized depreciation allowances; Du Pont told *The Wall Street Journal* in 1987 that it had spent a cool billion developing Kevlar over a twenty-five-year span and marketing it into a break-even position. As inflation escalated construction and operating costs, managements relying on depreciation to recapture the inflated cost of plant replacement found themselves left with only enough to cover nominal down payments on new plant. The alternative was to float new debt for new plant on top of debt previously floated to finance plant made obsolete, but not paid for, and/or to water stock not supported by earning power. During the buildup to the 1929 crash, the rate of unavoidable investment was catching up with, and crowding out, the rate of competitive return.

Capital idled in Treasury bills not only was reaping more reward than in productive work but avoiding any risk.

Veblen identified further power shifts within corporate structures that were inviting the entire managerial corps to develop steadily larger batteries of costly outside advisers. Soon consultants on retainer were duplicating the salaried bureaucrats, and vice versa. But all were specialized miniversions of trained ignorance, and from 1929 onward, all contributed to widening the gap between financial headquarters in New York and divisional operations scattered around the hinterlands. In the 1980s, fees to consultants rose with corporate debt, thanks to the antics subsidized by the manipulations of the takeover artists. So, therefore, did the incentive for corporate consultants, led by the investment bankers and the legal specialists in conglomeratization, to devise new methods of inflating corporate debt.

So deep into the system did Veblen's vision reach, and so far into the future, that his satiric tirades against parasitic overheads bloating the cost of doing business applied to the excesses of the 1980s even more than to the 1920s. By then, corporate "departments of public affairs"—and all corporations of consequence were saddled with them—were entrenched as the most vibrant centers of corporate expansion and the most conspicuous centers of waste. Of all the ironical consequences of the technological revolution, none spread more frustration than the stultification of its creative potential into the tawdry favor grubbing of the lobbyists' trade. Accounting and legal departments ran lobbying departments a close second in the inflation of unproductive overhead costs.

America bore no defense burden in 1929. The inflation of overhead and marketing costs, which Veblen identified even then as insupportable, was limited to her commercial economy. In the 1980s, accounting and lobbying had blossomed into the acknowledged profit centers of the defense contracting establishment. Engineering had been demoted to an adjunct of accounting. Even in 1987, a *New York Times* breakdown of the typical advanced technological corporate structure depicted the CEO flanked by departments of finance, production, and marketing, as well as a swarm of engineering advisers, but with no links connecting these advisers to the finance executives, and no place at all for real engineering, that is, research and development.

Built-in overheads, including hourly time sheets to protect the system against larceny, inflated the cost of procurement for even a single spare toilet seat into three figures, and toiletry is indispensable to missilry. Overhead followed from "specs." The more specialized the specs, the more the product cost. The faster the specs changed, the faster costs escalated. As products aborted, overheads took off.

While the technological revolution was still in its inspirational phase, it stirred exaggerated hopes that it would evolve into a money machine spraying enough purchasing power across the economic landscape to guarantee continuous demand for the productive capacities of the system. It ended up working in the exact opposite way in 1929 and again in the 1980s, and Veblen anticipated the cross up. The strains it put on the system turned out to be financial, not commercial. Demand for inputs of capital for future output proved continual. But calls for volume delivery of output proved sporadic. Price increases needed to keep the system running ran it down instead. Vietnam demonstrated that defense spending had lost its power to stimulate the economy.

As the teacher who sent his protégés off on the trail of business-cycle behavior, Veblen was aware of the historical ability of the system to pause for periodic reliquification, which was automatic in the days when business was free to pay its debts down whenever market conditions forced it to lower its prices. But as the student of the technological revolution who tracked the reciprocal inflationary impact of the laboratory on the marketplace and of the marketplace on the laboratory, Veblen saw how the recurrent stops and starts of the business cycle would keep the increasingly capital-intensive system underfinanced and overburdened with fixed costs, reflecting step-ups in overhead costs in general and in selling and interest costs in particular.

When Keynes's time came, he emphasized the high risk of productive investment, arguing that capitalists were always speculating on an unknowable future. But Veblen was the first to explain that new productive assets become obsolete before they earn their keep, so that the only way to keep balance sheets solvent is by continuous investment. Keynes had counseled monetary management of the system by enlightened governments responsibly advised to alternate monetary spurs and holdbacks to business activity. But Veblen cautioned that the technological revolution was inflating the entire

American system into a macroversion of its most advanced, capital-intensive plants, which were engineered to run around the clock or not at all. He saw big government inviting big businesses to grow beyond their ability to manage or to finance, and he foresaw how big government, under the prod of demand for weaponry, would outgrow its capabilities too—with neither government nor business knowing what the other was doing, yet with each looking to the other for a bailout.

A provocative parallel connects the messages of Veblen and his revered contemporary, Justice Louis D. Brandeis, the fountainhead of the stream of wisdom that flowed into the New Deal from Harvard. Superficially, no connection could have seemed more unlikely. Brandeis always had an eye on the main chance at the bar of justice, whether on the bottom line or pro bono publico. Rich and autocratic behind a façade of austerity, he operated at the apex of the power structure and, after a career punctured with controversy, ended up respected everywhere. Veblen never had a power base or even a title, and always lived in squalor. The annals reveal only one meeting, an awkward abortive audience in Brandeis's chambers on the occasion of Veblen's publication of *The Nature of Peace*. No doubt the venerated elder reacted with impatience to any suggestion that he might benefit from help on any subject, let alone share his undercover pipelines to publicity with an interloper who contrived to magnetize the media on his own. So no sparks flew, nor were any private tidbits left for history to chew on.

On scrutiny, however, an arresting interplay is discernible, centering around the fire each of these giants turned against bigness beyond responsible knowledge or behavior. Brandeis, typically moralistic, branded bigness as inherently bad. Veblen, typically sardonic, caricatured its operatives as incompetents. FDR, who had a nickname for everyone in his ken, dubbed Brandeis "the Prophet Isaiah." Whether or not this name was apt, Brandeis was indeed a prophet honored in his time. Veblen, by contrast, though hailed as a prophet from time to time, was never honored as one. Both shared the status of ethnics rejected by the WASP establishment—Brandeis more obviously so as a Jew, though less stigmatized because born in Kentucky and harvested from an early crop of immigrants; Veblen, at the height of his career, despite his biting English prose style, still suffering indignities at the hands of oafish colleagues, such as the one who asked deferentially whether his latest book were a translation from the Norwegian!

Brandeis taunted the new conglomerates of his day as dinosaurs, unable to perform as profitably as smaller independent specialists and constitutionally vulnerable to hardening of the corporate arteries. Veblen lambasted investment bankers for indulging insupportable inflation in the ratio of sales and administration costs to barebones production costs and, consequently, in the cost of living and of doing business, even while labor and materials costs were stable, as they were all through the 1920s. Each of these dissenters from the status quo provoked criticism as an iconoclast before winning posthumous confirmation as a realist.

But the parallel did not stop with the warnings Brandeis and Veblen issued against the deficiencies in the growing system. Both roamed far afield from their perceived pastures of expertise. In their respective ways, they broadened and cross-pollinated their constituencies. Both opened the restricted precincts of economic analysis to popular debate; they filled new forums with new audiences. (Brandeis introduced statistical exhibits into legal brief and opinion writing.) Each also, in his individual, eloquent, and controversial way, blended his vision of a restructured economy with a blueprint for peace.

Wisdom on statesmanship is not expected from experts on economics. Nor are statesmen expected to know an interest rate from a dividend rate. Extracurricular excursions of this sort, and the controversies they stir up, provide further illustrations of the provocative effects of improper behavior. Veblen and Brandeis were the first to draw speculation about peace into the orbit of economics. When Wilson proclaimed his Fourteen Points for peace in 1918, he established himself as spokesman for the world, but he failed to mention economics, like all his great predecessors in the history of political philosophy.

The two giants, after spending their lives swimming in separate channels, were swept together by the currents of history to germinate the principal intellectual influence on Roosevelt's presidency. Both thinkers were clear in their anticipations of a double challenge to the crisis presidency: for domestic institutional reform and for resolute

mobilization of resources to deal with the scissorslike threat represented by imperial Japan and Germany en route to nazification.

Veblen, "that visitor from another world," in Wesley Mitchell's
phrase, never left it. He was content to let his ideas travel to this
world as his ambassador. They serve as practical guides through the
maze of complications he anticipated. Their passports are subject to
no expiration dates.

5

HOBSON'S HEIRS
Economic Man in Political Custody

Inquisitors risk infection with the germ of the very heresy they target for extermination. The elders of Athens sentenced Socrates to a ceremonial suicide as punishment for "corrupting the youth" by asking too many questions and then proceeded to commit political suicide themselves with their everlasting arguments about the answers. When the inquisitors forced a recantation from Galileo, they had no way of knowing that, a mere four centuries later (no time at all as the theological crow flies), many Catholic educational institutions would sport observatories among their crown jewels. When Stalin's inquisitors imposed his simplistic Marxist orthodoxy in Moscow, they set up every victim of the KGB as a potential folk hero a generation later.

If victims of inquisitions do ultimately infect their inquisitors, Veblen was Typhoid Mary: the American system adopted his barbs as its reforms. Ironically, his attack on capitalist orthodoxy misled the proprietors of Communist orthodoxy into welcoming him as an ally when they took over in Moscow. They would have been confused to discover that he put the encrusted beliefs of Marxism on the autopsy table too. Nevertheless, Lenin, soon after taking over, sought out Veblen as the American oracle denouncing the corporate establishment, without realizing that Veblen was not even a socialist, just an uninhibited realist.

Lenin revered Marx as the last word in modern thought and believed as an article of faith that after Marx, only commentators were needed. Indeed, Marx had brought the broadest view yet seen in his day to economic analysis. But Marx sold himself short when he described his system as an amalgam of English economics, French politics, and German philosophy. He certainly earned the right to add his own incomparable mastery of history, and his analytical creativity, to the product mix. For the sake of a phrase, he narrowed these three diverse national cultures into specialized miniatures.

Veblen pulled this neat thinking apart, demonstrating the absurdity of these stereotyped national attributes. In Marx's time, Germany had already demonstrated its proficiency in pursuits far more material than philosophy and more practical than English economics. Veblen noted that Europe was hardly spoken for by Marx's "Big Three"; and that world politics, after Marx, had outgrown European politics. Even more urgently for Russia as well as for economics, Veblen's vision embraced the emergent engineering arts, whose impact went completely unnoticed by every other economist, except Schumpeter but including Marx (although Marx's focus was fastened more on the exploitation of raw labor than on the uses to which capital could put engineering). Marx used his interdisciplinary approach to forecast victorious revolution in Germany—with no success. Veblen used his interdisciplinary approach to search out the conditions for peace—with prophetic success. In the process, Veblen hit upon the ambiguity built into Russia's backwardness, which made it ripe for revolution but unappetizing to invasion. Veblen saw this backwardness for what it was: a crippling economic liability, but a decisive military asset.

Lenin misread the scenario unfolding for Russia that Veblen got right. Lenin had no doubt that the Marxist prophecy would prevail, but was petrified that his revolution would fail. He made the most noise from his perch in exile insisting on strict observance of enshrined text as decreed by Marx, never more so than when he was rewriting Marx himself. Reactionary ideologues provoked his contempt. But while he was writing about how to change the system, his former socialist comrades were beginning to change it by winning elections all over Europe. The administrative responsibilities they took over left them with no time to bicker over slogans to which no unchained wage slaves were about to march. Lenin in his garret had nothing to do except carp at compromises forced on socialist officeholders by the pressures of power sharing. His dogmatism remained unshaken by the remarkable resilience of the capitalist system in recovering from its recurrent slumps in defiance of the death knell Marx had sounded for it. Marx had proclaimed the imminence of revolution in 1848. Yet the turn of the century found capitalism going stronger than ever, apparently thriving on the better living conditions it delivered to the members of its socialist opposition— with a powerful assist from the efforts of the socialists themselves to reform the system. Veblen was the one realistic observer of these crosscurrents, who foresaw at the very outset of the Russian Revolution that both Washington and Moscow would grow stronger as they clashed, each system evolving toward the other while professing fear of the other.

Marx left a hallowed legend of his days in the British Museum researching the genesis and conceptualizing the shape of the revolution fated never to come off for him. His fiery polemics diverted attention from his temperamental unfitness to lead one. He was incapable of organizing a two-carriage funeral, let alone a country delivered to him on a platter by its enemies, as imperial Germany delivered Russia to Lenin in 1917. The kaiser's generals perpetrated one of history's most lunatic flukes by sending Lenin on a sealed train through the front lines into Russia as their guided missile for dismantling the czarist regime in order to free their own troops for massive redeployment to the western front.

To defend Marx, Lenin employed a simple technique: *dialectic,* which viewed every regime as setting the stage for its opposite. Accordingly, he insisted that every historical development which had proved Marx wrong was about to prove Marx right. He proposed to accomplish his defense methodically, subject by subject, citing one non-Marxist text after another for "objective" support. This plodding, literal minded dedication of Lenin's exposed him to the influence of advanced, non-Marxist economic thinking in England and the United States. Once Lenin took over in Moscow, his interest in this thinking, in turn, added a new dimension of impor-

tance to these economic thinkers, especially the one who led him to
Veblen.

Of the intellectual mentors Lenin found useful in validating his
master's work, by far the most important was John A. Hobson.
Hobson stands in the great tradition of independent British writers,
typified by Locke and Hume, who survived rejection by the academic
establishment and won acceptance from the emergent democracy of
literate, opinion-making, action-generating readers at large. I recall
with veneration and gratitude his Sunday evening "at homes" in
Hampstead for students on the loose in search of an intellectual
haven from the ravages of depression and the clang of rearmament.
The memorable history lectures of R. H. Tawney, his son-in-law, and
Eileen Power, at the London School of Economics, provided the
most receptive recruits.

Hobson developed a twin set of observations on the visible
change capitalism had undergone in the industrial countries of
Europe after Marx's death. His first perception centered on the
endemic difficulty of capitalism: the failure of domestic purchasing
power to keep pace with productivity. The second—imperialism—
gave Lenin his alibi for Marx's failure to anticipate the success of
capitalism in spreading wealth as well as accumulating it. Hobson
identified imperialism as the engine of war, and Lenin blasted the
socialist leaders as accessories to it.

Lenin treated Hobson's classic critique, *Imperialism*, published
in 1904, as the unimpeachable sourcebook he drew on for his own
tract of the same title, published in 1917, which became the bible of
Bolshevism. It supported Lenin's central premise that the apparent
growth of the system was building up to a greater fall than even
revolutionary critics had anticipated. Lenin had no inhibitions
against paying his intellectual debts. Like Marx before him, he
turned his sources into intellectual references, bragging about them
as assets.

The principal American independent economic thinker whose
work intrigued Lenin was a mentor of Hobson's, to whom Hobson
had looked for guidance for as long as Lenin had looked to Hobson.
It was none other than the ubiquitous Veblen, roaming the world's
fertile intellectual underground. He and Hobson shared, among

other qualities, a common resentment against the economics establishment for rejecting them.

Lenin, as the pamphleteer in exile, had seized on Hobson, the voice of documented and conscientious dissent, as his authority for pursuing his argument against capitalism. As the revolutionary in power, he tried to use Veblen for the briefing he needed on America. Dorfman, in his biography of Veblen, described a meeting shrouded in mystery between Veblen and Ludwig C. A. K. Martens, an engineer stationed in New York by a Russian company, whom Lenin had just named the Soviet Union's representative to the United States. The fact that Lenin took less than two years after the Bolshevik takeover to approach Veblen showed how much importance he attached to Veblen's thinking, considering how chaotic Moscow was, how difficult its overseas communications were, and how determined Washington was, having rejected the League of Nations, not to deal with the Bolshevik regime. Martens and Veblen were brought together by an anonymous intermediary who reported to Dorfman that the lunch started out as a silent disaster and culminated with "one keen, orderly mind excavating another." Dorfman took the identity of his mysterious source with him to the grave.

Veblen's message about Russia was that the country's backwardness left it immune to the post-Armistice military intervention aimed at overthrowing the Bolshevik regime: a prophetic insight into Vietnam's subsequent immunity to technologically advanced military intervention for the same reason. His message about America was that the technological revolution had put her fate in the hands of the engineers, who were solidly and safely on the side of their superiors. Moscow, in the early days of the revolution, was on edge awaiting word that the revolution was on its way in Europe, beginning of course in Germany. Veblen threw cold water on such naïveté, explaining that America would refinance Europe and pull it out of its postwar crisis.

Whether or not Lenin learned about Veblen from reading Hobson is not clear. Certainly, Hobson wrote enough about Veblen, and as a devotee, to have made it possible. True, as an earnest British dissenter within the evangelical tradition of reform, Hobson confessed to being troubled by the peculiarly American irony of Veblen's style in digging up the social roots of injustice, as if profundity were always expected to express itself with solemnity. Hobson grieved, believing that Veblen's stylistic preference for the needle over the hammer "postponed and even damaged his legitimate reputation as the keenest social thinker of our time."

But Hobson was very clear that the pragmatic thrust of Veblen's thinking centered on a premise of vital importance, and not only to every strategist in the Kremlin. Veblen's vision fastened on the distinctiveness of America's evolution into the power she was becoming in the world. As Hobson put it, "Veblen had his eyes set on a distinctively national economy for the United States." The Kremlin's decision makers have lived with this hypothesis ever since Bukharin signed his own death sentence by formulating his thesis of "American expansionism." America's economic ideologues have yet to fathom it.

Tragically, America's economic ideologues do not understand the workings of the system they are responsible for managing. Pathetically, Russia's economic managers do not understand their own history. No doubt they have an alibi: they were brought up on Stalin's criminal distortion of it. To their credit, they have been trying to tear off the blinders with which the Russian people view the Stalin legacy. But Gorbachev and his closest advisers still look back on the Lenin legacy with blinders, unaware that Lenin, the holy father of the Soviet Revolution, was practical enough to perceive that Veblen's message for America offered a by-product message for Russia under his rule. Abel Aganbegyan, Gorbachev's economic spokesman, makes no reference to Veblen in his American book, offered as an exercise in *glasnost* on *perestroika.*

Yet for decades, every schoolchild in America was reared without being told so on the evolution of the very distinctiveness that Veblen identified and that Hobson noted. Henry Clay asserted it in his proclamation of "the American system," in the days when canals were as pivotal as airports are today. Professor Merrill D. Peterson, of the University of Virginia, in his monumental study of *The Great Triumvirate: Webster, Clay and Calhoun,* defines Clay's American system as "the Hamiltonian system enlarged" to accommodate America's turbulent coming of age, and Clay himself as the father of the Monroe Doctrine. Lincoln proclaimed himself the disciple of Clay. Hobson stumbled onto Veblen's economics as the formulation it represented of the main line of America's political development.

Clay, arguably the most creative of all American master politicians, entirely innocent of the precepts of classical economics, was practical enough to seize upon the single feature that was to set off the American economy from all other industrial powers: its ability to prosper without dependence on exports. Neither the classical texts, the neoclassical texts (including Keynes), nor the socialist and communist texts made allowance for the possibility that the most important of the industrial powers would not be export-dependent. Veblen never wrote a full-fledged economics text and, as we know, was derided as a mere sociologist. If the politicians and economists in vogue during the late 1980s had given themselves the benefit of studying the mainstream of American history, or of sampling the writings of Veblen, they would have thought twice before turning the country upside down by devaluing the dollar with the idea of exporting America back to prosperity.

In the 1930s, Hobson, as Veblen's interpreter overseas, conveyed Veblen's theme of American distinctiveness with the special authority he commanded among the communist intelligentsia as Lenin's acknowledged non-Communist mentor on imperialism. History has lost track of the prestige carried by Hobson's name in those years of revolutionary turmoil. Lenin's self-appointed mission was to teach the living; he admitted learning only from the dead, with Hobson the notable exception. Hobson, insulated from any political intrigue, enjoying the run of the dissenting intellectual establishment (with familial ties to the *Manchester Guardian*), was the one eminence exempted from Lenin's strictures against capitalist apologists.

The remarkable influence of Hobson's opinions went beyond his teachings on imperialism, which Lenin absorbed, to his explanations of Veblen's insights encapsulated in his biography of Veblen. Hobson's decision to choose Veblen as the subject of his own swan song qualified Veblen for posthumous legitimization in the great tradition of creative intellectual emancipation. The line of Veblen's analytical descent overseas is clear: in one direction, through Hobson to Lenin, who knew it and boasted of it; and, in the other, to Keynes, who evaded the responsibility, though he finally acknowledged Hobson.

The publication of Hobson's *Veblen* in 1936 coincided with Keynes's belated discovery of Hobson as his lifelong ally in Britain. This was the very year in which Keynes published his definitive work, *The*

General Theory of Employment, Interest and Money. Yet Keynes lavished his praise on Hobson without recognizing Veblen as his own predecessor in Washington. By that time, Keynes had superseded Hobson and Veblen as the most persuasive voice for communist ideologues to heed as the advocate of capitalist economic policies aimed at a middle course between laissez faire and socialism.

Keynes's title is a misnomer, because the book was meant as a specific monetary antidote, desperately needed at the time, to the stagnation of the world economy after the Depression but before the war. No doubt Keynes intended *The General Theory* as an exploration of the theoretical foundations of the entire system, and its contributions did go as far as any purely monetary theory could go. *The General Theory* gave Keynes his start toward endowing the postwar world with its monetary system and gave Keynes the right to preside over the birth of the system at Bretton Woods. But Veblen's enormous impact on the New Deal helps explain the failure of Keynes's initial foray into Roosevelt's Washington. Keynes's subsequent adoption by the New Dealers confirmed his captivity to Veblen's ghost.

The publication of *The General Theory* represented an exercise in redemption for Keynes. By his own account, he had spent the six critical pre-Depression years, from 1924 to 1930, perfecting his *Treatise on Money,* only to publish it as an admitted exercise in failure, on two counts: it neither headed off the depression he anticipated nor clarified his views on money to his own satisfaction. When the crisis did strike and Roosevelt's activism took over from Hoover's do-nothingism, he nonetheless regarded himself as the obvious choice to advise Roosevelt. Justice Felix Frankfurter (as he was soon to be), the broker-plenipotentiary of the New Deal, obliged by arranging a momentous audience in the Oval Office that moved from deceptive warmth into a glacial chill.

Keynes had spent much of his career chiding traditionally gaited politicians who plunged into stagnant economic pools. He had endured the frustration of watching Winston Churchill wallow in them: witness his classic polemic, "The Economic Consequences of Mr. Churchill," provoked by Churchill's blunder, as chancellor of the Exchequer, in rushing England back onto the gold standard just in time for the credit squeeze this move precipitated to provoke the traumatically divisive General Strike of 1926. Consequently, he welcomed FDR's muscular response to financial crises tolerated by conventional operators, and his many expressions of enthusiasm for FDR have encouraged the impression that Roosevelt welcomed his initiatives and embraced his views.

Sir Roy Harrod, Keynes's faithful disciple whom his elder brother, the eminent surgeon and bibliophile, authorized to write Maynard's biography, documented this belief by publishing a typical FDR one-liner, dashed off to Frankfurter immediately. "I had a grand talk with K. and like him immensely," it said, and, frustratingly, that was all it said. Frankfurter, who was a past master at the art of reading volumes into grunts, let alone into thank-yous, used FDR's note as a calling card for Keynes, without the slightest concern for where, in the diffused and divided New Deal talent sprawl, Keynes might pick up any abiding influence. Harrod admitted that his own clumsy polling about Keynes's White House visit left him entirely in the dark. Frankfurter's prominence and his knack of playing the busybody encouraged his reputation as a serious student of the economy, though in fact he was entirely ignorant of its workings. But his role on the Court encouraged his vocation, which was that of a talent scout—especially for his always active English clientele from Oxbridge via the City of London, or, in Keynes's case, Bloomsbury as well.

"Felix's English connection" was a standing joke at FDR's floating poker game, though FDR was vigilant in keeping Frankfurter's flesh-brokering under control. FDR was a past master at the delicate game of stopping payment on blank checks before they could be cashed. He chose his well-meaning but ineffective secretary of labor, "Madam" Frances Perkins, whose sophistication in economic matters was on a par with Frankfurter's, to pass the word that Keynes was strictly on his own in New Deal Washington. She recalls in *The Roosevelt I Knew* that he told her after he saw Keynes, "I saw your friend Keynes. He left a whole rigmarole of figures. He must be a mathematician rather than a political economist."

The unmistakable Rooseveltian jab of the needle showed in this reference. No newcomers to the inner circle were routed by way of Fanny Perkins, and she was patient in her role on display, rarely for use. As for Keynes, he fell off more political horses during his interwar years in the wilderness than the Prince of Wales toppled from saddles between bouts with the bottle. But he learned enough about the turf that economists and politicians share to enrich us with the dictum that the ultimate test of the wisdom of an economist is in the quality of the political clients he attracts. (It could be argued that the bottom line is how successfully an economist takes his own advice. By this test, Keynes ended well. Veblen failed miserably, not merely because he died before the crash, but because if he had lived to see it confirm his forebodings, he would have had neither the inclination nor the money to clean up by selling the market short.)

Back in Washington, FDR knew what he was doing. He could not follow the lingo of economists, and he had no intention of trying. His way of dealing with the fraternity was to play the field. But the least practiced eye could have seen in those early years that Roosevelt, fortified as the beneficiary of Veblen's work, had taken office free to take advice or leave it. To what extent, even Roosevelt was not aware: he took his favors from economists where he found them, and ignored intellectual copyrights. He certainly had no intention of learning how to govern by reading Veblen's writings or anything else not written about himself (though his conservative critics harbored the deepest suspicions that he was an avid reader of the "English socialist," Keynes).

Keynes himself was naïve enough about the strange workings of American politics to believe that he had to know the president to influence the White House or that he could influence the president by impressing his ideas on a member of the Cabinet. He labored under the illusion that its members participated in presidential power sharing in the same collegial fashion enjoyed by members of foreign cabinets with independent parliamentary power bases. There's no doubt that Keynes did his best to sell his Depression remedy in Washington on its merits, but that the American political turf was too tricky for him; how the unlikely corridor of access for Keynes to political Washington led through the back door, but not of the White House, is described in Chapter 10.

By the same token, Keynes did persuade the New Deal economists, who had come to Washington indoctrinated with Veblenism, to adopt the specific monetary remedies he proposed for the Roosevelt Recession of 1938. But they greeted him as a fellow outsider and dissident. They, too, were frustrated by their own political ineffectiveness in Roosevelt's Washington after the New Deal had lost its initial enthusiasm and before the war inspired the dedication to victory. By that time, Roosevelt had lost his successive fights to pack the Supreme Court and to purge Congress of "reactionary" Democrats. While he sulked over his loss of control over Congress, and the economy stagnated, programmatic initiatives from his economic advisers seemed pointless.

Yet through it all, Roosevelt showed a native flair for anticipating the future he unknowingly inherited from Veblen via his Brains Trust; together, they politicized economics. Roosevelt himself marked a milestone for the history of econometrics after his time, as well as for the practice of economics in his time, with his casual expression of frustration with Keynes for *not* being "the political economist" his situation called for. With this negative throwaway line, he demonstrated how much the most gifted of economists can learn from the most untutored and impatient of their political clients.

A hint of mystery, revealing Keynes's own uncertainty in place of his normal assurance, clouds his treatment of Hobson. His earlier passing recognition of Hobson in the *Treatise* was reluctant to the point of being grudging. The contrast with his later treatment in *The General Theory* could not have been more striking. He saluted the English biographer of Veblen with reverence: "For nearly fifty years Mr. Hobson has flung himself with unflagging . . . ardor and courage against the ranks of orthodoxy." But he never mentioned Hobson's connection with Veblen.

In 1936, Keynes hailed Hobson's long-forgotten *Physiology of Industry* as "an epoch in economic thought," noting that its heterodoxy had frustrated Hobson's hope of an academic career. He did not volunteer an explanation of why he himself did not use his enormous clout at Cambridge to get Hobson recognition there. In *The General Theory*, however, Keynes embraced this pioneering work by Hobson as the first reasoned and systematic attack against the fetish of saving for the sake of saving. With this acknowledgment, he paved the way for the thesis he subsequently developed himself: oversaving is the cause of underconsumption and of the investment lags that bring sags in employment.

But this mystery around Keynes's long delay in recognizing his indebtedness to Hobson, and his fulsomeness when he did, is the least of it. His contradictory treatment of Hobson does not explain his unawareness of Veblen, though his belated tribute to Hobson does include passing references to marginal, mercifully forgotten, crank spokesmen for underconsumption theories, which always gain currency in hard times. Keynes, in all fairness and without the malice shown by Schumpeter, may have had a plausible reason for ignoring Veblen—beyond the temptation to jealousy, if indeed Keynes had been aware of Veblen's influence in America.

Though he lived to see Hiroshima, Keynes's professional preoccupation was entirely with monetary methods. The levers he trusted to move the world were financial, not technological. True, no one could have put greater emphasis on the monetary impact of the investment function than Keynes. Nor has anyone else developed a comparably devastating evaluation of the escalating costs to economic society from any compounded shortfall in the arithmetic of investment. Of all the criticisms hurled at Keynes at the height of his controversial career, none distorted his achievements more grossly than the know-nothing caricature of him as a "spender" insensitive to the investment process.

The realistic criticism is not that Keynes underestimated, or even ignored, the investment process; it is that he did not extend his scrutiny to how much work investment dollars would do, beginning in the sectors of the economy where they were put to work, and radiating throughout the system. But this criticism is not limited to Keynes alone. The thinkers who came after Veblen were inclined to be specialized; Keynes's area of specialization did not extend to the sociology of the production process as Veblen respectablized it for economic analysis. Veblen saw that the laboratory, once unleashed on the production process, develops a momentum of its own, exactly as markets do.

In the end, comedy always wins over callousness. Keynes may have been serious in thinking that he had the last word in the economic policy debate of his time. But the war was to make it moot. He was introduced to the Washington scene too late to have a chance at the last laugh. Veblen's ghost enjoyed it.

Keynes was in a downbeat mood at the end of *The General Theory,* when he delivered himself of his famous bid for proprietary influence over the pattern of ideas embedded in the structure of society. "The ideas of economists and political philosophers," he wrote, both when they are right and when they are wrong, are more powerful than is commonly understood. Indeed the world is ruled by little else. Practical men, who believe themselves to be quite exempt from any intellectual influences, are usually the slaves of some defunct economist. Madmen in authority, who hear voices in the air, are distilling their frenzy from some academic scribbler of a few years back.

This lament was no doubt meant to be a boast about his own influence; it proved to be Keynes's last hurrah. But he was not aware of the fact that some no longer audible voice had already preempted his preachings. Veblen was the forgotten "academic scribbler of a few years back," already enshrined as the shaper of the New Deal that Keynes himself aspired to be. Keynes had never been in the running to be drafted as the architect of the New Deal. Moreover, the analytical tools Veblen left behind him were doing double duty not only on the financial side, which turned out to be merely supportive, but on the technological side, which proved formative. This combination of functions was a distinctive American phenomenon, portending America's status as the military superpower with the malfunctioning supereconomy. Throughout, as Keynes did not anticipate but as Veblen did, it was the technological factor which forced the investment pace, independently of both consumer demand and financial conditions, and not the financial factor.

Politicians at the end of the twentieth century, confident they had absorbed Keynes's message, as Nixon bragged he did, still failed to grasp Veblen's earlier message that technology was driving the system into debt faster than they could lead it there. But well before Veblen died in 1929, he spelled out the plain meaning of economic development for the rest of the century. Engineering, harnessed to transportation and communications, would provide the infrastructure for interrelated commercial and military functions of the economy, superseding decision making by personal motivation. Enough of Veblen's insights won institutionalization during his lifetime, and by the New Deal, to obscure this central theme of his thinking from posterity.

The reason for the oversight was built into the culture of the interwar years. Hoover in the White House personified the distinction between Veblen's impact as a teacher and his neglect as a prophet. Publicity exploiting Veblen's discovery of the engineer landed the Great Engineer in the White House, but the challenge of policy-making revealed Hoover's innocence of technological reality and guaranteed his political failure. He saw no threat to America's security brewing in Germany and Japan, as Veblen had a decade earlier.

Although Hoover was quick to express dismay over Japan's invasion of Manchuria in 1929, he based his censure on morality, not national security. So did his secretary of state, Henry L. Stimson; in his political resurrection as the secretary of war Truman inherited from FDR, it was he who made the recommendation to drop the atom bomb on Japan in 1945. By then, however, the war had vindicated Veblen's thesis that technology—especially in transportation and communications—sets the pace of military and economic development. Reflecting this thesis, consciously or otherwise, Stimson defended his initiative against the outcry at the time on the grounds of technological uncertainty over whether the ultimate weapon would work. In 1929, both Hoover and Stimson had taken false comfort from the spectacle of Japanese foot troops slogging in the opposite direction from Pearl Harbor.

Where Keynes had satirized politicians as the mindless instruments of "defunct economists," Veblen had shown their hands forced by the blueprints of contemporary engineers. In the Britain of 1929, Keynes was mounting his polemical offensive for a spending program to counter the slump guaranteed when America stopped subsidizing the Versailles structure of reparations and debts. This calamitous American decision, pronounced inevitable by Keynes himself, gave Hitler's demagoguery the ring of legitimacy that sent him on a beeline to power. Nevertheless, the British debate over domestic spending droned on, oblivious to Hitler's threat.

In Washington, as late as 1939, Roosevelt indulged the political luxury of keeping an articulately isolationist former governor of Kansas on display as secretary of war. I remember vividly what passed for his general staff: two lieutenant colonels awaiting retirement, sharing an office with no staff or even files, lolling over the stock tables in the newspapers. When Keynes won his case for spending with the New Deal, he made it, not on military grounds but, on the contrary, as a civilized and humane alternative. The result of his belated policy victory proved irrelevant to economic performance but critical to political competition. Efforts to increase spending contributed significantly to Roosevelt's narrow third-term election victory in 1940; he campaigned claiming credit because, in his eighth year, unemployment was less than 15 million instead of more. The subsequent history of the twentieth century stamped Keynes's argument for spending as a sociological specimen. Technology deflated spending as the solution to stagnation.

Historians have forgotten the flamboyant political theory of continuous communist revolution advanced by Trotsky and outlawed by Stalin. But they have come to take for granted the continuous technological revolution that was still gathering momentum a century after Veblen's birth. They will do well to remember that Veblen, writing in the twilight of the era of the candle and the horse-drawn carriage, anticipated its pace, its scope and, above all, its explosive, recurrently disruptive financial impact. All thinkers in the great tradition of economic inquiry have recognized the triple distinction among transactions in consumer goods, in capital goods, and in financial instruments. But Veblen was the first to note the particular capital-intensiveness of the modern transportation and communications industries, as well as their continuous, open-ended exposure to the financial demands generated by the technological revolution.

The transportation and communications industries provide the infrastructure of superpower military reach, as well as of modern economic behavior. They also determine the definition of money because they move it. To paraphrase the old adage, money is as money does. It works only as fast as it is able to move, but it does work that fast. All parties to the monetary argument have agreed from its outset, when money was still minted, that the supply of money receives an extra boost with each step-up in the speed of its turnover: an arresting exception to the law of supply and demand, which decrees a drop in price for each increase in the supply of any product. The market for money is the basic market of the system. Nevertheless, during the early 1980s, the cost of money rose with its supply. Aberrations in the money market revealed distortions not accounted for by any theories or regulatory practices. Consistent with this reversal, and confirming the perversity of market behavior under crisis conditions, in 1985 debts rose while prices—including the price of money—began to fall.

Although the speed of money turnover varies with credit demands, its movement is limited by the technology of transportation and communications. As fast as the technology of transportation has managed to carry money, as fast as the technology of communications has taken over its transmission, the value of the work any unit of money can do has multiplied, and its visible supply simultaneously available in all money markets has multiplied, too. Money limited to the speed of canal boats or stagecoaches, and loaded onto them in strongboxes, did less, and in fewer places, than money received by instant push-button transmission across oceans and time zones, working three or four shifts a day around the world, and breeding more on each jump without losing any residues in transit.

In 1919, Veblen wrote that handfuls of technicians, rather than masses of laborers, as Marx and Lenin had supposed, control the nerve centers of modern communities. Since then, the financial transactions of these communities have come under the same centralized switching-equipment control as round-the-clock operations of basic utilities. The "invisible hand" envisioned by Adam Smith as orchestrating ongoing auctions between buyers and sellers asserts its grip on computerized transactions in a more literal sense than he could possibly have imagined: manipulating electronic brains and moving push-button money. Schumpeter expressed astonishment because Hume, himself a formidable explorer into the mysteries of markets, took "as late as 1752" to stumble onto this new invention of paper money. Push-button money has superseded paper.

Relatively few of the millions of people on payrolls are aware that the checks they write are routinely wafted into their bank accounts with the speed of light by electronic carriers as invisible as angels were ever believed to be. Even fewer of the more sophisticated cross section owning U.S. Treasury bills suspect that their huge flows, continually expanding and turning over, exist in the moneyed world as so many master-computer entries. No owners of Treasury bills, smug and secure in their havens, can finger this ultimate symbol of safety with earning power built into it and claim it as a tangible asset—like a dollar bill. Let some deranged technician, unable to reach for a nuclear missile control switch, erase the combined memory of the batteries of computer systems sorting out the Treasury bill float. Any such financial Hiroshima would visit more instant nonviolent, extralegal expropriation on the unsuspecting money owners of the world than Marx ever imagined in his angriest exhortations.

Altogether, Veblen demonstrated how the system fluctuated before the technological revolution changed it, and anticipated how the technological revolution would change it. Specifically, he identified the transportation and communications industries—along with the defense industries interconnected with them and based on them—as the carriers of the seed of the technological revolution. All three sectors of the American economy accentuate its distinctiveness. They operate in the private sector; yet, despite their depreciation allowances, each of them accumulates capital needs faster and more regularly than swings of the business cycle permit investable reserves to accumulate, whether from earnings during recoveries or holdbacks during recessions.

In America, compulsive investment input outruns any hope of competitive cash return: not merely for industries directly related to defense, but for defense-support work throughout the infrastructure of economic society (particularly to do with transportation and communications). Under the accelerating impact of their escalating capital needs, the technological revolution has turned business-cycle recessions into anachronisms that the system can no longer afford or, therefore, tolerate without suffering inescapable structural damage. Fixed charges mount up in these capital-intensive industries even when revenues fall off; and the charges, like the cash needs that go with them, cannot be avoided because sales are lost.

Economists who still welcome recessions as automatic corrections of excesses in the system are mired in its monetary history. Despite the transformation the technological revolution has engineered in present-day financial practices, the economic consensus has been as slow to grasp the economic consequences of the technological revolution as it was to accept the commercial consequences of the business cycle in the first place. Nor, in conventional terms of business-cycle analysis, have they learned to trace the chain reaction of interest rate inflation. It starts with the escalation of continuous investment demand that is unavoidable. It results in a deflationary backlash on the competitive attractiveness of the returns that capital-intensive corporations can pay. Their engineers routinely ready new generations of products for activation, or for modification and, almost immediately afterward, for scrapping as obsolete, with no sensitivity to considerations of cost or return.

Yet the dependence of industry on relative handfuls of engineers—in research as well as in operational and maintenance categories—has grown just as critical as the dependence of society on them. So has the sensitivity of social opinion, of legislative proceedings, of public health, and of military security, to the transportation and communications bottlenecks and breakdowns that tax the limits of tolerability whenever investment in either sector is slowed down or sped up. Road or elevator repairs, to take a common example, stir up as much irritation as the failure to make them.

The special category awaiting investment in regulated industries whose bottlenecking modern society cannot permit escaped the scrutiny of Veblen's predecessors for a simple reason: railroads and utilities enjoyed high visibility, which led them to be taken for granted. Capital investment in these industries accounted for such high ratios of total capital investment for so long that the parts seemed indistinguishable from the whole. The importance of these industries accentuated another source of structural distinctiveness for the American economy. Other countries, accustomed to coddling their capitalists, socialized their losses from the exponential inflation of investment requirements in transportation (especially airlines) and communications technology; even though these nominally capitalist countries cut back their historically influential defense industries from workhorses in their domestic economies to racehorses for their export operations.

The crossing of the curves between inescapable investment outlays and affordable investment returns has forced governments to assume responsibility for increasing portions of the investment burden—even in countries without America's defense burden, yet saddled with onerous unemployment costs and with irreducible import costs for materials and food when exports fail. Only America has relied on rigidly regulated corporations to finance ongoing investment essential to the functioning of the infrastructure. Only America has been saddled with continual defense burdens escalating independently of the earning power of the economy.

Veblen found the economic fraternity a bastard progeny of morals and psychology, groping for generalizations. He left it to the insensitive mercies of engineers, especially engineers involved with military projects, all dedicated to perpetual technological progress unrestrained by the exorbitant cost of the obsolescence they were constantly accelerating.

With a few notable exceptions, the economic establishment never thanked him for giving them a prestigious and profitable craft in the form of business-cycle econometrics. Nor did the establishment ever agree on the nature of its grievance against Veblen. The profession worshiped the cult of Economic Man, which Veblen demolished with his savagely satirical perspective on economic behavior in the broader context of social institutions and political power. Conventional economists are conditioned to track the business cycle as if they were playing bridge, by rules that never change. Veblen identified institutional changes—social as well as political—as the source of business-cycle change. Section II lays out the distinctive characteristics responsible for the performance of America's economic society and political economy.

America's
Distinctive
Assets

6

ROBIN HOOD'S REPLY

The Assets America Starts With

America is the first continental power in history to combine the capabilities of size and range, of markets and productivity, of money and muscle, and to extend her reach as she shed colonial appendages: the exact opposite of standard imperial practice. Historically, size and sprawl have hindered empires. Competitively, the race for national wealth has gone to nimble predators raiding the preserves of inbred dynasties. Strategically, islands have developed military leverage over the continents they flank: first England over Europe, then Japan over Asia. But England could not afford to keep its colonies, and Japan lost its colonies. America's geographical and sociological strength, by contrast, equipped her to do better without colonies than other powers ever did with them.

The phenomenon of superpowerdom was unknown before America found herself playing the part. She first rejected Theodore Roosevelt's legacy of colonialism; then, a generation later, grasped at Franklin Roosevelt's legacy of superpowerdom. By the time she was thrust into the role, FDR was gone. At Hiroshima, America asserted the prerogative of her position but compromised the responsibility. Only then did the hazy concept of superpowerdom dawn on a world that accepted it without formulating it. It meant more than being number one worldwide, as Rome and England were in the worlds of their eras. In the world of 1945, it meant being number one unchallenged by any rival or combination of rivals by any test—military, economic, financial, or political. (Only Marxism—or, as it then was, Stalinism—presented an ideological challenge, which, however, was muted by universal acknowledgment that American power had saved its leader's wartime regime.) Superpowerdom assumed leadership inspired and effective enough to manage the role with continuity.

Beginning with Korea in 1950, America's leadership did not live up to this assumption. Notwithstanding her successive miscalculations, however, America retained the single attribute of superpowerdom that she cannot surrender and that no challenger can match: the diversity of resources inherent in her geography and her sociology. Her ongoing status as the only diversified superpower illustrates the principle that the whole is greater than the sum of its parts. This distinctive diversity explains why any challengers exploiting America's managerial failures are bound to be limited because specialized. The distinctiveness of her geography and sociology also explains why any competitor outdoing America in selling cars, attracting bank deposits, or deploying tanks still remains obliged to deal with America on her terms, on the occasions when she herself figures out what those terms are.

America's distinctive sociology has evolved from her distinctive geography. These two basics formed American history, and they continue to mold America's policy options and opportunities more than any theory ever did. The combination of the two powered the country's historic waves of expansion by people on the move into new land and into new cities: commonplaces of American life, but unknown elsewhere.

Older countries had won their places by making the most of the geography they inherited and the sociology they created; England and Holland, for example, asserted themselves as maritime powers. Then the Germans, after their fashion, created a compound word—geopolitics—to describe the political impact of mobilizing national resources, especially people, to maximize geographical position: in other words, making the most of geography and sociology. Karl von Clausewitz, the German military strategist, defined war as "an extension of politics by other means." *Geopolitics* may well be defined as an extension of war by other means, including the show of force, the use of token force, and the mobilization of national economic resources for purposes of political bargaining. Germany's corruption by nazism gave geopolitics a nasty ring. Hitler's degenerate methods of redressing two of Germany's legitimate interwar complaints— "Germany is a 'have not' nation," and "Germany must export to live"—tainted the term. America is innocent of any of the ugly implications of Nazi Germany's claim to geopolitical domination by aggression and persecution. But America's geopolitical blunders since 1950 have put her on notice to make a belated effort at geopolitical effectiveness. The unprecedented risk of nuclear disaster adds urgency to the warning.

America's geopolitical clout gravitated to her; she did not acquire it by conquest. She allowed herself to be drawn into the twentieth century's two great wars only after provocation in the case of the first, and attack in the second. Far from collecting tribute from either victory, moreover, she volunteered to pick up the tab for each. Her ingrained isolationist tradition, reinforced by her economic maturity, led her to ignore her geopolitical advantage. During her heyday, she was too proud to present bills for services rendered. Her gradual decline atrophied her powers of calculation.

Throughout her history, America has shown various combinations of brawn and brains, always distinctive, reflecting her geopolitical uniqueness. The success of Harriet Beecher Stowe's pre–Civil War abolitionist tract, *Uncle Tom's Cabin,* in establishing a black slave girl as an American legend, symbolizes the country's rise to geopolitical preeminence in the generation following the Civil War. Topsy was born in a cabbage patch. America, too, sprang up out of open fields. Like Topsy, she just "growed." In her capacity as the world's lushest cabbage patch, the one most sought after despite her preference for isolation, she certainly did not plan the strange new role of superpowerdom in the world awaiting her. When she was younger and weaker, she thought more clearly and therefore acted with more geopolitical shrewdness—the Louisiana Purchase from Napoleon, for example—than she later did as the superpower in charge.

America's focus was still trained inward when the twentieth century opened on her debut as the up-and-coming power in the world. Despite her periodic pre-1914 panics, America had put her expanding cycles of imports—of capital, of machinery, and of labor—to more responsible, profitable use than any of Europe's other customers ever had. Because European capital funneled the largest portion of its investment flows into American transcontinental transportation and communications, the world grew smaller and America grew richer just as the 1914–18 war broke Europe and left every country in it dependent on America. Consequently, she was primed for a trial run at her future international role when the 1914–18 war brought her a bonanza. The older European powers rushed to tap her bulging reservoir of financial reserves when their hostilities emptied their own. Where else could they go?

The bloodbath of 1914–18 turned all of Europe into captive buyers of anything and everything America could ship and sent the American economy into a five-year inflationary boom. The stalemate in Europe's trenches transferred the power of resolution to America without her knowledge or consent, but with her money, her munitions, and finally, her manpower. She was as unprepared for victory as she had been for crisis. Nevertheless, during the interwar years of the 1920s and 1930s, America commanded the geographical and sociological advantages—the money, as well as the transatlantic remoteness—to buy time for Europe and subsidize Europe's unworkable peace treaty that first pauperized Germany and then drove it to target France and England for revenge. In 1928, when America decided that she had run out of money for Europe, Europe ran out of time and Hitler walked into power. America's decision to cut off Europe's dollar allowance had an immediate boomerang effect on America as well: Europe ran out of high-priced pin money to lend back as margin credit to the market makers on Wall Street, who promptly ran out of high-priced money for their customers too.

The 1939–45 war established America as the unique continental power in the world armed with a major economy, backed by a teeming and affluent population, simultaneously supporting and stimulated by a major military establishment, while still subsidizing a network of competing countries. She remained smart even after she recovered from the demoralizing shock at Pearl Harbor. In planning her war victory, she relied mainly on factories, not just on bodies, a priority that certainly put her in a class by herself in the history books. No other power had ever won a war of production. Moreover, during her postwar ascendancy, between her overkill at Hiroshima and her entrapment in Korea, America freed her principal client countries in Europe and Asia from the age-old needs that had governed their strategies for survival: to raise food for subsistence and to raise armies. Throughout the Third World, however, she achieved a more dubious freedom for her dependent clients: allowing them to raise armies and loans without raising food.

Before the stalemate in Korea, America managed better on momentum than any other power ever had before by calculation. Between her lunge into Korea and her costly turn into Vietnam, America held her own as a superpower stuck in a rut: a triumph of geography over policy, of sociology over strategy, of places and people over political shrewdness, economic wisdom, or financial discipline. No other country was as dumb or as lucky. After America stumbled into the Cold War with Russia, she achieved a victory of brawn over brains: of power over policy. Policy bungles based on geopolitical miscalculations are standard operating performance for America. Muddling along with them as her albatross, America substituted her ongoing omnipresence for the omnipotence originally attributed to her in most world capitals.

By the early 1980s, America had lost almost if not quite all of the familiar advantages she once held, most conspicuously, but by no means solely, in manufacturing. Even her presumably protected defense industries were overrun and undercut by imports; with the Pentagon as its customer, arms emerged as the best business in Europe, better even than the hotel business. America's momentum bogged down in a series of geopolitical entrapments: taking OPEC at face value; embracing the shah of Iran as democracy's tribune in the Middle Eastern oil belt; conceiving the lunatic scheme, revealed by Irangate, to siphon money from the mullahs to the contras.

Nevertheless, through the worst of America's ordeals, her geopolitical strengths had insured her against her managerial failures. After her erratic loss of leadership throughout the second half of the twentieth century, her geopolitical failures spread damage among her allies, as might have been expected, and fear among her enemies, as could hardly have been expected. The spectacle of America down but not out, and her competitors ahead but panicked, recalls the sportswriter's interview with a professional wrestling bum after a particularly devastating defeat. "Well, Gorilla, how come you got beat up so bad?" the sportswriter asked. Spitting teeth, squinting through half-shut eyes, with half an ear missing, the champ replied, "Well, I got that other bum on top of me, and I wouldn't let him up." America's stubborn formidability in the face of her dismal performance testified to her size, rather than her strategy (or lack of it): proof that her power was geopolitical, rather than political; a sociological attribute rather than a policy achievement; and hers to keep only because her leaders could not dissipate it.

By any geographical test, America is clearly in a class by herself among all the world's countries dominating continents. Geography has provided the power base on which her sociology has sharpened the difference between specialized national power and bona fide superpowerdom. Russia and China, continental powers themselves, live with the frustration of starting with even greater size, but ending with even less utilization of their resources. Both admit floundering, each in its different way, for lack of an economy able to meet modern standards of subsistence, let alone comfort, for its population. The Soviet powers respect America as the only continental power in the world to have activated her resources on the scale invited by her size, as if she had. Both admit dependence, in one form or another, on America's strength. Russia does when Kremlin spokesmen broadcast their fear of America, and China does with equally revealing ignorance when it sets out to copy America. For America has what each of them needs, and both know it. Only America does not.

Of the other continental aggregates in the world, neither Canada nor Australia is an economic power commanding authority commensurate with its size, much less a political or military force (despite an emerging Australian streak for strutting as "the superpower of the southwest Pacific"). Two obstacles stand in the way of each country: sparse population and excessive dependence on bulk exports of primary commodities into shrinking cash markets.

Geographically, the European Common Market could pass muster as a subcontinent. If the countries in it were to hang together, they would qualify for economic superpower status by the tests of population, wealth, and ability to produce it. In fact they haggle separately. All Europe is limited in scope by the Soviet presence in what the German geopolitical theorists and propagandists used to call its "Heartland." The Common Market at the peak of its prestige never was a geopolitical power in the sense that only a unified economic and military sovereignty can be. Jean Monnet was venerated as a prophet in his time for having conceived of the Common Market to cleanse the Rhine of blood. Hindsight revealed the fatal flaw of the EEC as geopolitical: the NATO alliance, far from standing as a unified transatlantic counterforce against the Red Army in Europe, activated the stunted ghost of the Holy Roman Empire. NATO's geopolitical inferiority was advertised by the withdrawal of France, the absence of Switzerland and Sweden, the dismaying obsolescence of its conventional weaponry, and the fantasy of eleven American divisions airborne overnight across the Atlantic to its rescue in the event of emergency.

In 1988, U.S. Air Force drill exercises revealed that no fewer than thirty thousand round-trips would be needed just to load the troops into the flying equivalent of cattle cars, with standing room only, the way their ancestors were brought over, but on suicide missions. The manuals made no allowance for the nightmare of C-141s passing the point of no return across the Atlantic only to learn that their landing fields had been destroyed along with their fuel supplies. Europe looked to its prosperous arms manufacturers for insurance against its ripped security umbrella. Their respective governments, understandably, made exports a priority. Europe evolved into the world's mercenary arsenal.

America, though an "also ran" in each of the four individual areas of national competition—military, political, economic, and financial—retained her distinctiveness as the world's only diversified superpower. But no challenger appeared on all four fronts. Paradoxically, therefore, America remained number one overall—by default!—while running behind in each test. Of the four national competitions, two sets overlapped: the military and the political in one functional framework, and the economic and the financial in the other. Russia was her only rival claimant to superpower status on the military and political fronts, where its nuisance value and propaganda impact are immense. Yet Russia's need to negotiate access to America's superior technology remained at least as powerful a restraint as was fear of precipitating a collision. Russia had no capability for follow-through on the related economic and financial fronts.

Japan on its tiny island base emerged as America's only rival claimant to superpower status on the economic and financial fronts. Once Washington demilitarized Japan in 1945, the Japanese diverted their disciplined social structure to the business of making war in the marketplace, which pays better and risks less than imperialism. But Japan did too well for its own good from its American dealings. America's slump into the deficit trap of the late 1980s tilted Tokyo's originally comfortable cost-benefit calculations into reverse as fast as its American markets shrank and, then, inescapably, Japan's lesser markets turned weak too. But to Washington's astonishment and dismay, dollar devaluation did not deter the Japanese steamroller.

When dollar devaluation failed as a defensive U.S. market ploy against Japan, Washington could see Moscow trading with Tokyo comfortably and from strength, even though Japan's industrial economy was strong where Russia's was weak. Beijing, with no industrial capabilities at all, traded from strength, too. Neither bilateral relationship with Japan depended on currency levels. Both backward continental economies were welcome to buy at will from Japan, but by using muscle rather than money. The contrast with America's craven beggar's stance was glaring in both cases.

So Japan accepted Russia's terms of trade; Russian military power gave it no choice. China's clout against Japan demonstrated continental spread and population pressure, plus the lure of an endless economic appetite. Russia and China between them showed that continental power can offset economic weakness with political leverage for bargaining against insular power, even when the island is rich and the continent poor. This was a reversal from England's naval heyday in the eighteenth century, when sail power harnessed the wealth accumulating in a tiny island base against immobilized, impoverished continental rivals. In today's world economy, a continental superpower that is rich, accessible to cash imports, and strong— as only America is—presents a posture of overwhelming strength to any client country, especially a dynamic but congested, land-poor, industrial island, which, like Japan, is dependent on a ready-cash, continental-scale export market.

In any assessment of the performance of Washington policy-making (or lack of it) by American public opinion, the behavior of Russia and Japan confirmed America's geopolitical default. The primacy Russia established over Japan, despite Russia's deficiencies in the marketplace and Japan's proficiencies there, accentuated America's distinctiveness as the only superpower whose diversification offered her a way to offset any and all of her weaknesses.

Yet despite America's disorientation, her diversification prevented Russia from dictating terms of trade with America in the high-handed way it did with Japan. The hardest line Russia could manage against America, economically and financially, was not to do business with her. But no fear of military confrontation with America inhibited Japan's chutzpah. Even China, though guaranteed to remain a nonstarter in superpower competition well into the twenty-first century, managed to translate its raw bulk of land and bodies into a stance of strength toward both Russia and Japan: a moral lost on America. This irony dramatized the troubles that America's policymakers have brought on her—as well as on a dependent world—by their failure to lead from her inherent geopolitical strength.

Without risking war or protectionism, war's economic equivalent, America was in a unique position to assert market power against Russia and muscle power against Japan; "hit 'em where they ain't," the oldest adage in baseball, applies to superpower competition as well. With just a little effort and not very much ingenuity, she also had the wherewithal to force market reciprocity from Japan, as well as military compromise from Russia. Yet America has never played the geopolitical hand given her by her distinctive status as the only diversified superpower in the world. Her loss of dominance left her dispirited, but failed to alert her to the need to deal with her felt by every other country with strengths necessarily specialized.

Every country outside America brandished some specialized form of geopolitical pressure on her in varying geographical and sociological mixes. All of them need something from America, and all of them, from the strongest to the weakest, have learned to use their specialized nuisance value to get whatever they ask for. Each of America's challengers, however, was hampered by an Achilles heel, stuck in its respective geography and sociology, that neutralized its specialized strength.

Russia could overrun any country in Europe, but could not finance the satellites it already had nor feed them in any better style than it could feed itself; in fact, Russia's imperium drew on satellite technology. Russia's massive military formations, organized to dominate broad landmasses, found no targets they could pinpoint in Afghanistan's mountain crags. Moreover, the Kremlin regarded Afghan Moslem militants as the advance guard, to use a favorite Marxist figure of speech, of an internal religioethnic menace to the Soviet regime more troublesome than any of the "counterrevolutionary plotters" Stalin chased.

Japan could no longer outfight China, but China could not begin to outproduce Japan. Korea was not about to erase Japan's marketplace lead, but its aggressive undercutting of Japan's subsidized dollar export bargains ate into Japan's profit margins. Korea's success intensified Seoul's demand for preferential treatment from Washington as a land base jutting into Soviet territory, topping Japan's bid as an island base. Israel could blast Libya, Iran, or Pakistan into nuclear ashes as decisively as it blitzed Iraq, but, unlike its Arab adversaries, it lived under constant pressure to avoid casualties and, therefore, has sidestepped major confrontations with massed Arab forces.

No one in the 1980s doubted which of the other powers was number one where it was number one: Russia by virtue of its military, China by body count, Switzerland and Bermuda (but no longer Hong Kong) as bank vaults adorned with their respective flags, Israel for missilry and incisiveness with the trigger, Japan for economic vigor and financial strength, Colombia for cocaine, America for nothing. But all of these top dogs in one or another national competition shared another built-in characteristic: an inability to control or neutralize unfriendly neighbors on any front but the specialized one. Each country was as strong as a bear in its own lair, but as slow lumbering into its neighbor's lane.

The world's intricate, interconnected structure of strengths and weaknesses duplicated willy-nilly the carefully contrived purpose written into the Constitution of the United States. Outside America, a worldwide system of checks and balances evolved, as unplanned as it was uncontrollable. It worked to keep the stronger countries strong but specialized and to give the weak countries bargaining power without, however, curing their weaknesses.

Financial power is as specialized as military power. Only America can mobilize both, reflecting her diversification. Specialized financial power can flourish in a sociological hothouse, insulated from military, political, and even economic power: witness Liechtenstein, Luxembourg, and the Cayman Islands. Even Bermuda and Hong Kong have qualified, without being countries. England used its financial expertise to engineer a comeback from its economic and political decline. Switzerland, because landlocked and congested, relied on financial power as a substitute for any temptation to use aggressive muscle power.

To complete the contrast, all of the top dogs in their specialized arenas around the world depended on above-average management to overcome their natural deficiencies. Japan certainly did economically, and Russia did militarily. America is the only power endowed with built-in advantages formidable enough to overcome the below-average managerial performance that had come to be expected of her by the late 1980s. But her smug reliance on her impressive reserves of resources led her to indulge her managerial failures and anesthetized her to the market losses and political dangers these failures inflicted.

By every sociological standard, the profile America presents is distinctive. Her institutionalized ability to consume at the world's highest rate has provided the productive segment of her economy with a rich bonus of domestic nonearning customers, both above and below the ages of employment. Of course, all demographic calculations are subject to a basic input-output ratio, balancing the birthrate against the death rate. But America's distinctive demographics have produced an unprecedented combination of variables—immigration to begin with, then mobility—that injected more dynamism into the American economy than her wars ever did, and these distinctive demographics continued doing so after she stopped making wars on a demographically significant scale.

Throughout, America's sociology, homegrown on its geographic base, has bred distinctive characters who embodied America's distinctive flair. Ben Franklin was the first in a long line of crafty, pushy personifications of America on the make. None played the role more flamboyantly than Huey Long, the populist governor of Louisiana (as well as senator and father of Russell Long, for many years the distinguished chairman of the Senate Finance Committee). During the Depression, the self-styled "Kingfish" emerged as the only potential rival who ever disturbed Franklin Roosevelt's sleep in the White House. By the time he descended on the New York scene, he had magnetized national attention by giving backward rural Louisiana the best roads, the best schools, and the best college football that money could buy, taking it from Standard Oil and, with no apologies, for himself.

This Robin Hood of the bayou organized his first encounter with New York's press sophisticates from the vantage point of his hotel bed, in which he held court after the manner of the Bourbon kings dispensing justice. Clowning to dramatize his extravagant promise to make "every man a king," he decked himself out in pajamas of royal purple, with a straw hat tilted on his head serving as a crown. In response to the obvious question about his ideology invited by his bizarre getup, he was ready with his nonchalant "aw shucks" answer: "Just call me sui generis," he chortled. If he had gone further in intriguing, outraging, and confusing the press, exercises that came naturally to him, he could have added, "The same goes for the whole damn country." America, as a country, encompasses a sociology all her own. In no other nation could an incumbent baron, such as Long, seize on the depths of a depression to promise more to most—and gain plausibility—without stirring up mob violence.

Schumpeter's definition of economics as historical sociology recognized the need for a broader environmental framework in which to view the economy and called for a corollary describing the supportive sociology. His concept evolved from the interplay between the geography that channels the energies of a country and the people who generate them. Geographically, for example, countries able to feed themselves are more prone to isolationist stances than countries that need to import to eat. Sociologically, countries offering growth opportunities to manufacturers and marketers of processed foods, as well as to restaurant operators, are likely to experience steady shifts in population from farm to urban living. "Elementary, my dear Janeway," Sherlock Holmes would have said.

All modern economic analysis ends by seeking a marketplace consensus, but begins with a government census, counting acres and heads one by one. Distinctions within the society are revealed by census results: more sociological than economic, because delineating the contour of the entire society nurturing the economy within. No high-speed computer printouts tell us more today about how to rate the ongoing performance of any society, or of one society against another, than this rudimentary head count did in the early days of economics. Any realistic analysis is bound to start at the same place, with a quantification of the productivity that "land and hands" provide in support of one another; in fact, as we shall see in Chapter 9, the title of the first practical economics tract was just that.

Population increases for any country, however, are a far cry from skills gained for its economy. Throughout the Third World in the twentieth century, population increases have been piling up socially unendurable liabilities, forcing experiments, so far fruitless, to limit birthrates. One of the basic dicta of economics establishes the distinction between population increases that enrich and those that impoverish: between demand and "effective demand"—bare want is not the criterion. Alone among the world's functional economies, America has been absorbing massive increases in her nonworking population. Even though America's demographic gains and educational losses have netted her more bodies than skills, they offer her a limitless challenge to strengthen her social structure and to reestablish her distinctive leadership by exploiting America's elastic potential for import absorption. The assets that America fails to use—beginning with the bargaining power rooted in her import capability—are as conspicuous as those she misuses.

Here are the accumulated sociological assets awaiting activation into economic productivity. No solution to America's geopolitical floundering can be brought into play until America learns how to minimize her social waste and maximize her human potential.

The drop in infant and juvenile mortality. The decrease in America's infant and juvenile mortality rate testified to her progress in perfecting the clinical arts, while making a start, however uneven, toward sanitizing the environment. True, the squeeze of the Reagan years brought a dismaying setback in health support for ghetto infants and juveniles, as well as in employer contributions to health insurance. The squeeze also inflamed calls for community support for child care into a burning political issue. But the setbacks were due to cutbacks in public funding; that is, while catastrophic in human terms, they counted as merely political irritants, readily reversible.

In clinical terms, America's combined public and private health capability reflected enormous breakthroughs in her ability to save significant numbers of babies born prematurely (or with defects), formerly given up for lost. Socially, these breakthroughs extended to the clinical ability to avoid premature deaths for youngsters in whom society and families had invested the cost of health, education, and upbringing. The American economy also accrued an ongoing dividend from public sector spending on social services and from private sector disbursements in fees, benefits, and gifts.

More than humane considerations ride on America's ability to sustain a favorable balance of births over deaths, and, in addition, to ensure a scheduled matriculation from education into employment. Over and above the economic axiom that an expanding, employable population is essential to an expanding economy, the entire Social Security system assumes a favorable balance of payments from those entering the work force to those leaving it. The system is based not on prefunding of future benefits, as life insurance is, but on the "revolving door" principle. It assumes that younger generations entering the work force will outnumber and outearn retirees leaving it, that is, that entries into the work force will be healthier and higher-skilled than previous generations.

The drop in the overall infant and juvenile mortality rate within the working economy was unforeseen when the Social Security system was legislated in 1937. Consequently, no measures were ever taken to harness the gushers of human energy this scientific achievement released. On the contrary, too many of the youngsters saved were left to the mercy of America's substandard public education system, or, worse still, the welfare system.

The extension of life expectancy. Between radical clinical break-throughs and humane social reforms, America managed to extend life expectancy as well as reduce infant mortality. In 1937, America started to count on the continuous collection of withholding taxes from a continually increasing number of young entries to the work force needed to support the simultaneous population explosion start-ing among retirees entitled to benefits. From the 1939–45 war until the oil war of the 1970s, this assumption of dynamism in the work force remained valid: incomes soared, with recurrent assists from bonuses and overtime. The vast improvement in clinical capability from the cradle to the grave, accentuated by the corresponding in-crease in social need, forecast an astronomic increase in claims for benefits from the Social Security trust funds.

But the improvement in the delivery of social services, especially health services, lengthened life expectancy—despite the deepening crisis of homelessness and hunger in the 1980s—faster than withhold-ing tax collections generated revenues. The unprecedented and dis-tinctive explosion in America's unearned incomes also helped extend longevity, despite the hardships caused by volatile interest rates. Retirees without investment income to fall back on suffered when interest rates rose; those with investment income suffered when inter-est rates fell. As the country's solvent geriatric base expanded, how-ever, the health services and the financial services industries, between them, sparked the dynamism in the service sector of the economy.

Financial insecurity and poverty rank with cancer and stroke as geriatric killers. A rising number of retirees and geriatrics (those over sixty-five) in America run out of money before they run out of energy. The enormous overload of geriatrics with no financial re-serves, or with marginal reserves rapidly depleting, imposed a new policy veto against the accepted assumption that recurrent recessions are healthy as well as normal.

No other country, let alone power, enjoyed a comparable injection of demographic vigor at both ends of the generation scale at once. More than incidentally, the extension of life expectancy well beyond the actuarial tables used by the life insurance companies deferred large reserves of cash from death benefit payments for institutional investment in stocks, bonds, real estate, and entrepreneuring. The John Hancock Life Insurance Company, to take just one example, was a founding investor in the American Research and Development Company, America's first venture capital enterprise, dedicated to exploring commercial uses for new technology developed during the 1939–45 war.

The drop in the ratio of the productively employed to the total population. This phenomenon is a normal symptom of shrinkage; first, the economy loses the increment of income, then the Treasury loses the increment of revenue. Unemployment is its familiar cause. America, however, enjoyed the benefits of a distinctively affluent drop in this pivotal ratio, thanks to the simultaneous growth in her nonworking population at both ends of the demographic spectrum, infant and geriatric. Where Russia was drafting bodies of all ages into the work force, regardless of their educational potential, America was generating a powerful and continuing human resource for her economic society with each drop in the ratio of the productively employed per total population: that is, more people consuming more on the demand side of her national economic equation. The Kremlin was forced to acknowledge the expansion of demand in Russia's marketplace equation as a national liability. America enjoyed it as an asset; though establishment economic thinking—supported articulately by the economic advisers to the 1988 Democratic presidential campaign—launched a crusade against "overconsumption" and advocated a return for the country to "living within its means." This thinking ignored America's reliance on expanded demand from future and past employables to lead her production: the exact reverse of the failed "quick fix" advertised as supply side economics, which bets on overproduction, although, in fact, overproduction precipitates recessions at best and depressions at worst. The supply-side rationale looks to people suffering losses of income to increase consumption!

America's falling ratio of the productively employed to the total population reflected her ability to increase her investment in her educational pipeline, to lengthen the training period in all professional fields, and to shorten the training period in all vocational and support fields. This distinctive demographic phenomenon has endowed America with an untapped pool of deactivated older people of experience and talent available to supplement community support for the growing portion of her infant and juvenile population no longer able to count on purely parental support to graduate into skilled employment.

The health-care industry. No other country, least of all Russia or Japan, can match America's. True, health insurance and socialized medicine have been institutionalized in Europe for generations. But comparisons don't apply. The services European systems offered came free or at nominal cost—and, in most cases, were worth the price. Moreover, the societies in question were relatively small and characterized by homogeneous and immobile populations, deprived of America's demographic dynamism.

By contrast, America's health industry was organized on a scale sufficient to support her accelerating social mobility and her exploding immigrant population on top of that. No other country could begin to match her ongoing annual scale of investment in health industry technology and training. Nor did any other country come close to the payback America earned on this investment in publicity on health. A medical journal joked, at the time of Reagan's surgery for colon cancer, that the technology behind his operation would end by saving more American lives than the money he was lavishing on his SDI program. Even though the health industry suffered contraction during the 1980s due to cuts in social welfare budgets and, even more, in employer-paid health insurance premiums, both setbacks are readily reversible. What matter are the size and quality of the health-care structure. The spread of congregate community care living and day-care centers in the late 1980s for geriatrics enjoying mobility but needing and affording quality support promised a new dimension of growth to the health-care industry and of solvency to geriatrics.

This prime American service industry asset is eminently exportable, especially to Russia, but only by government-to-government negotiation. By the late 1980s, Gorbachev identified the failure of Russian health care as a priority for *glasnost* and *perestroika*. He admitted that the rise in infant and maternal mortality was spreading fear at one end of Russian society; the drop in life expectancy was doing the same at the other. But *glasnost* did not extend to advertising the military implications of Russia's health industry deficiencies; instead, Russia advertises the overwhelming strength of its standing army. Two-thirds of Russia's armed forces, however, are reserves (nearly 30 percent non-native-Russian–speaking Moslems), who need continuous medical scrutiny and health care to avoid spreading infection in the work force and to guarantee performance when recalled. Russia's staggering health-care needs offered America an opportunity, not often found between antagonists, to sell her services.

A footnote on progress: In the seventeenth century, John Locke was an intellectual and scientific explorer for all seasons. A physician by training, and a philosopher by vocation, he scored a number of notable "firsts." He was the first to perform a surgical procedure inserting a rubber tube into an abscessed stomach for the purpose of siphoning out the pus, reversing the age-old custom of bleeding, which sealed the pus in and let the blood out. As an economist, he was the first to relate the rate of interest to the quantity of lendable money. As a philosopher, he was the first to assert that babies are born tabula rasa: without any preconceived ideas of God or of morals, with their brains responsive to environmental impressions.

As a political philosopher, Locke drafted the first constitution ever written to provide for due process as a guarantee of civil liberties. He also formulated the original version of the noble language immortalized by his disciple, Jefferson, in the Declaration of Independence: "Life, Liberty, and the Pursuit of Happiness." But Locke's pioneering libertarianism, reflecting his medical training, was more paternalistic than Jefferson's. In his tract, "Of Civil Government," he called for "Life, Health, Liberty, or Possessions": documentary proof that our own philosophical and constitutional history is rooted in the primacy of health among the governing arts, as the precondition to the enjoyment of liberty that it is. Locke knew something that we have forgotten. So did the ancient Chinese, who paid their physicians only when "patients" were well.

The economic liberation of women.   Though the politics and rhetoric of feminism remained controversial into the 1980s, the principle of women's economic participation, free from sexual discrimination or restraint, won mainstream acceptance fortified by legal approval. This remained a far cry from acceptance in practice, but it was a necessary precondition. Women steadily freed themselves on a mass scale from traditional obligations that had frozen them out of employment and educational opportunities to qualify for high-earning jobs. Each crop of young women entering the higher education system established new beachheads for professional acceptance. Women physicians first entered the field, then broadened their horizons—from pediatrics to surgery, to take one example. Contrary to dire predictions that the mass entry of women into the work force would result in stagnation of the birthrate, the number of single parents has jumped as the army of women at work has grown.

None of the other powers offered women comparable openings. Quite the opposite. Of the specialized superpowers, Russia reduced Marx's revolutionary call for sexual equality to a mockery; Japan provoked revulsion from independent-minded women against its severe macho discipline. Even England, despite its noble literature of emancipation and the achievements of great women, remained a male bastion.

Most other economies have remained ghettos for women attempting to pursue career incentives in business and the professions. No single lift to the American economy has been more powerful during America's upturns, or more stabilizing during her downturns, than the women's segment of her work force. The sustained breakaway growth in its net size, earning power, and professional status testified to this achievement, despite the incalculable handicap to the economy from the absence of child-care support. Continual increases in the number of women getting educations—including adult, professional, and vocational educations—guaranteed increases in their employment, their status, their earnings, and their independent investment capabilities. At the outset of the 1988 season for political fisticuffs, the Senate recognized the representative bipartisan character of the call from the economy for community support for the bulging population of single parents. It attached a child-care rider to the Welfare Reform Bill it passed by a vote of 95 to 3.

The immigration explosion. Immigration was America's distinctive institution before slavery became her "peculiar institution." Immigrants built the country: no country ever rewarded immigrants more generously, despite indignities endured on their way up. During long stretches in the nineteenth century, the demographic vigor of immigration into her centers of population furnished all the expansive thrust the American economy needed, even more than free land across the West. America demonstrated the force of the tautology on which classical economics is based: that more land use by more people adds up to economic expansion—that is, more equals more.

In the last third of the twentieth century, America emerged as the only world power enjoying major revitalization through immigration. Russia, Japan, and China all outlawed it. While the "German economic miracle" was still on in full force during the postwar boom, West Germany banked massive benefits from its army of "guest workers" (all male), but sent them back where they came from the moment its economy felt the pinch. England invited a considerable influx of immigrants from Commonwealth countries before restricting it in acknowledgment of the resulting social strains.

The distinctive feature of America's late-twentieth-century immigration flood was its racial and ethnic mix, led by Asians, Hispanics, blacks, and, increasingly, Middle Easterners. Their presence provided America with a broadened base from which to reassert her leadership throughout the nonwhite world as a positive alternative to Russia's pronounced white racist bias, sharpened by Moscow's ban on immigration. Productive immigrants from outside Europe enacted the same saga as their European predecessors, mobilizing the family work ethic as they took over large segments of metropolitan area retail trade and services, astounding their community establishments with their high rates of savings, capital accumulation, and qualification of their offspring for admission to universities for advanced technological training. Cuban exiles led the way in Miami during the generation following Castro's takeover, yielding America an economic windfall from the overextension of Soviet power. In her Chinese communities, brokerage firms were astonished and embarrassed during the stock market boom by the floods of anonymous cash that poured in, identified with none of the bank account or Social Security numbers required by regulations.

Free commodity markets. In the 1970s, Treasury Secretary William Simon defined *markets* "as people"; that is, as a sociological phenomenon, like the institution of marriage or education. The degree of market regulation varies with custom, reflecting national and concerned group interests. Commodity prices, as well as supplies, are intimately involved with the vital interests of the countries producing and consuming them. America maintains commodity markets from which the rest of the trading world projects its operating incomes and its living costs—no questions asked about the passports carried by the money being turned over. No other country could begin to provide the trading facilities America offers. Nor could any country challenge this American monopoly for at least one reason: most commodities, notably oil and gold, are quoted internationally in dollars. Therefore, America serves the world as its natural auction block.

Sotheby's exacts top dollar for the service it provides. The catch is that America has yet to exploit this priceless proprietary sociological asset—liquid public commodity markets—for reciprocal bargaining purposes, if only to prevent undisclosed manipulation by futures traders, including governments, not just, but mainly, Russians (in the commodity markets) operating against America's interests. Any other host country would long since have squeezed corresponding advantages for the services rendered or threatened to restrict access to them. No foreign power, for example, would be allowed to set up an exchange inside Chile to manipulate the future export price of copper up, or inside Egypt to manipulate the future import price of wheat down.

Commodity producers and consumers operate under pressure to use commodity futures markets to hedge themselves against price changes. Traditionally, therefore, futures prices followed current transactions markets. But America has allowed speculators to use futures markets to lead current transactions prices, frustrating commodity producers and consumers alike in their traditional pursuit of price insurance. She has invited foreign governments to manipulate her commodity futures markets to overcome their geographical limitations: commodity markets reflect geopolitical strength or weakness. She has also invited the commodification of her currency, defeating her own devaluation strategy and complicating commodity hedging operations with the need for a double hedge against currency fluctuations; never before has commodification undone devaluation.

Russia, in particular, has found this convenience profitable, contriving to politicize the commodities America has put on the auction block, especially the oil, gold, platinum, and nickel Russia sells and the grain it buys. The Brazilians make enough money bouncing coffee futures up and down in the Chicago trading pits to pay their interest bills and bring back the 5 cent cup of coffee. None of this incremental profit seeps back into the Brazils of the world.

But America has persisted in her inherited sociological anachronism that markets settle dealings between countries, and that countries settle their dealings in cash. So she habitually mistakes her liabilities for assets and her assets for liabilities, particularly with the oil she hopes to buy and the grain she needs to sell. We shall see in Chapter 14 how to realize the earnings potential of these assets.

America's past ability to raise her standards of living along with her productivity freed her from the scenario decreed by the Puritan ethic: to suffer in order to prosper, to plan interruptions of prosperity in order to enjoy its resumption. Economists pressing for domestic deflation as the price of restoring world competitiveness ignored the distinctive economic advantage rooted in the size of America's domestic market. This unique endowment of domestic consuming power had offered her embattled industries the economies of mass production: built-in cost savings on heavy volume, instead of the foreign practice of ruinous subsidies to finance price cutting in quest of volume too small to pay for itself. The advantage of size also enabled America to absorb massive overspills of production from the rest of the world, astounding her competitors by showering them with dollars.

America gave the Marshall Plan to Europe as a gift. She can earn the wherewithal for a Marshall Plan for herself as she puts it to work. The responsibility America owed the world was the same as the responsibility she owed herself: to harness all her resources. Nothing less could iron out the fiscal chaos reflected in the need of her government to borrow between $100 billion and $200 billion a year—the first jumbo fiscal stimulus to fail to spur the economy. This besetting American failure to use her geopolitical asset base as an international bargaining tool permitted her parallel $100 billion–$200 billion trade deficit to siphon out of her income stream every cent, and then some, of the purchasing power created by the budget deficit.

The rest of the world can't use the geographical assets that America can't give away and doesn't want the sociological assets that

America won't give away. The trouble is that America has forfeited control over other assets that only she can manage, but that the rest of the world can't manage. Foreign incursions have damaged some of these assets; only America can repair them.

7

CORCORAN'S CALCULATIONS

The Assets America Developed

As late as 1980, even the shrewdest American molders of events, hardened by experience and sharpened by study, regarded America as secure in her control over the proprietary assets that she had developed: her dollar power, her import power, her agripower, and her technological power. Not even cynics speculated whether America might lose control over any of these bedrock assets, let alone what the cost would be if she lost control over all of them. Yet mid-century assertions of "The American Century" by America's then most influential press magnate, Henry Luce, had long since been refuted by successive reversals—in Korea, Cuba, Vietnam, and Iran—that blasted the reelection hopes (and in one case, even the life) of four presidents—Truman, Kennedy, Johnson, and Carter. After her Irangate fiasco, while "standing tall" to the slogans of the Reagan revolution, America accelerated the decline she had tolerated for a generation. She managed to give away control over these four domestic assets that the rest of the world could manipulate but not manage.

*   *   *

Among senior event makers from Roosevelt's time to Carter's, none was more perspicacious or more pragmatic than Thomas G. Corcoran. Corcoran had served as after-hours entertainer in the presidential living quarters, where he charmed Roosevelt and his confidants with the folk songs he played on the accordion. FDR's office wife, Marguerite ("Missy") LeHand, officially his secretary, relied on Corcoran to handle missions, delicate and rough alike, unsuitable for presidential fingerprints. By day, Corcoran had functioned as the emissary between the White House and the power centers connected to it. Sometimes he just pretended to speak for the president; at others he flashed his magnetic grin and murmured, "Frankie likes to lie." The New Deal's emergency lending agency and the Justice Department were his special preserves; he appointed the New Deal's judges and prosecuting attorneys (making sure that they were young enough to enjoy long tenure or old enough to offer quick replacement opportunities). But behind his swagger and his charm, Corcoran brought a sense of national power and grand strategy to bear on the course of history.

"I'll take the 1980s on consignment," Corcoran laughed when he and I philosophized over his eightieth birthday at the turn of the decade. He did not like what he saw and was dead by 1983, while hospitalized with an infection picked up, ironically, on his last tour of the modern Asian trade routes he had helped to open. Corcoran was formidable as a modern Marco Polo; history was deprived of a treasure trove because he did not travel with a tape recorder. He was the figure in America's oral history who singled out the four distinctive assets America couldn't afford to give away, but failed to employ to her advantage.

"Tommy is a street fighter," Joe Kennedy used to say about him during the Roosevelt years. Kennedy, FDR's wartime ambassador to London, in contrast to the heroism of his sons, had made a record as a coward during the London blitz, when he pressed Roosevelt to send a destroyer to evacuate him and his family. Accordingly, he expressed admiration mixed with frustration as he said it. Corcoran was a muscular thinker, that is, both a thinker and muscular. He and his political partner in the management of the New Deal, Benjamin V. Cohen, ranked with the most important movers and shakers in cabinet history, though the pair were mere janissaries. Corcoran won respect for his instinct for the kill at close quarters, typified by his celebrated one-liner: "The way to handle a businessman who comes after you to say thanks for saving him is with an open razor."

Here is how Corcoran assessed the whiplash America suffered in the 1970s from her loss of the oil war under the Kissinger regime. "We are big enough to shrug off defeat in any jungle war," I recall his ruminating,

> and we can certainly sit out a deadlock against any other power or coalition of powers. We can spread more tension across the table just by sitting at it than any antagonist can make us feel by banging it. But the loss of our chips in the Middle East, and our invitation to Russia to pick them up, spells more trouble for us than we're equipped to handle. Our economy established us as a force in the world without our having to fire a shot. It's been the source of our power and it's becoming the source of our vulnerability.

He and I fought side by side with very spotty success during the 1939–45 war to establish a U.S. economic presence along the Middle Eastern oil link between Europe and Asia. Therefore, this complaint of his echoes in my mind as having ranged backward as well as forward across time. It illustrates the way the frustrations born of the experience of one generation again and again crystallize into prophetic wisdom for the next.

The conclusion that Corcoran drew from his forebodings suggested the need for analytical priorities in any effort to retrieve America's losses. "The only assets we have left," he went on to say, with his peppery pragmatism, "are the assets that we haven't managed to give away—the dollar, agripower, technology, and our mass market for imports. The reason that we haven't is that we can't." For all his realism, he underestimated the genius America demonstrated for ingenuity fortified by perseverance in giving away control over these geopolitical equivalents of crown jewels in less than a decade.

Corcoran identified these four proprietary assets as direct outgrowths from the two native assets—geography and sociology—that

America can't give away. These four are few enough to be manageable. They also make a formidable enough package to be negotiable
to America's advantage even by policymakers without very much
talent for the negotiating game. Moreover, America's competitors,
beginning with her rival superpowers, have made a free gift of a fifth
asset unique in the annals of political and economic confrontations:
This fifth asset is the dependence shared by Russia, Japan, and the
others on America to do well in order that they may do still better
in their dealings with her. America's ability to regain control over
these four proprietary assets is the precondition of her freedom to
expand and exploit this fifth asset.

America has forfeited control over each and all of these proprietary strategic assets. No power, let alone superpower, that has lost
position inside its own home base can regain competitiveness by
setting out, as America did with her devaluation, to undercut other
powers in their own territories. America's route back to competitiveness leads, not via lower living standards for her people, as her
economists of both parties have insisted, and as dollar devaluation
guarantees, but along the political high road that connects Moscow
and Tokyo with Washington. No price maneuvers—beginning with
devaluation—can redress marketplace grievances directed against
foreign competitors subsidized by their governments. Only straight
political talk, government to government, can.

Early in the Cold War, America was suckered into her "also-ran"
role. Truman took Stalin's bait and challenged Russia on the Kremlin's surrogate military turf in Korea. In Korea, America found no
centralized target against which to aim her industrial artillery, as she
had against Japan and Germany. Therefore, her invasion floundered
and failed, just as her subsequent invasion of Vietnam did. America
will always find sole reliance on military power uncongenial to her
sociology and wasteful to her economy. The only terrain on which
America can retrieve the control she has defaulted over these four
assets, and the bargaining power they offer overseas, is political, and
the signposts are economic, as they always have been for America.

But America has never maneuvered Russia or Japan into head-
to-head political bargaining on her economic and financial turf. Moscow and Washington have indulged alternating whims: refusing to
bargain and accusing the other of refusing to bargain. Yet America

has enough economic benefits to put on the table with Russia to wring and wheedle political concessions from Russia; the cost to America would be next to nothing, and the paybacks rich. Moscow's negotiators have been realistic in eying Washington as a richer source of loot than any economic hub under its control; no one can count the cash booty Moscow's traders have taken out of America's commodity futures markets. Once Washington learns how to sell goods and services to Russia in exchange for trustworthy relief from military burdens, Russia's dependence on America's economic reservoir would intensify, not lessen. Washington has never considered this a fair trade. It's not. America would be the clear winner on all counts: business done plus missilry saved and lethal risk avoided.

Tokyo has no political defenses against Washington, only economic and financial offensive power that, against America, boils down to nuisance value. Accordingly, Japan's motive for wishing America more prosperity rather than less is incomparably simpler, because uncomplicated by calculations related to power politics or ideological claims. An old farm adage describes it: fat cows give rich milk.

The rest of the world, including even Russia, remains too small, notwithstanding all its economic expansion, to manage for itself without ready access to America's economy. Every other domestic economy in the world is so small, relative to America, and so export-dependent, that all of them are obliged to treasure their American markets as pivotal. Moreover, the more other countries expand at America's expense, the more dependent they become on income from America. But the distinction between access to America's markets (including the right of access as investors) and control over them suggests that between extending hospitality and giving away the silverware. Until America reestablishes control over her rudderless economy and demonstrates again that she can manage her stated priorities, the rest of the world will remain powerless to mount a challenge to Soviet power, or, what comes to the same thing, to take advantage of Soviet weakness.

Soviet weakness is particularly dependent on the American economy to guide the Kremlin across its admittedly thorny transition, by way of *glasnost,* to *perestroika.* This transition represents a planned exercise in contradiction that comes naturally to a younger generation of armchair revolutionaries disciplined to think in terms of the Marxist dialectic; it is an exercise in the market equivalent of guided liberty or—as George Orwell might have said—an invocation of the spirit of 1776 preparatory to a trip through 1984. In its advance (or, perhaps, retreat) to capitalism, the Kremlin's strategists cannot manage by taking economies as disparate as Sweden's or Japan's or West Germany's as their model. Only another continental economy will do; and the only other continental economy in the world happens to run on precisely the same structural foundations and fundamental mechanisms that Russia needs: primarily domestic, diversified by a large agricultural sector, richly endowed with raw materials, complicated by an expensive defense nucleus, and offering less advantaged competitors a limitless import market.

America's four proprietary assets fit into one package for negotiating purposes. Here it is.

America's dollar asset. The world economy needs a common monetary standard on which to operate, and every government and central bank in the world knows it. The dollar meets the need, and no other instrument does. Certainly, none of America's competitors could substitute its currency for the dollar, and none of them would dream of trying—especially not the Japanese, who owed so much of their success to their bold use of the dollar standard. All of America's competitors recognize that the world is on the dollar standard to stay, and all are glad that it is. Their reliance on the dollar to serve as the money they use in foreign dealing has lured all of them into speculating on its price fluctuations.

After 1945, America overplayed the hand she dealt herself with her nuclear head start and was traumatized to discover herself obliged to share with Russia the nuclear power she believed to have been her monopoly. But she underplayed the distinctive monopoly her dollar power gave her as the custodian of the world's money, and she ignored the opportunity it gave her to bargain from strength. Nevertheless, America has retained this monopoly for the simple reason that no challenger loomed; not even the contrived collapse of the dollar nominated an alternative. Ironically the hysteria drummed up in behalf of "a return to American competitiveness in today's global economy" ignored the world's dependence on America's dollar monopoly in clearing its transactions. Gold offers no substitute.

The world has outgrown the limitations a return to the gold standard would put on its need for money. Moreover, gold has lost its traditional stability. The gold markets fluctuate even more violently than the dollar market does.

America's managerial default of her dollar monopoly turned out to be at least as distinctive as the fact that she had it. She lost control over her own monopoly without ever having exploited it; yet, in spite of herself, she retained the option to reclaim it. The dollar power that was thrust on America in midcentury combined a rare trusteeship held for the benefit of the world with a historic profit opportunity for herself. But America dressed herself up in the monetary mantle of empire without finding a workable political role to support it.

Claims to assert power for the benefit of mankind made by any government on the rise normally provoke hoots of derision. Such posturing is hypocritical when the strutting results from an imperialistic grab. America had indeed been guilty of an adolescent infatuation with Manifest Destiny early in the twentieth century, but her escapades in Cuba, Panama, and the Philippines brought her more liabilities than loot. In her maturity after 1945, when economic power was embodied in the dollar standard, however, she failed to safeguard control over it for her own benefit, let alone anyone else's.

In the 1950s, a distinctive new dimension broadened the dollar standard: the Eurodollar market. Neither Washington nor Wall Street could claim paternity. Russia did: the London branch of the Moscow Novodny Bank startled the stodgy banking fraternity by unveiling the first transaction. America's military rival superpower has no financial resources, let alone muscle, although Russia's financial technicians are unmatched in their worldwide dealings within the dollar economy. Their invention of the Eurodollar was hardly meant as a contribution to America's wealth or power, or as a policy to further a grand strategy hatched by the financial illiterates in the Kremlin.

The incentive that prompted Russia's banking bureaucrats to open the Eurodollar market to money flashing any passport, or none, reflected their own immediate need to find dollars. The solution the Soviet technicians improvised was the simplest one imaginable: they created dollars by depositing their own currency as collateral to begin with and, from then on, out of thin air. When America did not object, Europe's banks took them.

Once Russia stumbled onto the Eurodollar standard, every other country followed. From that time on, Russia managed steadily better in the world as it financed its aggressive stance, and America managed steadily worse as she permitted cloning of dollars on demand for every other country's financial operations. America's default of responsibility to retain control over her own currency in dealing with the world at the crest of her power echoed her start-up failure to exercise this same basic responsibility in dealing with the states. The first failure prevented America from managing for the states as they relied on her to do. The latter-day failure prevented her from managing for the world economy as it relied on her to do.

Never before had the world of international finance conceived of, let alone relied on, a money standard so loose: the dollar domiciled under the American flag, and at the same time available for issuance in foreign financial centers as "stateless money," an unprecedented financial phenomenon, inherently unstable, as no monetary standard can be and still remain functional. All foreign banks held the bulk of their reserves in dollars. Any foreign bank could create dollars by the simple device of making loans in dollars cloned from its own deposits in local currency and crediting borrowers with deposits. In the simple bygone world of the gold and silver standard, bullion—which was money—became "coin of the realm" only after it was minted by the sovereign. But the prevalence of this stateless money in the form of Eurodollars announced a momentous reversal of all previous practice and theory. It worked to the advantage of the issuers, and against America's.

The growth of the Eurodollar market did more than make financial history and remake the financial structure of the world. It also established two new political axioms of invaluable importance to manage the dollar standard as the asset it was for America and the necessity it was for the world. Both would have delighted Marx because each acted out his mischievous but acute vision of the dynamics of change alternately reversing and confirming an original purpose.

The first axiom: Russia's dependence on America. Moscow confirmed the dramatic extent of this change in the Soviet party line long before Gorbachev popularized *glasnost.* The improvisation of the Eurodollar market marked Moscow's switch from plotting the destruction of the capitalist system to taking part in its continued expansion.

The second axiom: the Eurodollar market's dependence on regulation from Washington. America's regulatory authorities shut their eyes to the danger the Eurodollar market spread with its Niagara of unrestrained dollar loans creating uninsured dollar deposits. Consequently, the inflation of the Eurodollar supply sprouted limitless liabilities for the dollar banking structure. As fast as these liabilities pyramided, however, they projected an offsetting new asset for America's dollar power: the universal reliance on the U.S. government as the lender of last resort for banks going bad on megadollar deposits. The threat of a collapse in the Eurodollar market, accelerating a chain reaction of worldwide bank failures amidst the increasing shakiness of America's banking system, confirmed a famous joke of Lenin's. When the time was ripe for revolution, he predicted, capitalism would sell the Communists the rope to hang itself. The Eurodollar market emerged as the shopping center where capitalism made the sale. Giveaways of "green stamps" kept it open and humming.

After America's monetary regulators failed to notice the initial growth of the Eurodollar market, they accepted its subsequent institutionalization as part of the scenery, even when Eurodollar interest rates started to lead domestic interest rates. As late as 1982, Federal Reserve Board Chairman Paul Volcker escaped challenge when Congress tolerated his assertion that the Fed had no jurisdiction over Eurodollar transactions between the home offices of U.S. banks and their overseas subsidiaries: evidence that he was more adept at cultivating his reputation for "toughness" as an inflation fighter than schooled in the historical rudiments of his own responsibilities.

As an inflation fighter on the domestic economic front, Volcker was empowered to tighten credit. But as an accessory to runaway inflation on the Eurodollar currency front, he claimed to be powerless to prevent the proliferation of Eurodollar bank credit faster than he tightened it in the domestic system tied to the Eurodollar pool. Accordingly, the Fed professed to be too busy fighting moderate inflation in the domestic dollar economy to be bothered with complaints about indulging rampant inflation in the Eurodollar economy. The board's statistics on transactions between bank home offices and their overseas branches have remained inexcusably fragmentary and belated, even for government statistics: support for Volcker's complacent view that these high-speed and high-powered transactions were for the markets to judge, not for the Federal Reserve to monitor, let alone bring within its jurisdiction for the regulation of credit.

Consequently, America repeated in her maturity the costly error that molded and marred her youth: she failed to extend the protective hand of federal scrutiny over the banks enjoying the power to create dollars by making loans and accepting deposits. This original sin of omission did irreparable damage to America during her formative decades in Washington's dealings with the states. It culminated in Washington's dealings across the oceans with the market chaos of 1987, triggered by the successive collapses of the dollar and the New York stock market. The "invisible hands" pumping dollars into the Eurodollar market shrugged off the cheapening of dollars held. The simple device of printing more dollars, and selling them short, made good the losses.

Traditionally, the country that offered its money for use in world markets set the terms on how it was used and where. When the world worked on the sterling standard, England routinely luxuriated in the triple comfort of a permanent floating debt (and no nonsense about paying it off!), a 3 percent interest rate on it set by market stability, and a balanced budget. London counted on this display of financial strength to support the subsidy it spotted its banking clients abroad—and every country of consequence was on the list—by tolerating and indeed planning on a trade deficit, which Whitehall and the Bank of England between them financed from Britain's banking, shipping, and investment profits.

From Waterloo to Sarajevo, England kept firm control over the credit it doled out to its clients—America always at the top of the list, especially during her periodic banking crises. To cover its expanding portfolio of loans to client countries, England fortified itself with renewable collateral in the form of annual income—the only trustworthy collateral—earned by imports from England's foreign debtors. England's domestic prosperity provided the import base on which the country built its world banking role. This was no happy accident. Free trade sloganized the policy blueprinted for the pioneering prime minister Sir Robert Peel by the second of England's great classical economists, David Ricardo, often popularly bracketed with Smith but immeasurably more experienced in the workings of the markets. (In fact, it was from Ricardo, more than Smith, that Marx took his leads in English economics.)

When Britannia ruled the waves, London ran the markets, because sterling was the world's money. Any country that banks the world develops a vital interest in giving foreign debtors a chance to earn their interest payments by taking their imports. London, in its day, made trade import deficits pay by earning "invisible" income from interest and dividends to build England's balance of payment surpluses. Trade deficits are part of the cost of doing business as the world's banker.

But England never let its free-trade policy interfere with its national interest, which called for continual negotiation with the debtors underwritten by its support. The Pax Britannica, while it worked, showed that trade deficits need not provoke crises but, on the contrary, will promote prosperity for a creditor power. America precipitated disaster a century later when she permitted herself to run the world's biggest import deficit while becoming the world's biggest debtor. A dominant power can either borrow more than it collects or buy more than it sells, but it cannot do both at once and still do well. America's loss of control over her own currency as well as over her dealings condemned her to do both at the most dangerous possible juncture of events: at a climax of a speculative bubble, just when Japan, swaggering with intoxication as the world's emergent economic and financial superpower, was inviting disaster by lending more than it collected and, simultaneously, selling more than it bought.

Japan's error as a creditor with a surplus compounded America's as a debtor with a deficit and confirmed Ricardo's classic dictum that a creditor power is on notice to operate at a deficit, and a debtor power to operate with a surplus. The risks swinging on the double blunder shared by Tokyo and Washington were magnified by Russia's forced entry into the world capital markets as a jumbo borrower. Gorbachev's emergency experiment in *glasnost* and *perestroika* had left his Kremlin desperate to finance an emergency step-up of equipment imports in an economic environment dominated by the painful deflation in the price of oil, and in a financial environment colored by the onerous failure of the price of gold, another major Russian source of export income, to rise in response to the collapse of the dollar.

Until America advertised her political plight by surrendering her sovereignty in the Eurodollar market, the most devastating and deserved criticism leveled at any failed power had come from Secretary of State Dean Acheson in 1962. He chided England for having "lost an empire without finding a role." Although Acheson was not primarily concerned with the financial sources of national power, he was a shrewd strategist with a keen sense of how to use national position as a negotiating lever. He had good reason to take America's power for granted in the immediate postwar world whose councils he dominated. He had conceived the master plan for the reconstruction of Europe, sponsored by his then chief, Secretary of State George Marshall (when Acheson was still under secretary).

From my recollections of Acheson as late as the mid-1960s, I am confident he had no inkling that his stern, avuncular admonition to England would apply to the United States as well, and with such shattering speed. Yet it did, and within a generation, not a long time as the fates of superpowers are reckoned. When America abandoned the dollar to the whims of the market, she lost a role without finding a policy. The implications of the default were lost on the new crops of money managers educated to regard money as just another commodity. When money is commodified, the commerce of the world is disrupted.

Just as a horse swinging its tail in the sun can't identify the flight plan of the flies stinging it, America's sophisticated global market makers were baffled by the apparent coincidence of America's domestic industry shrinking as her Eurodollar float expanded. In the fateful summer of 1987, *The Wall Street Journal* put the markets on notice that the volume of dollar futures trading in the Eurodollar futures market had overtaken daily trading, not only in the Chicago bond futures market but also in the Treasury bond market in New York—$70 billion a day and exploding on this once austere and stable investment market faster than computers could track the traffic. The *Investors Chronicle* of London followed up with a sober calculation that bona fide dealings in actual goods entering into world commerce accounted for no more than 10 percent of this daily deluge of hot paper.

Both the megabond market and the Eurodollar futures market were undreamed-of during the 1920s. In the 1980s, America's latter-day Candides rationalized the disaster escalating in both markets with clichés hailing America's evolution from "an economy of smokestacks" into "an economy of services," buoyed by the global dollar economy. In fact, the backwash from the Eurodollar market was choking the entire domestic economy. The rate at which distress spread from America's farm economy to her industrial economy, and from her industrial economy to her service sector, measured the political devaluation of the dollar from a prime asset to an exposed risk.

Defeat invariably provokes a devil theory. America's conventional thinkers, echoed by the Treasury, blamed her two galloping deficits on the high exchange value of the dollar. They mistook the price of the asset for the power embodied in the asset. They relied on devaluation to spur the currency markets to provide the policy they lacked. Instead, the hysteria that seized the markets precipitated the collapse of the dollar. Subsequently, the dollar's partial recovery, manipulated by the central banks acting in concert to trap shorts, accentuated its volatile dependence on the erraticism of interest rate futures and, therefore, proved self-defeating.

Given that the dollar standard is here to stay, its successful functioning is subject to one condition: that America negotiate the terms of its use by the world. In dealings among nations, a popular adage—"There's no such thing as a free lunch"—indeed applies (although the Joe Kennedys of the world, like J. P. Morgan before them and Donald Trump after them, manage their coups and their raids without picking up the tabs). While the exchange markets bounced the dollar up and down, and all the markets signaled their satisfaction with the simplistic devaluation cure-all supplied to assuage their fears, the operators (some of them governmental) who ran the currency futures markets enjoyed a picnic, without prejudice to whether they kicked the dollar quotes up or down.

Consequently, in the name of free markets seeking equilibrium, the dollar was subjected to disequilibrium more demoralizing than a mere rate change or a mere trade deficit could force. These operators kept the supply of their own currencies tight, to control credit in their own countries, and created dollars for their borrowings. In addition, operators playing hopscotch by the billions on thin margins at the banks between currencies whose bills and bonds paid low returns and those whose instruments paid high returns necessarily borrowed dollars to make each turn, long or short, in every other currency. When any power, let alone the world's only diversified superpower, unwittingly gives away control over key assets, the resultant imbalances pile up into crises.

America's cash import market. It remained the one big, steady source of legal cash income to the export-dependent world, which enjoyed no dependable cash market remotely comparable anywhere else, not even foreign aid or the cash take from dope traffic. The steady but spectacular rise in single-family households, reflecting in large part the economic emancipation of women, provided a built-in stabilizer that delayed the onset of hard times: a powerful reminder to economists with conventional stances of the creative, constructive role that social change plays in the progress of economic society (a reminder as well that influential economic thinkers need not necessarily be sociological reactionaries; John Stuart Mill raised one of the nineteenth century's first articulate voices in behalf of the extension of democracy to women as well as to capitalism).

By the time the crisis of 1987 exploded on Wall Street, the Washington trade policy pendulum had swung from the extreme of paranoia in 1973, when Nixon belied his brag of expertise in world affairs by rushing into his ruinous soybean export embargo against Japan, to an extreme of complacence. But by 1987, the size of America's trade deficit had flung her onto the defensive, and her presumed power to pick and choose customers for her exports on her own terms was forgotten. Not only was she scrambling to find export customers on any terms, but her ability to compete against imports had surfaced as an overriding problem. Yet the *terms* of debate were limited to access to the domestic American market. The *right* of foreign access to American markets was never called into question— at least not as a commercial issue: the embargoes against Libya and Iran were imposed as reprisals against terrorism, an expedient short of an act of war, and against South Africa as an expression of moral disapproval.

A double miscalculation provoked the national policy crisis

over imports. The Reagan White House was responsible for the first error as an innocent expression of its economic philosophy, which called for trust in markets to set the policies that only policies can set for markets. The governments subsidizing the import dumping were responsible for the second miscalculation: their unanimous political belief in Reagan's power to keep America's markets open without any negotiation of terms.

Both miscalculations reckoned without the early discovery by one of America's pioneering merchant princes, R. H. Macy, that "the customer is always right." His dictum on the distinctive workings of the American system applies to politics as well as to business and, particularly, to the stormy politics of import business. Until the market crisis of 1987, the "customers" shared by politicians on Election Day and by businessmen on shopping days had counted on easy access to the best of both pocketbook worlds. They had expected to earn more at home and to buy more from abroad. So they borrowed more to cover their present buying in anticipation of their future earnings.

Throughout 1987, however, the Reagan White House, while professing sensitivity to the signals the markets flashed, clung to its clichés and avoided addressing the issue of negotiating terms for America's imports. Its complacence accelerated import dumping into America, and dulled the American appetite for everything— including even imports. The resultant shrinkage in the American market, sharpened by the foreign fear of protectionism, deepened the universal need to accelerate dumping into America. "Devil take the hindmost" was the watchword in every capital rushing to make the most out of the American market while the making was good.

Nevertheless, the Reagan White House believed that "the markets know best," even after the dollar's plunge forced the administration to waste billions intervening to stabilize the dollar. The administration's incumbent intellectuals took the behavior of the markets at face value. Their dogmatism blinded them to the foreign governments underwriting immense corporate losses in order to subsidize imports into the American markets: proof that governments around the world, beginning with Japan, had developed an effective riposte against protectionist measures calculated to cut their imports, either by raising tariffs or devaluing currencies. More costly still, they were paying for the privilege of buying profitless prosperity by financing exports with their credit. With every such marketplace victory for their imports into America, however, they were advertising America's enormous unused bargaining power.

But the collapse of the dollar in time initiated its own cure. First, the deflation of America's import markets intensified the dependence of America's import sources on their American customers. Then, the backlash from the profit squeeze on America's importers provided a consolation. It bolstered the hope that the day Washington announced its switch from one-way free trade to two-way fair trade, the terms America set for foreign access to the American market would be met.

America's agripower asset.   America is the only superpower commanding a food arsenal, not simply crops for export. The distinctive characteristics of this arsenal are less important than the fact that America is the superpower that has it. For better or worse, American agriculture is married to American industry as agripower. The basis of America's agripower is geopolitical; it draws on her economic and financial power, but is more than a simple extension of either.

Both other superpowers are food "have-nots." Russia is choking on its vast stretches of primitive, parched land isolated from its population centers by an antiquated transportation infrastructure. Japan is hungry and paying through the nose for lack of an agricultural base adequate to support its industrial development (and, more than incidentally, to check its epidemic of stomach cancer resulting from dietary insufficiencies).

All the farm powers per se—Argentina, Australia, Brazil, Canada, the Common Market, and increasingly China—are miniatures of America's agripower. But none of them commands food processing services interwoven with an industrial establishment remotely comparable to America's. Nor is any of them in a position to develop bargaining leverage against Russia or Japan—or, for that matter, against America if once she resolves to use her leverage in her own behalf. Japan's foreign trade operation is disciplined to use its food buying needs as a reciprocal weapon. In America, any such stance is denounced as protectionist, and worse still, inhumane.

In the case of Russia, none of America's crop export competitors can manuever to convert commercial transactions with agencies of the Soviet government into political negotiations with the Kremlin. Only America can; her failure to devise methods has obscured her distinctive ability to do so. America can bargain from strength because of the food weakness that limits Russia's formidability as her rival military superpower of the twentieth century. (America can do the same with agricultural China as, and when, it evolves into military superpower status in the twenty-first century. China shares Russia's woeful inadequacy of food-processing capability, as well as of sanitation, storage, and transportation.)

Nevertheless, America has invited Russia's experienced grain market arbitrageurs to exploit her futures markets. Year after year, they have planted stories touting bumper Russian crops at planting time. Routinely, the susceptible U.S. futures markets have responded with routine springtime plunges, which Russian traders in the American trading pits have seized on. By the time prices surged again in response to the autumn news that Russia had suffered another crop failure, the Russians wound up harvesting not breadstuffs but cash from profits on the sale of futures contracts picked up cheap by their out-of-season claims of bumper crops to come. Thanks to this routine manipulation of the Chicago grain pits, the Soviets have counted on obtaining cheap or even free grain.

The futures markets do not deal in real-life edibles, just in gossip about them, and in pieces of paper. America's agripower need not be subject to manipulated fluctuations in futures markets. It would not be, if brought into play as a negotiating weapon for transactions five years at a time. The Russians are brought up on five-year plans that don't work. America could feed them one that would work for both countries, but has never tried! America's distinctive diversification is not rooted in the games played in her futures markets. On the contrary, America would retain every bit of it if Washington were to bring the futures markets under responsible regulatory management. The key to the leverage her agripower gives her over other countries turns on the industrial, financial, and political base behind her agriculture.

Thanks to the distinctive combination of these assets, as well as to their magnitude, America can mobilize her distinctive resources to develop, process, sanitize, freeze, package, move, store, and finance finished food products, facilitated by her capacity for producing them abroad, instead of just offering raw crops to the speculative markets for manipulation by her competitors in the futures pits. America has resources in depth to put at the disposal of her agripower, transforming the agricultural segment of her economy from a black hole to a paying proposition and a contribution to human welfare as well. Welcome relief from the ghastly, grisly spectacle of America bleeding herself dry paying lip service to "foreign aid" dissipated on backward, hunger-ridden rural ghettos, none of them wanting for cash to lavish on arms, the number-one import for all of them, notwithstanding the conspicuous absence of foreign sources of insecurity.

America's Libyan crisis illustrated the contrast between agriculture as the world knows it and agripower as only America can use it. In the very month in 1986 that Reagan imposed sanctions against Libya and confronted Qaddafi at missile-point, *The New York Times* published a revealing report from Tripoli, detailing the absence of food and rudimentary sanitation facilities in that tropical country. Clearly, the limitation was not financial—even with Libya's minuscule population and oil then bringing in less than $10 a barrel—but physical. Libya did not know how to use the money it already had to buy a modern diet and to preserve food against contamination. Buying raw foodstuffs at random was not the answer; only American technology could furnish a comprehensive one. American agripower provided a model in the vast irrigation engineering system that her sociology had provided in response to the challenge her geography presented. America has always taken her irrigation capabilities and sanitation requirements for granted but has never recognized them as the strategic sources of export leverage they are. Both may prove a more decisive weapon in dealing with Libya, and with Russia as well, than her entire missile arsenal.

Meanwhile, the agony of the farm depression America was suffering in the mid-1980s raised a basic policy question. Her industrial heartland started to rust when she permitted her farm belt to go bad early in the 1980s. The collision of the farm depression with the oil depression no doubt seemed a coincidence to the unwary at the time. But the compounding of the two, on top of the simultaneous manufacturing depression, diversified the jeopardy when real estate values collapsed—by no means in just the farm belt, the oil patch, or the "Rustville" sprawl. The export revival sighted in late

1987 and early 1988 raised hopes of relief that, however, soon collided with the manipulated recovery of the dollar and the steady shrinkage of cash markets abroad for American products. Reagan's Washington offered none of the governmental assistance indispensable in arranging substantial barter flows.

From the narrow, price-oriented standpoint that dominated Washington during the Reagan years, America's farm economy seemed the most infectious carrier of depression for the 1980s—especially as financial support was withdrawn from it. The political marketplace has always been ready to judge agriculture a welfare client because it has always been first to beg for help. Instead, the collapse of American agriculture, and the spread of stagnation throughout the system while the stock market was making record highs, demonstrated that agriculture was not the weakest link in the American economy; it was just the first. The industrial economy and the oil economy followed it into the dumps. Only the stock market retained vitality after anemia infected the working economy—a direct reversal of the 1929 pattern, when the economy survived the market crash for the better part of a full year.

Agriculture is the one major export-dependent sector of the American economy. American industry is not. It remains too large, despite its startling scrappage of capacity during the 1980s, to prosper, let alone expand, on cash exports. Either American agriculture and industry will starve together or, thanks to American agripower, both will boom together. Washington can do better for American agriculture at the bargaining table and can open the way for American agriculture to do better for American industry than by waiting for any cash crop export windfall. By the same token, when American agripower is brought to bear on America's food export opportunities, it creates new domestic earning power in place of the will-o'-the-wisp spoken of as American cash industrial exports. American agripower can resume its historic contribution to American prosperity only when it helps the world.

To be sure, the economic consensus seized on the manufacturing export revival of early 1988 as proof that devaluation was working the magic claimed for it in the markets, however belatedly. Smug speculations are always the riskiest, never more so than in the case of the consensus bet that Rustville was exporting America back to a stable prosperity. As if all the increase represented goods finding bona fide customers abroad. In fact, the opportunity to brandish lower dollar price tags abroad sent American manufacturers scurrying to flood their overseas affiliates with inventories. (When *The New York Times* put this distinction I made between sales and shipments to the export experts at the Department of Commerce, they admitted that they had never considered it.)

This bet reckoned without the need of American industry and its work force for a continental-scale domestic market, which all the world's fragmented markets put together could not duplicate, even if each and every one of them were not itself export-dependent. It rested on the evangelical academic case for overnight investment of megabillions to convert American industry from mass-production dependence on recurrent volume to small-scale, multiproduct operations cut back to match the size of available export markets: a surrender of America's distinctive economies of scale responsible for her lost ascendancy in her own markets. Finally, it ignored the realities in every cash export market supposedly eager to snap up America's bargain offerings: the economic reality of overseas shrinkage in an already uncomfortable environment of job loss; the commercial reality of traditional barter and countertrade to minimize further job insecurity; and the political reality of naked, unashamed, positively popular protectionism.

America's technological assets. America's most distinctive technological assets have been military, particularly, but by no means only, the ones aimed at space. They are the advanced systems that every country outside Russia wanted America to have, but that all other countries outside Russia counted on America to keep proprietary and competitive with Russia for their own protection.

Providentially, the American technological assets of greatest potential value in retrieving her lost world leadership were best suited to political negotiation, not for sale in the commercial marketplace. They fell into the same basic categories as those dominating her budget disputes: military (including communications), health, and agriculture. In all three categories, America owed the technological impression she made on the outside world to her size, rather than her performance. America's competitive performance suffered even in the government-controlled market for U.S. military systems, presumably the strongest segment of the economy. Under cover of the doctrinaire insistence of the Reagan White House on "free-market" forces, defense import prices undercut them. Hidden subsidies made the difference. America's failure to compete even in her own military procurement market, while her own government was mouthing platitudes against the bulge in the trade deficit, dramatized her fecklessness.

Lester Thurow, dean of MIT's prestigious Sloan School of Business Management, emerged as the most articulate spokesman promoting the lost cause of price competition in free markets as a nonpartisan national policy. He won fame as the presumptive adviser to the Democratic party, and indeed, the Democratic presidential candidates who lost in the 1970s and 1980s did adopt his recommendations. As the acknowledged mentor of failed Democratic nominees, Thurow called on them to abandon their political base among the country's working people and, instead, press to bring domestic costs and living standards down to whatever level might be needed to accomplish the elusive purpose of undercutting subsidized competitors abroad. (Democratic congressional campaign committees, closer than national candidacies to the victims of import subsidizing, are gaited to win elections; they have no patience with this political exercise in the economics of austerity in pursuit of foredoomed campaigns.)

Thurow made no allowance for how high foreign barriers against imports would go to counteract low U.S. export prices. Nor did he anticipate how poorly the American economy would perform after the Reagan administration he opposed imposed the sweeping devaluation he advocated. He also failed to distinguish between advantages preserved by competitors abroad through subsidized dumping of commercial products in world markets and the proprietary political advantage America enjoyed through space and other advanced military technology. Japan was not about to build nuclear submarines; neither were the members of the European Common Market, nor Brazil or Mexico. America has no need to export this kind of proprietary asset and is better off not trying. American industry would do well if it became competitive again in just its own home markets. Making the most of an advantage is better business than losing blood in a rat race.

*   *   *

One simple word summarizes America's distinctive fund of technological resources: *quantity.* America has more of every source of
technological formidability—from skills to customers—than any
other country. Many other nations have better skills and put them
to better use, but very few are big enough to offer avenues of opportunity to many entries in all advanced technological fields. America is
so big, so bountiful, and so full of skills able to attract partners
and/or money that she does not need to be better, or even to do better
in world markets. America is one country where, unlike Russia,
Japan, or Germany, being fired or going under is no bar to a second
chance.

Just as a failed haberdasher could succeed in winning a presidential election against a "slick New Yorker" (as Truman characterized Tom Dewey), so can the perennial American inventor still hit
the jackpot; though no longer as a diligent and faithful cog in a
corporate machine. The financial traps America's mismanaged corporate conglomerates set for themselves freed her pool of technological skills to bubble to the surface, borne by independent currents. In
the 1980s, America saw more new business starts than ever. Recon-
firming Veblen's blast against the "trained ignorance" of her "captains of industry," her corporate managements and their financial
advisers accounted for the most burdensome liability she carried
during the takeover boom; its one constructive feature took the form
of divisional managements' buying freedom, however flimsily financed, from corporate parents. The skilled employees who flew the
corporate coop and went on their own as start-up entrepreneurs
remained her decisive asset. Skilled entrepreneurs starting on shoe-
strings, demonstrating the knack of survival, enjoy the feel of money.
They go after it with flair and zest that no expense accounts or
bureaucratized managements or venture capitalists can match. No
other country could equal them in numbers, diversity, independence,
or success. By contrast with the stratified limitations on financial
opportunity in every other country, America's ongoing population
explosion of investor-entrepreneurs combined technical knowledge
with financial decision making.

8

BARUCH'S BOOTBLACK
The Assets America Hides

The official dimensions of the American economy are as misleading as legends of Victorian morality. They understate the scope and momentum of the entire domestic economic mechanism because they ignore two major underground elements: moonlighting (including refugee cash stowed away in foreign tax sanctuaries) and criminality. Both forces buoy economic activity by injecting huge supplies of liquidity into the money stream but deprive the Treasury of any revenue share from both activities. Consequently, the Treasury's computers consistently overestimate the level of federal tax receipts and underestimate the level of federal borrowing needs—a costly statistical double distortion.

Gambling, historically a third element, is the biggest and best business in troubled societies abroad. Its legalization accepts popular preference but is a portent of fiscal distress, monetary disturbance, economic squeeze, and social malaise. In America, legalized gambling has spread along with these symptoms. Its institutionalization has acknowledged the magnitude of the cash float outside the reach of the tax collector. At the state and local level, the incorporation

of gambling into the tax structure represents a systematic effort to
tap a new source of tax revenue, while indulging a popular sport. No
growth business itemized in the statistics runs gambling a close
second—especially during the football season, lengthened in response to fan demand. Incentives to mix pleasure with profits whet-
ted the popular appetite.

In the autumn of 1987, my masseur gave me a graphic illustration of how respectable gambling had become, and how broad a base
it was providing for the affluence fed by the financial markets. He
told me of his determination to put the proceeds of his next lottery
lucky strike into the stock market and asked me for a tip. The
incident reminded me of Bernard Baruch's legendary bootblack, who
offered him a stock tip in October 1929. Speculative fever at that low
a bracket satisfied the professionals that the end had come, and
Baruch spent the rest of his life saluting the bootblack as his adviser.
But my masseur's vow showed how much healthier the Wall Street
environment had been in the climactic autumn of 1929 than it was
in the anticlimactic autumn of 1987: Baruch's bootblack already had
the money for his margin.

Cash flows from the moonlighting economy, windfalls from out
of the blue, have an energizing effect. Criminality showers the economy with cash, too, but the toll it takes of people is crippling. In both
cases, the money flowing in and out of these underground extensions
of the legal economy is too elusive to be measured with precision. But
the accelerating rush of refugee money into the underground economy has revealed leakages in the legal economy that are large enough
to explain its sluggish tendencies. Unmistakable evidence of dynamism outside the legal economy, especially at entry-level jobs, confirmed unsettling criticisms of creeping governmental incapacity,
administrative as well as managerial. Nevertheless, America's bipartisan obsession with market behavior abroad distracted her policymakers from examining the bargaining power of her government
domestically. The police power is popularly seen as the most effective
asset she commands for reversing the unfavorable cash flows that her
legal economy has been running into the underground economy. Yet
the bargaining power of her government remains her neglected asset,
not only with the outside world but, even more, with her own flour-
ishing underground economy as well. Washington has failed to bring
it into play.

By December 1983, extralegal domestic cash flows from the legal economy won belated official recognition as too important to ignore. The Joint Economic Committee of Congress made the first try at a guesstimate. With apologies for the inherent chanciness of the data available, its study suggested annual variations adding between 5 and 20 percent to GNP—some margin of error! This calculation, though high enough at the time to discourage follow-on probes, turned out to be too low. Subsequent revelations showed that the committee had underestimated new highs ahead in illegal immigration and employment off the books, as well as trafficking in drugs and stolen goods. Between 1983 and 1987, this extra- and illegal supplement to GNP mushroomed, while the official GNP stagnated.

The elasticity of the underground economy has frustrated the consensus of economic expectations, which are conditioned to statistical performance. True, the official count of domestic money supply was jumping by leaps and bounds during the precrisis years of the early 1980s; the monetarists were vocal in their complaints against this "permissiveness," as they called it. But they were also confused, if less vocally, by the failure of these huge increases in the money supply to prevent the creep of deflation into the price structure by way of the earnings stream.

This important polemical issue aside, however, the accelerating inflation of the underground economy covered up an ominous deflation in the official count of the money supply. It opened a cash gap too big and too elusive to be tracked between the artificial world of computerized statistics and the real world of consummated transactions. No one knows how much of this cash disappearing into the underground economy ever comes back to the legal economy, though much of it eventually is forced to do so. Money hiding out is always on the lookout for attractive offers. A consumer economy, running on credit cards and showering discounts indiscriminately, is at the mercy of cash buyers. No one knows how many recorded transactions—including bank deposits by legally identifiable depositors—originate outside the legal economy.

But everyone who stops to think about it knows that moonlighting and criminality provide powerful and continuous support for the economy, not identifiable in the statistics on income or spending (or, for that matter, on savings either). At one and the same time, the combined effect of moonlighting and criminality leaves the income statistics understated but endows the spending/savings figures with an unmeasurable add-on. Nevertheless, the statistics on spending/ savings, though benefiting from these invisible illegal boosters, show a decided lag behind those on employment. Any sophisticated reading of the statistics on both sides of the income-spending ledger flashes a reminder that the country has a great deal more money at its disposal than the government or the markets think it has, but is doing a great deal less with it than the government or the markets think it is.

Personal income is by far the largest component of GNP. By 1986, the growth of the underground economy became so conspicuous that the statistical policymakers felt obliged to legitimize a head count of its moonlighting component. They did not dare to try to claim a hidden "inflator" from the underground economy in the retail sales figures; though all dollars scattered throughout the underground economy to pay for work, supplies, and services either go into the mattress or flow back into visible channels of consumption. Despite the undeniable lift to both retail sales and savings, each remained too stagnant to invite tampering.

Employment was the statistical series bubbling over with synthetic energy. Accordingly, the guardians of this statistical presence factored a plausible and convenient moonlighting "inflator" into the employment numbers just when the routine monthly countdown on inflation petered out, but before hard evidence of income deflation and employment uncertainty accentuated the need to bring the underground economy within the reach of statistical measurement.

This same evidence pressed the statistical authorities to extend the parameters of the work force to include "aliens" (explicitly identified as Hispanics) in underground employment—as if all aliens were Hispanics, as if each legal full-timer had only one job, and as if only aliens engaged in the common American pastime of getting paid off the books, as if the Immigration Act of 1986 had not granted wholesale citizenship to millions of non-Hispanic aliens with roots in the illegal economy. One fact of life statistically verified by the Census Bureau did, however, provide a commonsensical basis for this gunshot approach; during the 1980s, the Hispanic population grew five times faster than the total. Just as the practice of counting marginal labor as full-time members of the work force exaggerated its dimensions and failed to account for the shortfall in spending powers, so the insistence on counting each job holder only once understated the parameters of the work force and its potential for producing earning power.

More glaring still, the new moonlighting inflator made no allowance for the routine reliance of shoestring businesses, old as well as new, on moonlighting help. A startling number of them, no longer concentrated in New York but spread representatively across the country, accounted for a boom in sweatshop labor, reflecting a distinct gravitational tug toward hourly rates paid across the Mexican border. Nor did the inflator allow for the growing practice among larger corporations (especially defense contractors, uncertain of continual government cash flows) of cutting payrolls and increasing dependence on outside contractors who rely on moonlighters. The Hispanic inflator did nothing more than provide window-dressing for official claims of dynamism in the legal work force.

Yet the fact of statistical recognition of the underground economy calls for a closer look at its anatomy, especially after the Immigration Act of 1986 snafued the official 1985 calculations of alien employment. "Real" GNP in 1985 (with inflation adjusted for it and/or squeezed out of it) amounted to just under $4 trillion. A conservative count, splitting the difference between the Joint Economic Committee's 5 to 20 percent spread, calls for assigning a 12.5 percent add-on for underground GNP. This median calculation of the underground inflator adds some $500 billion of untaxed income to the official 1985 total of $4 trillion for GNP.

Being conservative, and making no allowance for continuing expansion of the underground economy, any appreciable portion of even $500 billion recouped by the Treasury each year would make a signal contribution to reducing its $100–$200 billion deficit to manageable proportions. Assuming the effectiveness of incentives to switch from underground to legal employment—which I will offer in Chapter 16—suppose even half of this conservative estimate of $500 billion, $250 billion, were induced to go legal. At the 20 percent tax rate, the Treasury would collect an annual bonus of $50 billion in new revenue from such a $250 billion legalized income float. No proposals to tighten the screws on present taxpayers come close to promising as much revenue. Tapping this new vein would help stabilize interest rates along with dollar exchange rates, ease anxieties about the deficit, and call off arguments about raising taxes. America has been down that road before: higher tax rates guarantee still more moonlighting; higher consumption taxes guarantee still more retail bootlegging. In 1987, self-defeating tax reforms did just this, especially in construction, the country's most diversified industry.

The size of this massive toll diverted from the Treasury to the underground economy reveals a frustrated and angry society imposing its own de facto withholding tax on the government. For working people, illegal moonlighting has been like booze during Prohibition: anything but furtive. Furthermore, legal moonlighting by money at work was churning another visible but anonymous Niagara of U.S. government and corporate dollar debt in the Eurodollar market. Interest earned there is untraceable, unreportable, and untaxable. In a society on the warpath against double standards, moonlighters who worked for their money were outside the law, while money moonlighting outside the country was within the law.

Here are the broad outlines of the structure of the American underground economy—domestic moonlighting by wage earners, foreign moonlighting by runaway money, and criminality, especially dope sales. Practices prevalent throughout the government, the economy, and the Eurodollar market have sprouted an exponential expansion of underground cash flows. The money involved adds up to a huge, untapped American asset, continually reproducing itself and neglected for policy-making purposes because morality rejects its recognition—let alone reclamation—as an asset.

Domestic moonlighting. The United States has emerged as the only country in the world able to stimulate a booming black market for labor and, at the same time, to support a continuing expansion, however selective, in her legal employment. It has been, for example, the only country where, despite the boom in moonlighting, legitimate retailing, subject to direct taxation, has held its own against the knocked-down cost of bootlegging merchandise in the streets.

Moonlighting opportunities broadened from Pearl Harbor through the 1980s with the dramatic expansion of the full-time employment base, thanks to increased participation by women, Hispanics, blacks, retirees, teenagers, and the handicapped. Gratitude, however, is not an economic motive. No sooner was the welcome mat put out for recruits into the labor force than their rebellion against withholding and "bracket creep" (as it was before the Tax Reform Act of 1986) sent them joining older hands at the game in the chase for more take-home pay. Each phenomenon fed on the other: resentment against the tax bite and the cost push contributed to the drop in productivity and in the quality of workmanship throughout the legal economy, which, in turn, fed the demand for moonlighting labor. What's more, encounters with moonlighting labor tended to elicit expressions of satisfaction with its quality, and therefore, to become habit-forming.

Black markets—for goods as well as for labor—are natural by-products of perceived overregulation, overtaxation, and under-compensation. They are expected to flourish during wartime as shortages pinch. But an underground boom in peacetime reveals social strain flawing economic performance and expressing political discontent. The moonlighting economy in America functions on a larger scale than in countries like Italy and England, where its growth has been striking. At first, this condition seems normal: everything in America is bigger than its counterpart anywhere in Europe. But comparison between the underground economy in America and elsewhere is misleading, while the contrast is commanding.

Elsewhere, at least until the slump of the 1980s, either the underground economy or the legal economy has enjoyed striking growth: not both together. In Japan, payoff costs are routinely structured as brutal price markups into the legal system; domestic consumption suffers. In Italy, the underground economy substitutes for a legal economy, thriving uninterruptedly during recurrent crises forcing suspension of the government. Moreover, Italy's underground economy benefits from easy export accommodation by the banks, which are presumably supervised or at least scrutinized and, certainly, shaken down for bureaucratic approval of routine banking practices.

By contrast, for nearly half a century, America's moonlighting economy has supplemented, not substituted for, a legal economy that commanded respect through all its fluctuations. The enormous momentum of America's legal economy inspired income expectations high enough to persuade people either to remain in the system despite their frustrations or to straddle both economies, often quite comfortably. Therefore, America's legal and extralegal employment grew together, the first accruing the fringe benefits, and the second the cash. The free choice of alternatives in every community in America put the legal system and the policymakers presiding over it on clear and stern notice that withholding had festered into a resented tax on legal employment.

The contribution of moonlighting to national productivity brought an overdue correction of the inferiority complex into which Americans allowed themselves to be bamboozled. Putdowns of American labor charged that American workers had lost their interest in working, as well as their dedication to working well. But in the moonlighting economy, American working people have shown themselves to be at least as skillful, as hardworking, and as conscientious with their tools and materials as working people anywhere— when their motive is to work for themselves: weekends, holidays, and nights. This went double for women and minorities.

The more complicated a society gets and the more capital-intensive, the more dependent it becomes on repairs, which are always labor-intensive and, therefore, the undisputed domain of moonlighters selling their skills. Under cover of the prevailing confidence in managerial precision and control throughout the expanding informational society, the microchip revolution spread the unworkability and inefficiency on which moonlighting mechanics fed. Moonlighters answering emergency calls for repairs named their prices, taking control of the modern marketplace with the air of authority asserted by lords of the manor in the days when labor was chained to the soil.

By the late 1980s, diversion of labor time to moonlighting accentuated the chronic shortage of skills in the legal economy, and not just for repairs. The tug of money earned from moonlighting brought a new deal to the work force, reversing the traditional expectation that higher status brings more pay. Instead, people sacrificed status to grab cash. Lower-down jobs in the moonlighting economy paid enough in cash to woo skilled workers away from higher-status jobs in the taxpaying economy. For a young single woman making $400 a week, subject to withholding, in New York City as a traffic manager with a prestige firm in the garment district, the incentive offered as a barmaid in a neighborhood Queens tavern for $500 a week take-home cash was irresistible, worth the sacrifice of prestige. The pay levels for New York City may have been unrepresentative, but the principle and the practice were common throughout the country.

At the outset of the 1988 presidential campaign, George Bush drew howls of bipartisan derision from the economic establishment when he was seen, as the jibe went, caught again with his "silver foot in his mouth." His promise to create 30 million jobs in eight years stirred up the hooting; his own partisans among the economic mandarins led the chorus. But Bush's blooper was one of definition, not of arithmetic. Apparently he did not realize that the great bulk of the jobs he promised already existed. If 117 million Americans were counted as employed within the legal withholding system, and a good 20 percent more (a conservative guesstimate, remember) in just the *extralegal* (not the outright illegal) economy, no fewer than 23.4 million Americans were already at work making hard cash outside the legal economy. No presidential feats of managerial virtuosity were needed to "create" a paltry 6.6 million jobs in eight years.

The prevailing attitude has been to wink at moonlighting as a utility and a waste. America has been shrugging off moonlighting by her work force as too widespread a problem for enforcement and, therefore, off-limits to policy. The moonlighting economy is indeed too big to police: so big that only policy can reach it, and certainly too large for America to neglect.

During the war years, when Reuther and I used to philosophize about ways and means of avoiding the normal cyclical postwar bust, he told me, "The guy on the plant floor relies on forty hours a week to make out, but looks to overtime to pay for his big ticket items." Under cover of the claims made by Reaganomics for job dynamism, overtime disappeared, diverting the incentive for overtime into the underground economy. When Michael Dukakis sounded his call for "good jobs at good wages" in his acceptance speech at the 1988 Democratic convention, he mistook American working people for members of a stratified foreign working class, content to settle into a rut. Perhaps he projected the theme of his appeal as the son of immigrants from the social to the economic arena. Lane Kirkland, the shrewd president of the AFL-CIO, came closer to the mark when he told me, "The working stiff wants to make a buck"; modernizing Sam Gompers's muscular call for "more," and transposing it from the checkbooks to the net worth of the American people; their status has been upgraded to that of people of property. The solution presented in Chapter 16 would guarantee the achievement of both objectives: kindling incentives to collect continual overtime subject to withholding (the bumper crop of revenue the Treasury needs!), and ensuring the work force a riskless ride on the market escalator.

Foreign moonlighting. American sophisticates, working their dollars in sanctuaries abroad, have created an asset entirely different from the cash earned by moonlighters working at home. In the Eurodollar market, where money has moonlighted in large lots, the game has been played outside the regulatory system, while still being entirely legal. Eurodollars have been as acceptable to the domestic banking system as domestic deposits, but the incentive to hold them has been increasingly offset by the risk of ongoing dollar devaluation. While interest paid on Eurodollars has been beyond the reach of U.S. taxation, Eurodollar deposits—including jumbo trust deposits— have also been beyond the reach of federal insurance protection. America's multinational banks thus enjoyed the best of two worlds: borrowing access to this bulging new reservoir and the domestic franchise to relend the dollars they imported at will. The proliferation of dollar deposits around the world has offered American moonlighters of the entrepreneurial, managerial, and professional classes abundant opportunities to move very considerable caches of cash into perfectly respectable dollar holdings abroad.

Chapter 7 examined the Eurodollar market as a substitute for the dollar. It has mushroomed under the umbrella of American sovereignty while staying beyond America's regulatory control. Here, the focus is on Americans with the money and the know-how to buy free rides in the Eurodollar economy. The bustling Eurodollar market has asked no questions, imposed no restrictions, and operated a continuous tax-free turnstile for moving undeclared currency holdings into negotiable securities earning market rates of interest.

So Americans found themselves free to double their participation in the dollar float on both sides of the world, by keeping their dollar bank and securities accounts open overseas and borrowing back their own money into domestic accounts. Therefore, Eurodollars, collecting interest on time deposits and investment securities in overseas sanctuaries, have performed the apparent miracle of seeming to be invested while generating new liquidity for laundering into tax-free repatriation. U.S. tax filers utilized tax-free Eurodollar accounts overseas to create taxable interest deductions at home and tax-free interest income abroad. Corporations issuing bonds in the

Eurodollar market took deductions on the interest they paid, while holders of the bonds who happened to be U.S. taxpayers collected this same interest tax-free in their overseas sanctuaries: a breach of the trade-off in the U.S. Tax Code justifying the deductibility of interest on bonds as "passive income." The Tax Reform Act of 1986 accentuated the limitations on passive income entitled to deductible interest. Despite its endless complications and ambiguities, however, this obvious loophole eluded the scrutiny of the tax reformers.

So respectable did refugee American moonlighting dollars become that, during the 1984 campaign, the Reagan administration gave them a safe-conduct pass. A sophisticated management at the U.S. Treasury would have recognized the Eurodollar market as the depository of proprietary American assets it could lure home. Instead, the Reagan Treasury created new liabilities by borrowing Eurodollars especially created to indulge refugee dollar holders in their overseas tax havens. Traditional wisdom counts America's deficits as weakness, curable only by America's acquiescence to terms dictated by her creditors.

In this case, unanticipated by the sacred texts, the terms were exacted by the presumably taxpaying owners of America's own dollar hothouse in exile. Her Eurodollar creditors duly delivered their ultimatum, and the Treasury, conditioned to observe the rules of the game being played against it in the Eurodollar market, promptly acquiesced. America's policymakers and their economic advisers advertised their naïveté by their surprise over the refusal of her deficits to drop. The stock market advertised its susceptibility to official claims with its blind buying of the recipe cooked up by the Reagan Treasury.

The White House then legalized tax avoidance by acquiescing to an ultimatum from Eurodollar traders for new bonds to be issued to anonymous bearers of Eurodollars. These traders were readily identifiable as managers of European branch offices of major Wall Street firms, and their demands were put to the U.S. Treasury by their home offices. Reagan's opponent, former senator Walter Mondale, received no alert from his advisers that past confrontations with foreign dollar lenders had regularly preceded America's historic financial crises.

Both 1984 presidential campaigns shared the prevalent ignorance of history. Neither recognized the danger in the bearer bond sham. The Treasury not only surrendered the bonanza of tax revenue from interest paid on government bonds: it also accepted the Eurodollar market claim to set the terms by which America could borrow its own currency abroad. The Treasury's failure to protect this prime American asset escalated its exposure to the 1987 chain reaction of market disasters that wrecked, first, the dollar market, then the U.S. bond market, and, finally, the stock market.

The run on the dollar triggered by its collapse was a first for a world reserve currency. When gold served that purpose, stability was its hallmark; it never collapsed. In 1931, when sterling collapsed, it was no longer the world's money; the dollar, then still impregnable, had taken over from it with America's assertion of creditor power. After 1985, however, the discount on the dollar sped up the worldwide chase for dollars in every back alley of the underground economy.

As officially mandated cheaper dollars poured into the float, the float masters in the Eurodollar market responded to desperate incentives to print more of them to stay even. Simultaneously, dollar leakage from America's domestic legal economy into her domestic extralegal underground accelerated and ran into the surrounding criminal channels of the world economy. The dope masters felt the same pressure from the collapsing dollar as the oil producers—to produce more in order to earn as much. Meanwhile, the Eurodollar market bought the idea of dollar devaluation as much as yuppies. From 1985 to 1987, Eurodollar flows into the stock market outpaced the market's meteoric rise. No one will ever know how much dirty money flowed in with it.

By no means for the first time, expatriate American money made too much of a good thing—just as Tory money did by sandbagging the Continental Congress in Hamilton's time, and as J. P. Morgan did in the name of his British clients by provoking the panic of 1893. History did more than repeat itself: it increased the stakes.

The criminality factor. America has always loved big talk, but bragging also invited scoffing as a response. Her failure to come to grips with domestic drug criminality, while preaching against it, illustrated this propensity. Over the years, she has tolerated the growth of a criminal economy of huge proportions, accumulating untold wealth, flouting the law, and wasting lives within the orbit of the vibrant resources at work in her economic society.

The drug economy has flourished on three fronts where America commanded formidable assets, which she neglected. America's unused bargaining power with all of the countries of drug origin and transit was the first. Her failure to track, let alone tap, the cash take of the drug economy was the second; history buffs recall that Al Capone, the czar of the bootleg era, was bagged not for murder, but for tax evasion. The victims in the aborted war against drugs were the third.

To take just one conspicuous example of America's failure to lay down the law to countries where her drug imports originated, Turkey is high up the list, though it is a member of NATO. It boasts the most formidable military force of any member in its area and of its size. The fact remains, however, that NATO has been allowed to dwindle into a dubious ally at best. So America's indulgence of Turkey's prominence in the drug traffic has left her with the worst of both worlds: drug blackmail from a marginal military client abroad and an epidemic of social decay at home. No doubt Istanbul engaged in a cynical ploy in the 1980s, tempting Washington to lavish a little more military patronage on it to induce more crackdowns on local drug trafficking. Istanbul was explicit in threatening the status of American bases in Turkey if a finger were put in the dike of Turkish steel imports into the United States. Washington invoked domestic budgetary austerity instead of buying Turkish police action.

Although losses in dollar terms were the most easily measured of the damage to America from criminality, the future cost, by definition incalculable, of neglecting her young people far outweighed any of the immediate cash considerations, from drug law enforcement to drug addiction treatment, in America's belated quest for priorities. The exposure of her educational system to contamination by imported drugs has followed from her neglect of her overseas bargaining power against countries of origin, as well as from her own obsolete methods of offshore policing against this imported epidemic. Moreover, the dollar cost to the Treasury, and to American society, of negotiating enforcement in countries producing and shipping drugs would produce staggering returns, sufficient to help finance

and man American military defenses now being paid for with borrowings or axed for lack of funds.

Terrorism expanded under the protective cover of Third World
governments—on both sides of the Curtain in on the drug traffic: the
money did triple duty, in backing for the terrorists, payoffs for governments, and markets for drugs. The Eurodollar market furnished
the ideal shelter for drug money—not merely as it paid for entry into
the American market, but immeasurably lusher as the ideal haven
for loot sneaking back out with tax-free anonymity. Exposés of cov-
ert U.S. government agencies arranging U.S. air base cover for soldiers of fortune flying in dope seemed sensational at the time, but
were entirely predictable and routine.

Economic forces, obvious but ignored, explained the embarrassment the Reagan administration brought on itself through its zealotry in unsupervised support of the contras. The moment the White
House undertook to subsidize the contras, it signaled an implicit
commitment to buy the cash crop of the region. The alternative was
just as plain: for years, the U.S. government has spent billions paying
U.S. farmers not to grow nutritious crops. Paying our freedom-
fighting allies not to grow dope would have been a mercy bought at
a bargain—certainly worth violating the precepts of "free-market"
economics.

Colombia, also high up on the drug source list, won congratula-
tions for joining with Exxon, a symbol of marketplace competence,
to squander a sheikh's ransom on investment in productive capacity
wasted on the market: a huge new coal mine starving for customers.
What was left for Colombia, besides coffee and bananas, but the
rewards of dope? In a rational world enjoying the decay it brought
on itself, as the modern world has been taught Rome did, Americans
might have expected to buy gratitude with their purchases of the raw
material for drugs. Instead, she has earned hostility within Colombia
as the marketplace in which the big markups are pocketed on the
manufactured product. As Colombians see it, the markups they take
are petty cash by comparison.

Inside the United States, Miami emerged as the first metropolitan hub to prosper on a firm foundation provided by dope, as the
Connecticut River mill towns had on cloth, and as the Pittsburgh
valleys had on steel. Thanks to the drive against the smuggling of

marijuana, a proper boom developed in the homegrown cash crop: an argument for protectionism against imports of legitimate products! Dope also invaded and helped corrupt the financial markets during their 1985–87 spree. Active trading of drugs for stocks demonstrated a receptiveness to free-market bartering of snorts for securities that left the Reagan administration looking old-fashioned in its resistance to government-to-government barter.

Early in the history of the Republic, the Logan Act outlawed any private dealings by American citizens with foreign governments in behalf of the United States or in lieu of direct dealings by the American government itself. Since its enactment in the first decade of the nineteenth century, no question about the constitutionality, legality, or practicality of the Logan Act has been raised; nor has any amendment been offered. It stands as a unique postscript legislated to the Constitution by our start-up experience as a nation, with no subsequent need for the process of amendment.

The reach of the Logan Act has gone unnoticed down through the years because the ban it imposes is indispensable to the exercise of sovereignty by any bona fide government, however young. When the need for the act won bipartisan recognition, surviving fathers of the Constitution, leading their warring factions, were still in charge of the ongoing congressional debate, so that the controversial question of "original intent," reopened by Judge Robert Bork in the memorable controversy over his failed confirmation as a Supreme Court Justice, was settled from the outset. Only one serious effort has ever been made to circumvent the Logan Act: the Irangate conspiracy, itself an undercover government operation syndicated out by unauthorized conspirators to mercenaries, conceived in ignorance and foredoomed. A case of the exception proving the rule.

Encrusted ideological delusions have blinded America to her stubborn and painful collision with the world she has nurtured and protected. Imported economic theories, never used as guideposts abroad, have led her to mistake foreign government operations for free-market forces. The pitiful figure of Don Quixote, armed with his lance and tilting at windmills, haunts her bipartisan economic texts, making them as ridiculous in their solemnity as he was.

Congress agreed on the Logan Act in an era when the powers of government were conceived of as police powers, purely political and military, though, even then, modern economic society was still evolving in local enclaves, limited by the logistics of sea- and horse-power. Long since, the rest of the world—its advanced as well as its backward and, certainly, its developing portions—has extended the orbit of continual activist governmental vigilance from policing domestic operations to competing, economically and financially, with other countries. Some compete conventionally with commercial products, others respectably in the arms business, some criminally in the drug business, some—the more backward—in arms and drugs both. All other governments with any capability for exporting arms routinely draft their heads of state, from monarchs down, to double as arms sales agents.

Europe tittered for years over the tale of the king of the Belgians, hustling to earn export sales commissions, on the phone with the shah, exclaiming, "We can close this deal, king to king." Margaret Thatcher undertook to close a Pentagon sale with Reagan, and Reagan spent his last years in office denying that he had with Khomeini. Of all heads of state in the world, only the president of the United States is free from this tawdry burden to double as a traveling salesman, thanks to his responsibility to bargain in behalf of the world for the United States. Reagan mixed up his two missions by allowing the markets to bargain for America and by volunteering to bargain arms for hostages. He did not understand the major challenge of the twentieth century: how to manage the merger of the police powers of government and the money power concentrated in big business.

Gorbachev, by contrast with Reagan, found himself in the same position as all the world's other heads of government. Russia is the world's number one exporter not only of oil, but of arms. Accordingly, Gorbachev found himself the world's number one arms peddler. Classical economics taught that international business and financial dealings are best left to participants and barred to governments. In the twentieth century, however, Lenin contributed more to confusion about capitalism and statism than Adam Smith. His venomous polemics were relentless in accusing European socialists of selling out to capitalism.

But the history of the century demonstrated that capitalism needed to buy protection from socialism at least as much as it pressed to win control over it. Certainly, as socialists formed governments, capitalists in every country relied on their marriages of convenience with the socialists in office to ensure protection of their foreign interests. Gorbachev's exercise in *glasnost* presented a novel challenge to the besetting modern problem confronting governments of all stripes: how to apply the police powers to the management of a modern political economy—by definition, a blend of capitalism and socialism, of markets and regulations. He challenged his co-heirs of Stalin to yield power and privilege without threatening them with the menacing methods of Stalinism.

America has abundant capabilities for policing the ramparts her government is responsible for watching. As her Constitution in all its marvelous precision and flexibility stipulates (and as the forgotten but indispensable Logan Act supplements it), only her government can deal with all the other governments targeting her for rich pickings. When the country was celebrating its fifteenth decade under the Constitution, FDR's New Deal was settling the vestigial controversy over the Commerce Clause of the Constitution.

FDR established the responsibility of the federal government for invoking the clause to rescue the domestic economy from depression and provide insurance against a return trip. But the American government did not wait for any argument, let alone for its first depression, to authorize the clause in its start-up powers to regulate foreign commerce. From the outset, it assumed direct oversight over all inward traffic, provoking controversy only over the methods—such as tariffs—used to distribute sectional benefits and penalties. Moreover, the Constitution reserved prior regulatory control over the dollar to the government. The market for money is the common denominator of all markets.

Like Keynes or not, he did lay down an axiom when he began his *Treatise on Money* with the declaration that government "for some four thousand years at least . . . claimed the right not only to enforce the dictionary but also to write the dictionary" in defining money. The Founders, Jeffersonians included, were conservative enough to understand this 150 years before Keynes explained it. The doomed world of 1930 had forgotten this rule when Keynes inveighed in vain to remind its opinion and decision makers of it. Reagan's managers, ignorant of every other rule in the geopolitical manual, did not know that the rule book began with the jurisdiction the Constitution thrust on the federal government for currency and commerce. "Stateless money" invites outlawing.

War is best conducted on enemy territory. As with war, the safest and shrewdest strategy a national political economy can adopt is to maneuver within the home ground of foreign competitors. Japan has done so by a series of latter-day Viking raids inside the American economy—first, through marketing, then by building, and, finally, by investing in property, in lending, and in speculating. Soviet military doctrine modernized this universal axiom with a characteristic blend of German and American thinking. It added the insistence of historic German doctrine on the offense as the winning defense to the priority American naval doctrine has always given to overseas "choke points," whether airports or sea-lanes.

Unlike authoritarian regimes, America neither has nor needs any economic, financial, or social doctrine. But she has overindulged her blind, historic faith in what Judge Thurman Arnold called the "Folklore of Capitalism." The economists of both parties preach the tenets of Smith and Ricardo as long-gone litanies. Their hypnotic incantations to market magic mourn, if economists but knew it, for America's neglected assets—outside her legal economy at home, and outside her borders in all foreign power centers, neglected by her government.

Under different names, and without modern complications, the problems created for the Treasury by the diversified underground economy of the late twentieth century were the same as those that Hamilton met: how to bring the alienated into the system; or, if the holdouts, rebels, and aliens themselves proved irreconcilable, then at least their money. The taxpaying economy is not likely to regain its lost vigor until its managers rediscover the relevance of the Hamiltonian approach to modern urgencies. Hamilton offered the affluent pockets of Tory opposition a buyout not only without prejudice, but with profit—a financial contract that he used his unique influence over his uniquely prestigious president, and his equally unique collaboration with Madison across party lines, to make the new government honor. The Tories whose money Hamilton invested to feed his starving Treasury had not known they were for sale until Hamilton bought them. The denizens of the modern American underground, whose money is needed to restore flexibility to America's deficit-ridden Treasury, need not know either. Modern complexities call for an updated American social contract, cemented with money channeled via our megamarkets, sufficiently attractive to the participants in the illegal economy to lure them en masse into the legal economy.

America's
Yardsticks

9

PETTY'S PATERNITY
Measuring
Economic Performance

The sheer size of modern society has dictated the substitution of statistics for experience; the distinctiveness of the American economy has necessitated the adaptation of correspondingly distinctive American statistical procedures. For better or worse, only the government can monitor economic society. Even if the economy were free to act out Smith's fable, and to function miraculously emancipated from government intervention, all its participants would depend on the continuing dissemination of statistics by the government to know whether any of them were being prudent or reckless at any given time. None of them can answer their own questions about what to do until the government tells them what the economy has been doing. The government itself cannot arrive at intelligent determinations of national economic policy without a realistic countdown of net savings and investment performance and availabilities. Only the government can tabulate these results. Only the government can give managements and investors the data on its own vast operations essential to decision making in the modern economy.

Stock market folklore beguiles us with the bromide that market action continually adjusts prices to reflect what everybody knows (the claim is that it always knows more than any individual or group can know). In practice, the statistics government publishes adds up the total of what everybody reports already having done: a quite different proposition. Because all participants in the economy are conditioned to trust the stock market to signal how the economy will perform, they are obliged to trust the government to clock how the economy has been performing. In practice, moreover, the stock market derives the authority with which it speaks to the economy from its own susceptibility to the statistics the government gathers far beyond the inbred haunts of Wall Streeters. Invariably, the same critics who dispute the responsibility of the government for managing the economy are the first to base their claims and their criticisms on the statistics it provides.

With indispensability goes responsibility. Statistical integrity is the first condition of government's ability to discharge its managerial as well as its reportorial functions; statistical competence is the other. Government in America has failed to meet both statistical requirements. It has fabricated the fiction of a "statistical economy" whose performance is supposedly in step with the working economy, but is in fact glaringly out of step with real-world transactions and trends. Nevertheless, economists of all points of view are conditioned to take official progress reports of the statistical economy at face value. So, therefore, are their political clients (as well as corporate contributors to political campaigns). In the autumn of 1987, for example, Republican presidential candidates were pointing to the employment statistics as evidence of prosperity blossoming, not realizing that the statistics were "cooked," and that claims based on them would provoke negative reactions.

Yet the markets vibrate to the behavior of the statistical economy and to continuing rumors anticipating the next round of releases. Statistical first impressions have survived recurrent adjustments of publicized data into line with the sobering evidence of hindsight. Habitually, therefore, the rumors come so quickly, and the adjustments follow so slowly, that they send markets off on false starts, either up or down. The markets expect the government to manage the economy by monitoring the statistics. But the markets do not suspect that the government moves the economy by manipulating the statistics.

Catchphrases remind us that politics boils down to economics, and that economics is subject to politics. But statistical method is the missing link between them. It was not always missing. Economists with political flair once made a functional connection between the national resources they quantified and the national political policies they then recommended. The purpose of this section, in its first chapter, is to ferret out from the folklore of capitalism the lost origins of government statistics that proved trustworthy in England's national interest; then, in Chapter 10, to unearth the hidden role of statistical pioneering in engineering America's emergence as the diversified superpower of the latter twentieth century; and, finally, in Chapter 11, to expose the statistical manipulations that covered up the American government's default of its management responsibilities.

The trail starts with an English Renaissance man in the era of the Stuarts and Cromwell. Sir William Petty is the name by which history reserved a slot for him. He set out on his adventures as a cabin boy apprenticed onto a merchant ship, and ended with the posthumous award of a baronetcy; he was the ancestor of Lord Shelburne, secretary of state under the elder Pitt, and of Lord Lansdowne, Victoria's last foreign secretary and one of her viceroys to India. He made his original reputation lecturing on medicine at Oxford as the favorite protégé of Sir William Harvey, and the most prominent protagonist of Harvey's then still controversial views. Harvey discovered the circulation of the blood and originated the measurement of biological processes; he was also Charles I's court physician.

Within a generation, Petty extended Harvey's technique to the statistical measurement of social and economic phenomena. He was the first to quantify the national resources needed to set national policy, and to fortify national policy with the pragmatic test of statistical measurement. With these achievements, he secured his claim as the father of economics. Just as Veblen is overdue for resurrection as the father of economic sociology, so Petty is as the father of political economy. Like Veblen after him, Petty's work was practical enough to achieve instant institutionalization, so that he was forgotten as his ideas were absorbed into the mainstream of the history Americans have been allowed to forget.

Petty established the statistical base for economic policy-making with two innovations. The first demonstrated the dependence of economic growth on population growth; he pioneered the enumeration of net population change (the calculation on which life insurance is based, and on which the irretrievable cost of war is assessed). The second identified the dependence of economic growth on titles to land; he laid out the first detailed and comprehensive land survey and used it as a basis for taking a census. In the process, he demonstrated the practical workings of the marginal theory in classical economics, reserving large tracts of low-grade and foreclosed land for himself as his fee, to his profit. In a genuinely expanding market, the lowest-grade entry always scores the greatest gain.

When Petty fashioned statistics into the decision-making tool of policymakers, he put its findings at the disposal of national strategy as well as private wealth; the usability of the fundings assumed their accuracy, and also their objectivity. His pioneering exercise in census taking, the private initiative he undertook, based on a sampling of London parishes, documented his call, bold in his day, for population increases and his controversial assurance that the resources of seventeenth-century England would support them. His title search of Ireland—the Down Survey, as it was called—provided an indispensable map of the country for taxation and land grabs. He activated a two-way trade between England and Ireland, on both sides of which England profited. The brain trusting Petty provided for Cromwell's annexation of Ireland established the foray as the first calculated exercise in commercial imperialism.

True, earlier civilizations had tried their hands at one or another rudimentary form of census taking. The Bible took a stab at the task with the Book of Numbers. The Romans counted households for tax purposes. One of Britain's earliest political documents was its Domesday Book. But although each of these respective head and land counts had a purpose, none of them developed a statistical method in the national interest for a political purpose against rival powers.

Four seminal figures in the history of economic thought— Smith, Marx, Schumpeter, and Keynes—agreed that Petty was the originator of modern economic analysis: formidable unanimity across the ideological spectrum. Disagreements among economists are always good for a laugh, though an uneasy one, loaded with frustration. But this notable four-way agreement on Petty's achievement has stood the test of time. Each of these four, in turn, dropped his clue as an aside—Smith in a typically sycophantic letter, drooling obsequiousness; Marx in his equally typical historical sorties; and Keynes, even more typically, in a chance remark dropped at a cocktail party to his distinguished medical brother, a bibliographer by avocation, numbering among his achievements, by a provocative coincidence, the bibliography of Petty. Schumpeter elaborated one of his gracefully erudite footnotes into a tantalizingly brief profile.

All four of these figures recognized Petty as a mentor; each of them shared his achievement of influencing economic decisions by realists in power. All four credited his paternity to his statistical pioneering. Petty's statistical innovations also explain their reticence in clarifying the reason for their indebtedness. All four aspired to recognition for their own enumeration of national economic wealth for a political purpose: witness Smith's best-selling title, *The Wealth of Nations*—not of private competitors for it.

Marx and Schumpeter provided conflicting definitions of Petty's innovative thinking. Marx credited Petty with originating the labor theory of value, reason enough for Marx's partiality to him. But Schumpeter was on solid ground, too, in describing Petty's innovation as "a land theory of value," reflecting his own awareness of the enormous influence that Petty exerted on the revolutionary political economics the Physiocrats were soon to unveil in eighteenth-century France.

Petty summed up his thinking in his dictum that "labor is the father . . . of wealth, as lands are the mother." This memorable declaration, as Schumpeter said, "put on their feet the two original factors of production of later theorists." The revolution in land theory that Petty started activated the ferment in French thought that culminated in reddening the guillotine with the blue blood of an absolutist Bourbon monarch. Petty's emphasis on the land as well as the labor components of the value equation, in addition to his direct influence on French economics, muddied Marx's neat characterization of Marxism two centuries later as simply an amalgam of English economics, French politics, and German philosophy. The cultures of nations do not fit into compartments labeled for the convenience of philosophical system builders.

The process that inflates land values merely by establishing land usability provoked endless controversial theorizing. Petty's insight turned appraisals into the incontrovertible start-up method of operating. When he abruptly quit his brilliant medical practice and enlisted with Cromwell as physician-general to the unpaid, disease-ridden army of occupation in Ireland, he identified claims to land and the fight over hands to work it as the driving force in history; *Land and Hands* was the title of his classic tract.

Not only did land emerge as the burning issue in the French Revolution one hundred years after his time; it also dominated the American Revolution. Washington, like Petty, was a surveyor; the Continental Congress, adopting Cromwell in Ireland as its mentor, took advantage of the opportunity opened by the new statistical profession of surveying, which stakes out land for quantification. Like Cromwell, it paid its troops in land, opening the west to a new army of free landholders. The land issue astounded the Marxists another one hundred years later by dominating the Russian Revolution; they had expected the labor issue to provoke it. It upset their simplistic premises again during the Chinese Revolution nearly half a century after that. The same fight over land use obstructed the modernization of the Third World and, consequently, contributed to the start of the debt depression of the mid-1980s. Statistical method holds the key to the problem of refinancing land debt in Third World countries.

Imperialism followed naturally from capitalism. But channeling the proceeds of imperialism called for more managerial skills than merely unleashing an army of invasion in search of booty. A number of interrelated national goals beckoned to Cromwell's England: to increase the population without running short of food; to emulate Holland as a maritime power able to counter the armies of France; to trade a standing army prone to civil war for a naval power guaranteeing domestic civil freedom; to provide English shipping with colonial ports of call; to export penniless nobles in need of land and income and, reciprocally, to import a steady supply of cheap labor; to ennoble England's emergent strata of political and commercial nouveaux riches without impinging on established privilege, while broadening the rental income base of its upper classes; and, finally, to control the supplemental food base needed to support the shift of

England's population from agriculture to industry. A full plate of policies, every one of them prophetic, every one practical.

But the key to all of these goals was title searches to the ownership of Irish land, and quantification of the crops and income the land would bring. A countervailing consideration was how much new imported food supply the English marketplace could take without deflating the income of English landowners. The service Petty rendered was to turn Cromwell's rape of Ireland into a going business by administering Ireland's principal resource, agriculture, as an adjunct of the English economy. This strategy, based on Petty's statistical pioneering, reduced Ireland to starvation and left it watching the fruits of its lush land ("green gold") being gobbled up on the other side of the Irish Sea, while England expanded, free from inflation. Not a pretty story, but the form progress takes seldom is.

Where Schumpeter gave Veblen short shrift, his profile of Petty is elegant and judicious: "Physician, surgeon, mathematician, theoretical engineer, member of Parliament, public servant, and businessman—one of those vital people who make a success of almost everything they touch, even of their failures." Petty's principal failure detoured him into a temporary stay in pauper's prison in the course of protracted litigation over his extensive Irish landholdings. But these low-grade holdings accounted for his smashing success and laid the foundation for his vast fortune. He died rich, ennobled, a legend in his lifetime, and the upstart aristocrat of talent who founded a long line of aristocrats of privilege, all powerful, a few of them also talented: a monument to the majesty that was England.

Schumpeter himself underestimated Petty's attainments. Petty was also a practical entrepreneurial innovator, the model of the engineer-businessman whom Veblen hailed as the carrier of progress. He invented carbon paper; duplication is the common denominator of modern record keeping, relied on by management as well as for the detection of mismanagement. As Gorbachev was fated to discover in 1987, when he committed the Soviet system to *glasnost* and *perestroika,* freedom for market forces presupposes freedom of information. But he found himself blocked from accomplishing his goal for the economy by the lock the KGB held on the copying function Petty had identified as critical.

The sea was Petty's first love, naval architecture one of his passions. He designed faster types of ships, including the catamaran. Indeed, in 1987, the U.S. Navy was still experimenting with his notion of the superior speed potential of a double-hulled vessel— albeit with engines and a flight deck. It also commissioned Japan's advanced, but idled, shipbuilding industry to build two tailor-made twin-hulled antisubmarine sonar detectors. The Defense Department's ban on a major Japanese machine builder for pirating antisubmarine sonar devices to Russia did not inhibit the decision. Recognition of the catamaran's potential as the last word in modern antisubmarine engineering presupposed that twin underwater sonar detectors are more likely than a single one to find a submarine. America could be seen putting Petty's engineering heritage to more practical use in her time of trouble than his economic heritage. No doubt because the former was more difficult to adapt than the latter, a symptom of the trouble America made for herself was reflected in her difficulty in doing things the easy way.

The foundation of economics was by no means the only monument Petty left behind him. He joined with Robert Boyle and Christopher Wren, who also studied with Harvey en route to their trailblazing careers in chemistry and architecture, as founders of the Royal Society. The concept for laying out the West End of London originated with Petty. After the Great Fire of 1666, he located the West End beyond the drift of the winds carrying polluted air from the Port of London on the Thames; he identified the ship-borne rats that had caused the plague of 1665 as the Typhoid Marys of the day. Petty acted out Bacon's dream of new sciences growing out of existing ones. (Late in the twentieth century, the transition from the air to the space age dramatized this process of discipline splitting across the spectrum of scientific inquiry and technological application, vindicating the practicality of Bacon's vision.)

Petty's claim to fame as a working Baconian rests on his coups in fathering not only economics through statistics, but epidemiology, demography, and surveying. In a wry throwaway line, he anticipated the need, from his hard-coin era, for modern banking: "When in need of cash, start a bank." Where the Italian Renaissance produced a flowering of art, the English Renaissance (with due respect to Shakespeare) orchestrated the flowering of the practical arts that Bacon had foreseen. Sciences, in Petty's hands, were not self-propelling procedures whose progress was to be recorded passively. Beginning with what we call economics, they were tools for the engineering of policies needed to restructure the society of his country. The new world of the seventeenth century was eager to measure anything and everything preparatory to using it. Petty was the first to transfer the passion of his contemporaries for measuring natural phenomena to measuring social phenomena, beginning with national population growth and land values, calculated as the basis for engineering political advantage for England.

Political Arithmetick, Petty's own inspirational, self-explanatory, pragmatically implemented phrase, the title of his last book, tells it all about economics. One of the most prominent of the literate political activists Petty recruited, an MP named Sir Charles Davenant, sealed Petty's claim to recognition as the father of economics with this pungent summary: "By Political Arithmetick we mean the art of reasoning by figures upon things relating to government." This dictum for all seasons puts the burden squarely on statistics to guide policy and policymakers.

It is not surprising that Marx, Keynes, and Schumpeter paid tribute to Petty as the father of economics for his Baconian practicality in basing economic theory on hard statistical evidence, ready for use in government policy-making. But it is revealing that Smith felt obliged to revere Petty's performance in the act of reversing him. Only Schumpeter has whiplashed Smith for the mischief he made with Petty's slogan. "The suggestive program wilted," Schumpeter charged, "in the wooden hands of the Scottish professor and was practically lost to most economists for 250 years"—devastating criticism from a ranking twentieth-century conservative, steeped more systematically than any other scholar or thinker in the history of the literature.

Schumpeter went on to expose Smith's hypocrisy: "Smith took the safe side that was so congenial to him when he declared that he placed not much faith in Political Arithmetick." Between the definition with which Petty baptized economics and the fable Smith subsequently promulgated that markets could be free, the conflict is irreconcilable on the merits. Sycophancy and venality inspired Smith to reconcile it. He boarded pupils for tutoring. In 1759, his showcase student was the son of Lord Shelburne, the Marquis of Lansdowne, heir to Petty's title. Smith supplemented one of his term bills with a fulsome letter volunteering that Petty was the fountainhead of the wisdom he made a show of applauding. Specifically, after paying tribute to the noble lord's patronage of "arts, industry and independency" in Ireland, "a miserable country, which has hitherto been a stranger to them all," Smith added, "Nothing, I have often imagined, would give more pleasure to Sir William Petty, Your Lordship's ever honoured ancestor, than to see his representative pursuing a Plan so suitable to his own Ideas which are generally equally wise and public spirited."

But Petty's insistence that England's emergent economic interests command top priority in her national decision making—heresy, according to Smith!—established him as the dominant policy-making figure in the society that reared Smith. The spectacle of Smith, paid to board and tutor a direct descendant of Petty, and feeling obliged to grovel before the privileged descendant while reversing the revered ancestor, illustrates the Pilgrim's Progress of John Bull from the muscular reach of the commercial revolution to the complicated calculations of the industrial revolution. This profile of life at the economic summit acted out Schumpeter's enigmatic but profound definition of economics as *historical* sociology.

Smith's style revealed how academic and abstract his approach to the drive of modern economic processes was. He began each chapter of *The Wealth of Nations* with elaborate illustrations drawn from the history of ancient Greece and ancient Rome. But both, being slave societies, were irrelevant to the labor market, which Smith nevertheless identified as the key to the free modern marketplace. He never admitted this basic contradiction between ancient and modern society. The form of Smith's rhetoric was for show; the substance of his logic was for advocacy. He trusted analogies with the slave societies to appeal to the educated classes reared on classical studies that no longer served any practical purpose, as Bacon had complained three centuries before Veblen's satires on conspicuous consumption and veneration of the antique. Smith was the culprit responsible for detouring modern economic thought away from the navigable channel that Petty dug for it in the age of commerce and that Veblen later extended into the age of technology. By contrast with Smith's, Petty's style was terse and skeletal, stripped down to calculations relevant to modern problems.

Smith invoked the fantasy of self-propelling market forces to relieve the government of its inescapable responsibility for setting national economic policy. In the process, he discredited the political component of Political Arithmetick. His stratagem simultaneously dispensed with the arithmetic component as well. He anticipated the modern fashion of substituting psychological responses to perceived market incentives for statistical observations of them. Belatedly but inescapably, however, the realities of twentieth-century life have reactivated Petty's concept. Common, if not technical, usage refers to political economy—or political economics, as Political Arithmetick came to be known after Petty was forgotten—as the universal instrument for setting national policy in the demonstrable national interest. Its reinstatement in fact calls for its resurrection in theory by students of history.

Petty was as political in his operations as he was pragmatic in his conceptualizing. His plunge into politics made him his fortune as the Cromwellian on the make who brokered the new deal waiting to be made between the exhausted Cromwellians and the inert Stuarts. Petty went with Cromwell until the time came to cut the deal bringing the Stuarts back. Then, he helped broker the Restoration; this explains why Charles II did not dare either trust or ignore him. England's choice was between a reversion to the anarchy of civil war or acquiescence in treason; both future Stuart kings were on the payroll of England's enemy, King Louis XIV of France, and Petty was by no means the only one who knew it. So central was Petty's work to the currents of his time that his practical intervention in applying the doctrine of sovereignty acted out the theory of his other great mentor, Thomas Hobbes; namely, that any sovereign is better than none at all. Petty served his apprenticeship when Hobbes was still immersed in medical studies. In fact, Hobbes gave Petty his microscope, one of the first in England.

At the outset of the breach, the issue between Charles I and Parliament flared up as one of divine versus civil rights. But Charles and Parliament were partners in crime before they declared themselves crusaders against each other on the battlefield. Their imperialistic fight was over loot, not law, as legend has it. Parliament, in a last-ditch effort to strike a deal with Charles, offered him Ireland as a joint venture in legalized land piracy. Petty, the one-man statistical bureau, gave Parliament the numerical ammunition to know what it was offering. Charles did not have the sense to accept. But Petty carved out his role by providing the land survey in the interest of the Cromwellian revolution, which evolved into the imperial interest and, incidentally, his own. The competing claimants for the land trusted Petty to use his surveying tools judiciously for the protection of all the participants.

The structure of capitalism, with its imperialistic outcropping, expanded with Petty's statistical operations. The English Crown began trying to overrun Ireland while the Spanish Crown was still trying to overrun England. But the famous expedition into Ireland that Elizabeth organized for Essex on his way to the executioner's ax and romantic immortality was merely a repeat performance of a classic Viking raid, with no economic rationale to justify it. Under Petty's influence, England began to fix not only national goals, but new imperial operations as well, by statistical interpretation of national resources, beginning with population, instead of by sovereign whim or need to stave off an invasion or bankruptcy or both. Not until Cromwell's conquest of Ireland did the budding capitalist world get its first glimpse of its imperialist future independent of monarchical auspices. Petty activated the economic rationale—the lure of the "green gold" of Irish land at a discount—with his proprietary statistical technique of appraisal, subsequently standardized. The evil that capitalism perpetrated sprang from the same seed, like Cain and Abel, as the wonders it worked. Petty quantified imperialism's rise just as Hobson did its demise. The bad name imperialism earned with losing wars has obscured its profitable beginnings in piratical marauding.

Buying shares in pirate ships was an old English custom respectabilized by the aristocracy and legitimized by the Crown. Cromwell's nouveau riche followers bought the idea along with their estates, their titles, and their wives. They were eager to throw money at any government-sponsored scheme to make a killing. The sweetener in the deal was the offer to takers to buy seized Irish lands at knocked-down prices. The catch was that, before Petty set to work, no land titles were staked out; no one knew who could buy which lands where. The discovery by the English raiders that the beleaguered Irish had no clear idea themselves of who owned what invited a three-way factional land scramble among the stay-at-home speculators, Cromwell's pirates, and the natives.

But Petty was wily enough to reduce Hobbes's thunderous warnings of anarchy from an arbitrary command to a negotiating process, subject to the direction in which statistics pointed the national interest—in England's case, toward maritime power, rather than involvement in France's land wars.

Because he foresaw England's maritime future, Petty put himself forward as a candidate to head up the Navy. Charles lavished honors on him, but reserved this one for James, his own brother and successor, a more abject tool of France than he himself was. Petty's most ambitious catamaran model, appropriately named the *Experiment,* was launched by the king himself but sank with all hands aboard in a storm violent enough to destroy the entire fleet of conventional design sailing with it. No doubt the Stuart brothers welcomed the disaster as a convenient pretext for sidetracking Petty's aspirations to redirect English naval policy against their paymaster in Paris. So much for the wages of kingmaking.

To identify Petty with the paternity of economics solely because of his statistical achievements is to focus on only the arithmetic component of Political Arithmetick. It is to lose sight of the political component, which he brought into play in engineering each of these landmark technical achievements. It is to lose the flavor and miss the thrust of the pungent definition of Political Arithmetick as the "art of reasoning by figures upon things relating to government." This art involves a continual two-way process: generalizing quantified diagnoses into national policies, and particularizing national policies into quantified decisions. Petty's approach to national policy-making has outlived and outperformed Smith's theories of laissez faire and free trade. It holds even more promise as a lever on the crisis that exploded on America during the Reagan regime than it did for England when Petty advanced it during the Stuart Restoration.

When Petty improvised his rudimentary head counts, everybody took government for granted as the dominant economic force in society by virtue of its being the dominant political force. Accordingly, the main impulse of economic life was to press government for help. When government was drafted to direct the emergent economy of the seventeenth century, the Crown was big by the standards of the time, but busted, and caught in a continuous circle, by no means always as vicious as the mythology of laissez faire has led us to believe. To meet voracious claims for subsidies from the young economy sprouting amidst the aged structure of sovereignty inherited from feudalism, the government put excessive claims on the economy to divert budding resources for the taxes it needed and the loans it forced from the rising business classes. The government and the economy took turns giving and getting.

In the seventeenth century, government was in a race to provide new services for the economy as quickly as the economy was demanding them, and the economy was in a race to provide new assets for the government as fast as the government needed them. Each needed to learn how to measure the evolution of the other in microcosm and in macrocosm. Petty developed both demonstrations.

In nineteenth-century America, statistics did not command either its original or its latter-day priority, as it did throughout Europe. Under cover of the cant about self-reliance and pioneering, the U.S. government did so much for business during the Civil War and afterward, when it opened the West to the railroad promoters, that neither party to the old deal felt the need to quantify economic performance—or dare to advertise it. The achievements of the economy spoke for themselves; its recurrent setbacks were suffered as natural growing pains, and not blamed on government, which was not looked to for relief. All anyone had to know about the economy, then or forever, was thought to be compressed into Smith's dicta, popularized in the secular sermons of Herbert Spencer. At the outset of the twentieth century, the American economy responded with vigor to the distinctively American technological theory of value that Veblen proposed. It became so busy catching up with technology and the markets technology blasted open that it had no time to bother with statistics. Accordingly, statistical practice looked to commercial transactions for its data, not to the impact of technology on them.

America lost her statistical integrity late in the twentieth century. In the process, she displayed amnesia about her own distinctive statistical innovations in the 1930s, which I will summarize in the next chapter. Unfortunately, the American search into history for clues to the present has gone the way of reading for pleasure and mind stretching. Policy-making has suffered. Legend has it that shipload after shipload of penniless settlers from England, Scotland, and

Wales arrived in America armed with their Bibles and their copies of Macaulay's *History of England.* Macaulay was partisan, he was florid, but he wrote history in the grand manner, and America was brought up to read him and many others, just so long as it was history. At the turn of the twentieth century, Harry Truman grew up in a small town in the "Little Dixie" section of Missouri on a diet of Civil War history, though he never sampled the increasingly questionable benefits of a college education. Not until the opening of movie houses anticipated the advent of the media society did the study of history start its long slide out of fashion.

In 1975 Professor Frank Freidel, of Harvard, the eminent historian of the Roosevelt era, complained in *The New York Times* in behalf of the Organization of American Historians of the neglect of historical studies in the educational system. As the light of the television screen illuminated the dismal intellectual decline from the 1960s to the 1980s, academic historians learned how to adjust to their subordinate status and the budgetary subordination it brought. Some abandoned the quest for intelligibility and concentrated on the computerized correlation of data, as if they were economists rushing into equally remote econometric extensions of advanced algebra.

Earlier in the century, statistics won acceptance as a prestigious tool while history was still in fashion. Uses for statistics in America were discovered by two notable figures: Louis Brandeis, who exploited statistical revelations that he uncovered at the turn of the century, and Herbert Hoover, who advertised the statistical services he thought up twenty years afterward. Brandeis retrieved the statistical art in the private sector; Hoover dissipated it in the public sector.

The use Brandeis made of statistics helped establish him as the most formidable trial lawyer in the country and as the pioneer of legal advocacy in the public interest as we know it today. He enlivened the practice of arid case history law by injecting flesh-and-blood experience encapsulated in irrefutable form from eyewitnesses, participants, and investigators. Brandeis sharpened statistics into the same kind of precise tool that Petty fashioned it to be, chiseling out sharply delineated slices of life. At first, lawyers of the old school were outraged by his introduction of evidence from the economy as admissible in legal proceedings, and they entered objections. By the time his work was done, however, statistics was embedded in the practice of the law: economic, accounting, and engineering experts were as essential to litigation as lawyers themselves; and government agencies were routinely asking questions of the kind he made headlines answering.

The intelligible heritage that Petty started, of trustworthy statistics compiled to set the course of national policy, was lost by Hoover. Hoover had made money for himself in London before 1914 as the chosen expert for sponsors of new mining issues to consult; the inflationary flames fated to burst out with the shooting were already sputtering. He was on the inside of the war investment boom that hit the American economy before America entered the war. His fortune made, he launched his second career as a public man, running the food assistance portion of the war spending program with great administrative distinction and dedication to humanitarian priorities. But with his experience of the wartime movement of money and goods and their abrupt postwar cutoff, he might have been expected to anticipate the precipitous economic collapse of 1919–21, and to be recommending remedies—certainly by the end of 1920, when he was named secretary of commerce. Yet he was inexcusably stunned, sharing the general confusion. Nor did he learn from this disaster the need to develop an overall statistical apparatus of national warning signals.

Hoover did make use of statistics, but only to establish himself in 1920 as Young Mister Clean and Safe Mister Progressive in a Republican party tainted as corrupt and reactionary. Where Veblen had urged economists to adopt the flexibility of engineering, Hoover brought to economics the rigidity that Veblen had also described as attaching engineers to their superiors as instruments at best and deadweight at worst.

Always unimpeachable as a public servant and always inept as a politician with a high profile, he was a nonstarter for the lost cause Democratic presidential nomination in 1920. He had name recognition, but no delegate base, nor any idea of how to put one together. The Ohio Gang behind Harding, however, were a sorry and sordid lot, and they needed fronts as imposing as the marble façade of the Department of Commerce. As its secretary, he was on display as the tame, respectable former Democrat who could be shown and even heard in public in place of the predatory secretary of the Treasury,

Andrew Mellon, who could not. Mellon was delighted to use Hoover as his cover, and Hoover reveled in making headlines for two.

Hoover was invented for the role, but he needed to invent a function. He was eager to play an important part on the national scene in order to pursue his great ambition. He was sensitive, however, to any criticism, especially from his senior colleague, Secretary Mellon, but also from Mellon's fellow moguls, suggesting that he was using the government to interfere with business.

Hoover found the ideal compromise in the exploitation of economic statistics. The image he cultivated as the engineer-businessman come to Washington was at once novel and familiar, always good for a colorful story projecting his colorless personality. Veblen's vogue invited Hoover's claims. Statistics certainly advertised a function for the Commerce Department to fill and, therefore, a justification for its palatial headquarters and army of drones to match the Department of Labor. Moreover, the very exercise of compiling statistics invited pompous headlines saluting achievements in the public interest by the Great Engineer during the early 1920s.

All this statistical activity created an atmosphere of bustle, all of it empty, involving the government in neither controversy nor innovation. The drumbeating in the media over the public service rendered by the Commerce Department's statistics entirely ignored the role of statistics in identifying the national interest in international dealings. Early in 1929, when Hoover ascended in an aura of wisdom from the Commerce Department to the White House, telltale climactic market speculation, financed by high-cost borrowings, was beginning to destabilize the transatlantic financial mechanism. Intergovernmental juggling of Europe's war debts, still unfunded and insupportable by earning power, had already overstretched global finance. The stock market crash merely formalized the deterioration that Washington, guided by Hoover's expertise, had ignored. Hoover was intoxicated with the statistical illusion he had certified.

Consequently, the Depression found the country not only demoralized but defenseless. Just about the only statistical tools at hand were useless Customs and Census records collected routinely all along, plus tabulations of bank failures, debt defaults, and bankruptcies— all too painfully clear at that point even without counting. The statistical arsenal at the government's disposal lacked any overview clocking the performance of the economy as a whole, or even its key segments and functions. To the bitter end, the statistics Hoover compiled could not persuade him that the system had collapsed. Therefore no rationale had been formulated for identifying the sources of the wreckage, let alone for quantifying them or the wreckage itself. Such innovations as were developed by private researchers and institutions—mainly by former students of Veblen, notably Wesley Mitchell and Paul Douglas, then at the apex of the academic structure—mainly focused on business-cycle performance, which, however, the Depression interrupted. The best that could be said of the Hoover experiment in statistical publicity is that it advertised the need for the function it failed to fill.

The worst that could be said was more serious. The Great Engineer, turned promoter of statistics, deprived the Commerce Department in the 1920s of the counsel of the economic savant responsible for establishing the pivotal role of the engineer-entrepreneur in modern economic society—even though Veblen was then laboring in the academic vineyard at Hoover's beloved Stanford. (Hoover did not even use his decisive influence to win appropriate senior professorial status for Veblen.) No doubt Hoover's limited vision prevented him from seeing the connection between Veblen's creative application of "historical sociology" and the statistical practice needed to activate business-cycle theory into a reliable tool for national policy-making. Veblen, toward the end of his career, was too possessive of his creative role as the dean of qualitative analysis, and also too preoccupied with overall national policy in its international ramifications (especially vis-à-vis vestigial militarism after 1919), to bother with statistical minutiae. He brought up Mitchell, Douglas, and the rest to be the craftsmen measuring the functions he differentiated. At bay, in the Oval Office, Hoover did not think to send for any of these ranking academic disciples of Veblen to measure the disaster that engulfed him, let alone to assess the need to take more remedial action than he did.

Anticipating Reagan, Hoover's indulgence in laissez-faire ideology inhibited him from calling on pragmaticians in a position to arm him with the statistical competence he needed to defend his presidency. The missing link between the opportunity Hoover lost and the Depression crisis that befell the world when America was its creditor and transatlantic intergovernmental debt broke the system, went back to Petty's seminal discovery of the policy-making uses of statistical method in the national interest. Governments groping for guidelines to policies need practical and trustworthy statistical methods to develop them. One of the ways governments now govern is by managing the economy. Private decision makers always take on faith statistical announcements by government. The responsibility is on government to adopt statistical methods that will work for its policy-making as well as throughout the economy.

10

KEYNES'S KEY
Measuring Governmental Pump Priming

Roosevelt continued to run against Hoover long after beating him. FDR found this easier than facing up to the dimensions of the crisis Hoover left. Government spending remained the core issue confronting the country. A bewildered and benumbed public still shrank from any as too much, and no expertise dared guess how much, targeted where, might prove enough to end the Depression brought on by the failure of business investment. From 1933 to 1937, the Roosevelt Recovery bought the president time, which he wasted exaggerating the mandate his 1936 reelection landslide would and did give him. But although the recovery did represent a vote of confidence in FDR's mystique, it failed to break the strike of capital and collapsed into the Roosevelt Recession of 1937–38.

FDR, his promises and claims grown dubious, was left with two alternatives: more spending on subsistence or a switch in emphasis to procedures inviting business investment (or, as a substitute, government investment in plant and equipment). Predictably, he tried the convenient one first. It called for a helping hand to consumers at the bottom of the heap in the hope that the ripple effect would trickle up into the marketplace (a reversal, defying gravity, of Hoover's promise that benefits would trickle down from the rich to the poor). With so many would-be consumers in distress, Harry Hopkins and Eleanor pushed for this approach as humane and, therefore, bound to prove helpful; the pair were old allies from his start as a brass-knuckles social worker to his wartime role as presidential emissary. FDR welcomed the opportunity to play Santa Claus for the needy as helpful politically in his feuding with Congress. Handouts for subsistence were the acknowledged lesser evil.

But not for long. Failed White House experiments in pump priming were increasingly recognized as the greater evil, increasingly intolerable. Clear evidence of frustration, spreading from the White House itself across the economic landscape and back into political Washington, encouraged congressional veterans with an instinct for political survival to dig in until the New Deal administration, aging by 1938 into lame-duck limbo, could be declared programmatically bankrupt. FDR's only chance of staging a comeback hinged on the ability of his new crop of Brain Trusters (as by this time his original Brains Trust was called) to come up with the inconvenient, unfamiliar alternative: investment procedures with a chance to work.

FDR's Brain Trusters sensed the opportunity and were ready with a novel alternative to sell to Congress that the economy would buy: a package of direct government investment and government-insured programs, patterned after the Federal Housing Administration, that they insisted private investment no longer could or would launch and that the economy could not do without. But because FDR's investment-minded advisers were realists, they saw that the key to the right answer was to ask the right question: How to count the idle money responsible for the idled men and machinery? Or put positively, in terms of basic statistical functionalism, how much idled money would need to be invested in what sectors to put how much machinery and how many people to productive work?

The advocates of this New Deal strategy of constructive desperation were clear that the dollar amount any new government-initiated investment might take would be less important than the form. They rated a dollar of investment money—whether financed by government, by industry, jointly, or with private money backed by government guarantees—as measurably more productive in employment than dollars doled out on subsistence. Their timing was right.

They could see that frustration left FDR ready to do something for show that did not commit him to do anything for keeps. They agreed that he would jump for a statistical demonstration, while resisting a budget for actual investment. Moreover, they bet on his receptivity to political alerts, which was as lively as his obtuseness to economic analysis was dense. They understood that government effectiveness begins with statistical realism.

The vehicle custom-built for the overdue official quantification of America's distinctive economic mechanism was called the *Temporary National Economic Committee* (TNEC). Any such vehicle at that point required dual sponsorship: from Congress as well as from an agency that had links to the White House while enjoying statutory independence. Roosevelt's breach with the Democratic Congress, sharpened by the collapse of the economy and the Senate's rejection of his Supreme Court–packing bill, put him on trial to win congressional participation in any such venture. So did the clear dependence of his administration on Congress for the new appropriations called for by the statistical demonstration soon to be unveiled—especially on the vast new multibillion-dollar scale projected for meaningful investment in plant, machinery, and, with luck, a return to overtime. Finally, congressional participation was essential to the success of the new committee because Democrats and Republicans alike were winning support for denying the administration its petty cash spending requests. Congress needed education even more than Roosevelt did on the need to resolve the 1938–39 crisis of investment stagnation with net government investment (beyond tax revenues).

The purpose of the TNEC was to provide a statistical alternative to the recession within the Depression that blighted the decade between the market boom no one had the stomach to recall and the war boom no one had the fortitude to anticipate. The TNEC was designed to sell opinion (beginning with FDR himself) on the idea of activating the government to break the deadlock between capital unwilling to look for work and industries unable to find it. Anticipating the radical implementation of the conservative credo later called Reaganomics, the New Deal proposed to recommend reliance on the supply side of the national economic equation to activate the demand side. The TNEC's sponsors did not propose to do anything programmatic themselves, but they were prepared to recommend a great deal.

But, first, they set out to define the cumulative shortfall in investment stemming from the crash and the resultant collapse in consumption, preparatory to measuring the investment revival needed to revive consumption.

Contrary to the myth of his sophistication about economic matters, FDR didn't have the intellectual stick-to-itiveness to follow a full-dress, analytical presentation, no matter how vividly diagrammed. Under the skin, Franklin Roosevelt and Ronald Reagan shared a stubborn aversion to the idea of government spending, as well as to the discipline of in-depth briefing. Yet both enjoyed uncommon latitude to indulge vast amounts of spending, thanks to their popularity, which was buttressed by their promises to stop "overspending" by unpopular predecessors.

The Temporary National Economic Committee was well named. It was disbanded once the educational revolution it started was institutionalized in the statistics that simultaneously totaled gross national product and broke it down into functional sectors. The congressional members of the committee entrusted the participating government departments and agencies with routinizing the TNEC's statistical innovations. Like Veblen's thinking, the TNEC fulfilled its mission and was institutionalized, its achievements its forgotten monument.

Fifty years later, three references to the TNEC brought it out of the archives, but misread its mission. First, Professor Ellis W. Hawley, of Princeton, produced a tract limiting its scope to the off-and-on antitrust politicking of the New Deal (which, indeed, is the direction it might have taken against the insurance industry, had the war not intervened and had the New Deal been forced to stage a struggle for its own wartime political survival, instead of being guaranteed survival by the struggle forced on the country). Then, Felix Rohatyn, of Lazard Frères, a perennial candidate for the secretaryship of the Treasury in a Democratic administration, advised Reagan in *The New York Review of Books* (a most unlikely medium for the purpose) to duplicate the experiment, under the misimpression that the TNEC was a bipartisan commission of inquiry run by Congress. Within weeks, Governor Mario Cuomo, of New York, the elusive noncandidate for the 1988 Democratic presidential nomination, used an impish interview with James Reston, the elder statesman of *The New York Times,* to second the motion.

Cuomo's veiled reference to the TNEC, repeated for emphasis, suggested that he had been listening to Rohatyn, and that he was just as unclear about what the TNEC did or even was. The TNEC was not a congressional commission, nor was it an inquiry; it was a demonstration of political leverage packaged for programmatic application in a depressed economy. Its sponsors had worked out an answer to the problem before they went public with their questions.

The cast of characters in the TNEC changed the popular perception of the American economy, as well as professional methods of measuring its functions, and, therefore, the way it worked. In the process of blueprinting a quick and peaceful end to the Depression, these men helped direct the economy toward preparations for the war America won and the worldwide postwar boom she financed with the economy she expanded. These players salvaged Roosevelt's second term and substantiated his claim to a third. No one can understand the subsequent "Keynesianization" of America without a knowledge of the people who brought the TNEC to life, of the motives that drove them, and the breakthrough they achieved by making the case for investment modernization before the war won their argument for them.

I knew all of them; I participated in the adventure as a confidant and ally; and I shared their frustration as they schemed to persuade Roosevelt to let them at least measure the magnitude of the task of achieving recovery and/or starting preparedness. Though they failed even to persuade him to regard each as a by-product of the other, they succeeded in establishing investment as a sound and constructive activity for government to engage in when private industry would not—especially when government could do so without either spending money or interfering with industry—by the already-tested expedient of government guarantees.

By that time, the activist spending approach had run into a dead end and out of time. In those prewar years, Roosevelt's do-nothing approach was exhausting patience on both fronts, domestic and foreign, as fast as it was accumulating risks. The international activists, many of them standpatters in the domestic argument over recovery, unwittingly resolved the domestic argument in favor of the economic activists. Intervention overseas inescapably began with domestic activism to bring the American economy into play as the decisive factor in the war, reversing the contest from an Axis blitz into a war of production that America, once declaring it, could not lose.

Once a respected "swing" base in Congress was found for government investment as a compromise alternative to the familiar dilemmas of whether to spend or not to spend and, if to spend, on subsistence or expansion, the advocates of the TNEC were freed from the pressure to continue selling this new approach to the president. They started to bring him their converts with suggestions calculated to repackage the idea of government-financed or guaranteed investment as a budget-respecting winner and to put his own proprietary brand of progressive conservatism on it.

The TNEC originated in the fertile brain and political maneuvering of Chairman William O. Douglas, of the SEC. He engineered the renewal of the partnership between the administration and its disaffected congressional supporters by persuading two of his closest connections on Capitol Hill, Senator Joseph O'Mahoney, of Wyoming, and Representative Carroll Reece, of Tennessee, to sponsor the TNEC legislation.

O'Mahoney was a prominent and attractive Democrat (habitually referred to in those bygone predialing days by Capitol switchboard operators with Southern drawls as "O-ma-honey"). A staunch New Dealer, he nevertheless had angered the White House by casting the vote that defeated Roosevelt's ill-conceived scheme to pack the Supreme Court. Douglas, always keeping his own lines open with the targets of FDR's peeves, drew on his credit balance with O'Mahoney to cement a reconciliation with Roosevelt; O'Mahoney's antitrust activism did provide a basis. Reece was a Republican of independent stance and impressive seniority. Douglas was careful to handpick both from power bases remote from the center of wisdom on the southern tip of Manhattan Island.

Another principal player was Marriner Eccles, the articulate chairman of the Federal Reserve Board. Thanks to his stature and alliances, Eccles enjoyed a variety of claims to respect across factional lines. He owed his chairmanship of the Fed to his ingenuity in operating the only banking chain in the country that kept its doors wide open without government help during the banking crisis of 1932–33. Banking by air was his innovative secret weapon; he chartered a DC-3 to fly bundles of cash from the Federal Reserve Bank in Denver to his tellers in a satellite of Salt Lake City, for morning display on his counters as protective visual magic against depositor runs.

Eccles, operating from his remote base, discovered Keynes's recipe for monetary expansion as the antidote to slump half a decade before Keynes even formulated the theory advocating it. In fact, a chance remark by Eccles sparked Douglas's improvisation of the TNEC as a vehicle for dramatizing the stagnation in the private savings reservoir. Describing the frustration of banks eager to place loans at interest rates of zero-plus but finding no takers, he complained, "It's like pushing on a string." Eccles and Douglas agreed that the trick was to sell industry on the idea of using government guarantees for investment in new productive capacity for whose output businessmen had no customers and could see none.

Yet Eccles was eyed with suspicion by hard-core New Deal activists on two counts. He was an independent advocate of public works, but opposed to restrictive union practices—certainly on the government construction jobs that promised to mushroom as the TNEC's advocacy produced action. More suspect still, he remained loyal to Secretary of the Treasury Henry Morgenthau, under whom he had originally served as assistant secretary; Morgenthau was anathema to the New Dealers because of his rigid resistance to spending, large or small, regardless of the merits of any project.

At the same time, Eccles emerged as the administration's committed spokesman for its innovative loan guarantee programs, typified by the Federal Housing Administration (FHA), of which he himself was the architect. The FHA brought more help to more people than any government spending exercise. The respectable Veterans Administration (VA) program—a far cry from memories of Hoover standing pat against the 1914–18 war veterans' Bonus Army and ordering General Douglas MacArthur to fire on their miserable winter encampments outside Washington—was a by-product. The FHA and VA built capital for people and harvested revenues for the government in the process, while freeing the administration from dependence on Congress for housing appropriations. Eccles's immense popularity and prestige expressed a vote of confidence from the people with common sense who flocked to pay the government its nominal FHA and VA guarantee fees: a healthy-minded contrast with the corporate managements that preferred to mumble about "government interference" and operate with plants throttled back.

Morgenthau was linked by the closest personal ties to the president: his own family purse strings. FDR was kept on a financial leash well into his presidency by his mother, "the Duchess"; she was stridently to the right of the *Social Register,* as it was then. Morgenthau's father had been FDR's first political angel, and FDR made sure that he inherited the son, whose taboo on spending did not extend to FDR's election campaigns. Franklin Roosevelt was obliged to rely on Morgenthau family largesse to supplement his boyhood allowance until he reached his embittered prime of life and the pinnacle of power. Even then, the perquisites of the presidency were so meager that he complained freely that he "was barely breaking even in office." So Morgenthau's intimate access to Roosevelt was a fact of life that his New Deal coterie grumbled about but coped with. (The frustration of the New Dealers with Morgenthau's fiscal penny-pinching led them to nickname him "Henny Penny," after the character in a children's story.)

Morgenthau's narrow fiscal conservatism, which Eccles always opposed, led to one of the more destructive paradoxes of the Roosevelt era. While Morgenthau retained the most direct line to FDR, the New Dealers operated as public tribunes; yet they were locked in continual combat with Morgenthau on the spending issue. Roosevelt had committed himself and his administration to achieve recovery by inflating consumer spending, as prescribed by the popular wisdom. Yet, at the worst of the Roosevelt Recession, Morgenthau's recipe for recovery was a sales tax on every item of $10 and over, with 18 percent admittedly unemployed in a work force of only 55 million people. The TNEC was the first official body to identify targeted capital investment, not random consumer spending, as the engine of recovery.

Morgenthau's presence advertised Roosevelt's secret strategy for keeping him at the Treasury. "Henny Penny's" absolute fidelity and stick-in-the-mud conservatism provided FDR with the cover he wanted and needed to flirt and flit with new ideas such as the TNEC until events he could not control might push him to make a move he could not finalize beforehand or evaluate afterward. Eccles was a principal source of such new ideas. His dissent from dull-as-dishwater banking orthodoxy intrigued FDR, who delighted in making mischief with the establishment and, incidentally, in keeping Morgenthau off-balance. FDR was quick to agree with Morgenthau's detractors that his money man was not the quickest. Morgenthau could always count on Eccles to defend what Eccles referred to as his "bruise-prone sensibilities," but to oppose him programmatically. Eccles supported the shift of the Budget Bureau out of the Treasury and into the White House, where it grew into the powerhouse it never could have been under Morgenthau. Life was never dull for FDR's Palace Guard.

Eccles held still another wild card at Roosevelt's poker table, over and above the fact that he was the only habitué who understood the problem of investment paralysis and the shrinkage it compounded. He had taken under his wing an unknown former Harvard economics instructor named Lauchlin Currie, whom he promoted to the protective obscurity of assistant director of research and statistics at the Fed. In an administration brimming with gifted, enterprising, scholarly, and articulate economists, all senior aides to ambitious national political figures, Currie was the star. He was slotted to be the key witness on the macroeconomy, whose contours he had already delineated, at the TNEC hearings.

Currie, when I came to know him in 1937, was the prototypical frustrated bureaucrat, a candidate for deft minor satire. The more recognition he won for the talents he demonstrated, the more resentment he harbored over the lack of recognition for other talents he claimed (which, as matters turned out, were better left unrecognized). He despaired of elevation to the summit of the profession as the first economist to be ensconced in the White House by the side of the president. He was convinced that an accident of marriage blocked it. What the Marxists call one's "social composition" soured his creative career as a statistical technician and drove him to compromise himself and disappear into exile.

His wife's sister was married to Phil La Follette, the charismatic but controversial governor of Wisconsin, on whom Roosevelt fixed one of the hates he enjoyed indulging. Phil followed his father to the governor's mansion after the patriarch moved on to national prominence and power in the Senate, where "Fighting Bob" reared his older son, Bob, Jr., from a page boy to the universally respected workhorse of the Senate Finance Committee. When Currie, hang-dog, confessed the connection to me, he swore me to secrecy, as if he had committed a crime, though I immediately assured him of my warm relations with both La Follette brothers and of my continued respect for him. From Currie's side of the marital relationship, he blamed his wife's family for blocking his advancement. From his wife's side came reminders that she had married beneath her political station.

Democratic economic advisers with pretensions to political realism have perpetuated the Republican political myth that Roosevelt was too ideologically committed and too politically contemptuous to cultivate intimacies in the "Club" on Capitol Hill. In fact, he was eager to build congressional alliances, and he always needed new ones because he always broke off old ones. Two considerations guided his choice of allies: he did not trust the Democratic congressional establishment, and he was aware of its backroom ties to the Republican congressional establishment. Consequently, he was always on the lookout for a connection outside the Democratic fold that he could use as a pivot for splitting the Republicans.

FDR fastened on Bobby La Follette, the lone Progressive, as his Senate confidant when he launched the New Deal. La Follette's base in the Republican cloakroom made him doubly attractive as the link to Republicans and a goad to Democrats. Bobby was the first to hear the siren song FDR played about the presidential succession. Senators of all points of view in both parties agreed that Bobby had assimilated more knowledge of the arcane procedures of the United States Senate than any living colleague or even staffer. All of them sought his counsel despite his formal status as a party of one. His wisdom was epitomized in a statement to me: "If progressive social programs are to have any future in this country, we'll need everyone on a payroll to ante up on the tax rolls for them." (See Chapter 16 for my effort to put this dictum of La Follette to practical use to resolve America's twin crisis of deficits and productivity.)

I spent hours trying to persuade Bobby to become a Democrat, warning him of the ambush that awaited his independent progressiveness in a Wisconsin Republican primary. His reply was always the same: that he wanted no part of the Democratic city machines. He died in a bathtub with his wrists cut, after Joe McCarthy beat him in a Republican primary.

Unfortunately for Currie's opinion of himself, he believed he had married into the wrong side of the family: Phil's. Even when Bobby was the White House ambassador to the Senate, at the height of the New Deal honeymoon, FDR harbored vocal suspicions of Phil as "an opportunist": a character analysis FDR was qualified to offer with authority. By 1938, when Currie's opportunity came to leapfrog over his despised director, now long forgotten, into the White House, the debate over isolation versus rearmament was on; Bobby had joined Phil on FDR's "smallpox" list. Currie was convinced that his career was at a dead end. He saw no hope of attracting professional rewards sufficient to finance alimony.

The TNEC gave Currie his magic carpet. While the project was germinating, Corcoran and Cohen sold Roosevelt the bold idea of seizing the offensive with Congress by demanding a White House staff recognized in the budget; the president of the United States had never had one before. This need was urgent, and the initiative belated. As with ideas that take hold, the 1938 Reorganization Act became part of the administrative scenery, and therefore, its importance was obscured by the stream of events. One of its provisions achieved a giant leap forward in political recognition for the economics profession: an economics adviser to the president—namely, Currie. The creation of the post anticipated the establishment of the President's Council of Economic Advisers. The political lure of the 1938 Reorganization Bill was a presidential offer of reconciliation with Congress, by inviting its committees to vote themselves staffs too—another novel idea. This act turned congressional staffs into the de facto, extra-Constitutional arm of the U.S. government they have since become.

Currie's practical economic judgment was as abysmal as his theoretical economic grasp was virtuoso. He was convinced that the growth of America's economy, tiny though it was by end-of-century standards, lay behind it in 1938, and that cushioning stagnation was its best hope for the future. FDR's inertia justified this hypothesis in that dreary chapter of history. Notwithstanding a term and a half spent running against the Hoover Depression, and a bold response to the innovative recommendation to designate an economic adviser, the president was sitting still instead of acting on the warning from that same economic adviser that the Hoover Depression could perpetuate itself, as it started to do with the Roosevelt Recession.

Currie argued that the roots of distress went too deep for any automatic business-cycle adjustments to alleviate it. He despaired of the government's ability to overcome political resistance to the idea of spending and to effect a rescue with enough force in concentrated enough form to revive investment in plant and equipment, even though Hitler's challenge was becoming worrisome. In the world of the 1930s, establishing a transatlantic logistical support base was a more ambitious military undertaking than the growth of the modern American infrastructure subsequently made it. It took Hitler's stunning military breakthrough across Europe to prove Currie wrong. Meanwhile, Roosevelt's preference for treating time as an ally instead of an enemy tempted him to go along with Morgenthau's resistance to the massive public investment effort that Currie advocated to reverse the Roosevelt Recession. The inertia in the White House matched and accepted the inertia in the economy.

I was one of the two outsiders from New York in on the discussions over who was to be appointed the president's economic adviser, and, just as important, who was to make the New Deal case for a recovery program of government investment at the TNEC hearings. The other was Beardsley Ruml, the treasurer of R. H. Macy, whom Eccles subsequently made chairman of the Federal Reserve Bank of New York. Ruml and I were great friends and neighbors in Redding, Connecticut, along with Stuart Chase, the Voltaire of the New Deal.

Ruml and I, from our separate points of view, agreed on recommending Currie for the key role in the TNEC hearings, and, therefore, for the advisory job in the White House; the lament among the New Dealers was that selling Currie's thesis to "the Boss" would be more difficult than persuading the business brass to buy it once the president did. Eccles, characteristically generous, made no effort to stand in Currie's way. He recognized that the White House, not the Federal Reserve Board, was the power center to sponsor the statistical case for government investment in targeted economic expansion. Currie agonized that Roosevelt would reject him for his dream job because of his family connection with the outlawed Phil La Follette. But FDR never mentioned it in accepting Currie—not even after Currie betrayed his trust.

Well before the TNEC surfaced as a political project, Currie had adapted Keynes's general propositions about financial flows to the distinctive features of the American economy. But precisely because of Currie's genius for reducing concepts to calculations, he was not the enthusiast who found for Keynes the audience awaiting Keynesianism. The intellectual gadfly of the New Deal did. Jerome Frank was his name. Jerry Frank had a first-class mind, fifth-class judgment, and second-class status in the New Deal hierarchy, begrudgingly accorded him as its most valuable recruit from Wall Street. He was the only participant in the New Deal with firsthand knowledge of how the power system worked. He regaled the New Dealers with tales of the techniques the big boys had been using to run the country. Recalling Mitchell's profile of Veblen, he fascinated them "as a visitor from another world," as indeed he was.

Frank was brought to New York from Chicago at the outset of the Depression by Paul Chadbourn, the most powerful Democratic corporation lawyer–lobbyist in the country, to brain trust the juggling of the vast corporate empire Chadbourn's firm controlled. Chadbourn "carried the bag" for New York's flamboyant reform governor, Al Smith, whom FDR had first hailed as the "happy warrior" and then knocked out of the presidential sweepstakes. The big money, with its instinct for misgauging sure things, was assuming that Smith would be president in 1933.

On the New Deal agenda, no issue took precedence over the exploitation of agriculture by industry in a society one-third of whose voters were agricultural. Chadbourn was the spider in whose web the crop-processing industries were interwoven. Frank learned from him how the business interests of the country collected "goodies" from their Republican favorites in Washington but bought protection from the regional Democratic power structure. In that Republican era, the Democrats were entrenched in a network of baronies in state houses and city halls, reaching into Congress. Sixty years later, the embers from the hot stoves Chadbourn tended were still fiery enough to flare into a successful injunction petition by Al Smith's geriatric son against publication of Chadbourn's memoirs.

A brief digression concerning a stormy chapter in Frank's history is in order. When he enlisted in the New Deal, his involvement in crop-processor politics routed him into the Department of Agriculture, whose innovative legislation he drafted. In 1935, an ambitious assistant of his wrote an order, which Frank signed, requiring the meat packers to open their books to the department: at the time a revolutionary intrusion, long since routinized, into corporate privacy. The packers opened fire, and Secretary Wallace sued for peace by firing Frank. The assistant was Alger Hiss. "Alger had the gall to ask me for a recommendation to the State Department," Frank later sputtered to me, "after I signed the opinion he wrote for which *I* got branded as a Communist!" Because Wallace let Frank down, Harold Ickes picked him up (Ickes and Frank had been fellow crusaders against the Chicago city machine). In 1937, Douglas, a staunch ally of Ickes, arranged for Frank's appointment as an SEC commissioner.

In the midst of the Roosevelt Recession, Brandeis retired. Frank, though deeply involved as Douglas's deputy in the TNEC project, nevertheless went off on a second crusade: to elevate "the redhead," his affectionate nickname for Douglas, to the Supreme Court. His first stop was consistent with the bipartisan spirit of the TNEC. He approached Senator William Borah, of Idaho, the ranking Republican member of the Senate Foreign Relations and Judiciary committees.

Borah was a powerful and, more than incidentally, a romantic figure in Washington of the late 1930s. Formidable on the floor and in committee, everyone who was anyone—not only in Washington, but even in the Idaho backwater of the old morality—knew that he fathered the daughter "Princess Alice" Roosevelt Longworth gave to the ineffectual Speaker of the House (I remember the child as pathetic; she eventually committed suicide). Alice had served as hostess for her father, Theodore Roosevelt, at the White House. Borah's conquest of her symbolized his towering role as a chieftain. Borah's charisma echoed half a century later when F. Forrester Church, the eloquent Unitarian minister son of Senator Frank Church, chairman of the Foreign Relations Committee during the Vietnam years, wrote reverentially of the Republican isolationist as the hero of his father, the Democratic internationalist.

Borah had two holds over Roosevelt. He was the spokesman of the West, which was up in arms over its long-standing claim to a seat on the Supreme Court. But Borah was the even more troublesome tribune of isolationism. Roosevelt was positively petrified of the challenge Borah's fiery rhetoric would stir up if FDR decided to run for a third term while Hitler was moving aggressively and forcing him to move to save England. Pearl Harbor was undreamed-of; slogans of isolationism, particularly west of the Ohio River, were aimed exclusively against U.S. involvement in another European war. In the desperate atmosphere created in the White House by the political surrender of England and France to Hitler at Munich, any nod from Borah was welcomed as a bouquet. Frank enjoyed a special welcome with Borah by virtue of being an outspoken isolationist himself (unusual for a Jew in the Hitler era), as well as the outstanding legal scholar and practitioner in the New Deal; neither Cohen nor Douglas had ever tried a case. Frank persuaded Borah to join the TNEC to give it prestige: another coup.

But another manipulative obstacle loomed. The vacancy on the Court was earmarked for a westerner by senatorial courtesy, an unwritten article of America's Constitution enjoying parity with the written ones. Douglas had indeed been born and educated in the West. But he had not returned since he went east to the Columbia Law School, and he had been appointed to the SEC from Connecticut. Under these awkward circumstances, Frank resolved to persuade his fellow isolationist to anoint Douglas as a fellow westerner. Frank's stroke of genius overrode the presidential appointment hanging on Douglas's wall at the SEC that designated him an easterner.

Nevertheless, to make sure, Frank touched a second base with Brandeis, taking Douglas along with him. Cohen, always wily and mindful of his fences at upper levels, had been careful not to cut his own ties with Brandeis when FDR did in resentment over Brandeis's role in beating back the scheme to pack the Court. Cohen, a trusted protégé of Brandeis since the drafting of the Balfour Declaration promising to create a Jewish national home in Palestine, did the honors in sponsoring Douglas with the old man. Once the word went out that the patriarch had embraced Douglas as the son he never had and proclaimed him the successor he wanted, Frank's success was all but official.

Frank's coup won him a reputation as a first-team player in the Washington game; his elevation to the chairmanship of the SEC established him as a figure in his own right, instead of simply Douglas's idea man. A rich episode in his private life opened up new vistas for the New Deal, the TNEC, and the future of statistical realism in America. He fell madly in love with one of the century's leading ladies. Rebecca West was a magnificent writer and a reigning beauty. Frank was an irrepressible conversationalist; even ecstatic intimacies, as he used to recall, did not silence him. It seems that he once punctuated a delicious interval of detachment from worldly pressures with one of his tirades against "Henny Penny." According to Jerry, his exhortations spurred her to sit up and exclaim, in the grand manner of the British hostess, "You must meet Maynard." Once Frank did, the whole New Deal did. None of the commentators on Keynes's career, especially those who made much of his homosexual joustings in Oxbridge and Bloomsbury, have had a clue about the circumstances, the timing, or the sponsorship of his emergence as a formidable figure in Washington.

No back-door—or bedroom—route into the corridors of Washington power could have been more unlikely for Keynes. Frank guided him to Currie, a bureaucrat too low on the chart to be allowed even to sit below the salt, let alone to be favored with letters of introduction from Frankfurter, the Supreme Court justice with the Oxbridge franchise but without any orientation to the political realities of economic policy-making.

Before the president himself had an economic adviser, all the New Deal's Cabinet secretaries and agency heads had economic advisers who formed an informal network. Their political champions, Corcoran and Cohen, usually showed up at their sessions and lent sympathetic ears to the advisers' complaints about their bosses. Their brainstorming about statistics invariably added up to discussions of where and how to get funding for petty cash projects. Corcoran's presence commanded attention because of his dual capacity as the president's hatchetman and the New Deal's ambassador to the Reconstruction Finance Corporation (RFC), the administration's major source of capital funds. Ickes made sure he had two representatives—not only Ben Cohen, who made his office with Ickes, but Abe Fortas (the future Justice), whom Ickes had picked up along with Frank from Wallace's purge—because Wallace, whom he loathed, had two representatives there, too, Ezekiel and Bean. Fortas knew nothing about the economy; Cohen was a successful professional investor, as none of the economic advisers were. The Treasury representative was Harry Dexter White. Currie was a charter member. Frank could not be kept away even though, as a presidential appointee, he outranked mere departmental economic advisers. Once Frank introduced Keynes into this lively network, "Maynard" was "in," accepted as the fountainhead of policy ex officio, as he had never been up to that point in Whitehall, where his status, though formidable, was that of dissenter.

Events dictated the scenario that assigned this cast of characters their roles. Douglas and Keynes stand out as unforgettable figures for history to evaluate; Frank and Currie were movers and shakers of the period since lost to view. Though Douglas and Keynes were contemporaries in wartime Washington, each at the height of his powers, no evidence has threaded them together. Frank, Douglas's lieutenant, was the intellectual intermediary between the two. Frank was an admirer of Eccles and a codiscoverer of Currie. Currie couldn't sell water in a desert. But the same eloquent enthusiasm that had established Frank as Veblen's latter-day evangelist among the Greenwich Village intelligentsia during the 1920s won him the most influential audience in the world in 1938 as he expounded the principles of Keynes Americanized à la Currie on the Washington cocktail circuit.

The administration was at a loss to justify how the American economy could have come to a standstill while Europe was doing relatively well—to be sure, under the stimulus of the war threat. Douglas supplied the missing link: he saw the need for a credible foreign excuse to justify a steep speedup in U.S. government–inspired investment. Luckily, Keynes was a foreigner qualified to cite first-hand evidence of the grim nature of the European economic stimulus. Providentially, Currie tailored Keynes's general European theory to distinctive prewar American needs in time for Frank, the most unlikely of political sponsors, to package it into a political product.

Across the years, I can still hear Frank echoing Keynes's lucid logic as he put the question to the receptive group of interagency economists: "Would we rather prime the pump now by building homes, or later by building tanks? By breaking ground for factories, or digging trenches to populate again after a war starts?" Keynes's way of posing alternatives brought the war that was rumbling over Europe within earshot of Washington. Not that home building would have increased the special-purpose industrial capacities needed to support the arms effort ahead. Nor would public works projects, sanitized from ideological contamination by fallback military uses and aimed to catch up with the cumulative stagnation of the Depression decade, have ignited an overnight revival of full employment with continual engineering breakthroughs that had been deferred by the Depression.

My strategy in working for the appointment of Currie, as the dry-as-dust technician he was, dominating the witness stand at the TNEC en route to the White House, reckoned on the pragmatic impact of his statistical virtuosity. His very defeatism about prospects for the American economy was guaranteed to convince the pragmatic political mind in the Oval Office of the urgency and massiveness of the investment needed from the public sector to activate the unreinvested accumulation idled in the private sector. Moreover, his incisiveness in converting Keynes's random spend-lend formula for recovery anywhere to fit the distinctive features of the American system met the one practical objection to the clamor for spending. Until Currie unveiled his calculations, none of the advocates of spending programs knew how much to ask for.

Professor Hyman Minsky, of Washington University, in his milestone of post-Keynesian analysis, *Stabilizing an Unstable Economy,* relied on academic sources to identify Professor Alvin Hansen, of Harvard, as the TNEC's star spokesman. But Hansen, who had just announced his conversion to simplistic Keynesianism, was enlisted as window dressing to put his prestige behind the New Deal case that capital had gone on strike; Hansen demonstrated that industry had not even been reinvesting its depreciation allowances. The TNEC followed Currie's script to the letter, using the calculations he developed to jolt the White House into action. Currie's model provided the answer to the exasperated question Tommy Corcoran asked about Hansen's popularization of Keynes's theory: "Why don't we just charter some DC-3s and drop a bunch of dollar bills over the cities?"

Currie's TNEC testimony marked the transition in the history of economic thought from Veblen's observations to Keynes's evangelism. Veblen had noted how the revolution in transportation and communications technology subjected the system to the compounding capital needs of the gargantuan new infrastructure the engineers created. Keynes was crusading, with understandable desperation born of urgency, simply to put the system back to work—in the absence of private investment, by public spending. Currie synthesized Keynes's case for emergency government spending with Veblen's case for continuous investment in transportation, communications, and utilities.

Currie also clocked the workings of the American economic mechanism inside the American political apparatus. His presentation came as a reminder that no general theory of economic behavior developed elsewhere can work in America, at least not without major adaptation. Currie, always the disciple conceptually, yet always arrogant analytically, demonstrated that he knew how America's distinctive federal system functioned, and that the distinguished author of *The General Theory* did not.

War signals from across both oceans fortified Currie's statistical case for modernizing the infrastructure, beginning with transportation. His method of identifying functions of the economy differed from Keynes's. By spending, Keynes meant government injections into the money stream. His references to investment, apart from public works as a last resort, assumed that entrepreneurial commitments to plant, machinery, and inventories would follow. Currie's innovation was to call on government to insure the financing of direct investment by industry in the transportation infrastructure, and without putting the government into the transportation business, shortcutting the wait until increased consumer spending might move enough trains to persuade railroad managements to start ordering equipment. Every dollar invested in transportation technology forces follow-up investment in communications technology and, to meet demand from both, in manufacturing plant and equipment; faster trains call for higher-powered signaling and switching equipment calling, in turn, for modernized facilities and tooling.

In the war-torn spring of 1939, FDR was too worried about attacks from America Firsters on his supposedly secret war plans—though in fact he had none—even to consider any government-insured investment required by preparedness. Not even the opportunity to pull off the kind of double play he loved—financing the recovery from his own recession as the prelude to preparedness he knew he needed—tempted him to break his cover of neutrality, although the RFC, inherited from Hoover, basked in congressional respectability and was authorized to issue loans for capital investment to businesses in the name of recovery.

Eric Larrabee, in his provocative study of Roosevelt as war leader, demonstrates that Washington, in continuous negotiation with Tokyo after having broken the Japanese communications code, was on full war alert through most of 1941. Preventive investment in state-of-the-art communications technology in 1938—even though radar had not yet been invented—would have transmitted the pre–Pearl Harbor diplomatic alert to the military.

In May 1939, when the TNEC took Currie public, it was careful to prevent him from explaining what he was revealing. The hearings presented Currie as just "a fact factory" (an arrestingly misleading description of his genius for assembling figures as novel spurs to analysis). Their sponsors showed a sharp political feel for the resistance to new spending shared by Capitol Hill and the White House. Accordingly, they disclaimed any interest on his part to advance "suggestions on policy." Currie projected the national income requirement (*after* depreciation) needed to reach for "full employment"—then still just a New Deal chimera—at $100 billion: petty cash by the inflationary standards of the 1980s, but utopian in the deflated world of 1939. Currie noted that, at the top of the past two cycles, of 1923–29 and again, briefly in 1937, investment from all sources (business, consumer, government, and foreign) had peaked just above 19 percent of national income; that is, even this high rate of investment contribution to the income stream had proved inadequate to sustain boom income levels.

At the 1929 top, net national income (an altogether new item, baptized for this solemn occasion by Currie) had come within statistical pennies of $90 billion. In 1937, the income flow had shrunk to $52 billion. By mid-1939, the numbers were not in for 1938. But under combined hammerblows from overspeculation in inventories and the unplanned switch of Treasury operation from pump priming to fiscal tightening (due to the first tax flows from Social Security), the economy was knocked below the level of political acceptability and, certainly, well below the targets Currie sighted as necessary to provide full employment plus a return to expansion. In Currie's breakdown by categories of investment, the railroad and utility category accounted for over 15 percent of the 1929 total, but for only 10 percent of the 1937 boomlet: translated into cash, a shortfall of $8 billion. (Adjustments for inflation from decade to decade remained a habit for the future, after inflation began.)

Anticipating Currie's testimony, and the austere disclaimer from the TNEC of any policy suggestions from it, I had published a series of articles calling for a $10 billion program of government-sponsored capital investment, mainly in railroad equipment modernization, as the easiest for government to finance for industry on a safely insured basis, and as the most urgently needed. Cohen reacted by complaining that the railroads had a standing government guarantee of all the 3 percent money they wanted for all the rolling stock they could buy, but wouldn't take it because their traffic loads were down. The railroad presidents of the day resisted emergency additions to their equipment fleets, even after shortages had bottlenecked defense production. I criticized them severely at the time—perhaps discriminatorily, in view of the resistance Roosevelt himself showed, as late as May 1940, to any Army buildup. (Ironically, Joe Kennedy, the most outspoken isolationist among Roosevelt's supporters, headed the Maritime Commission, the one preparedness operation FDR did launch—perhaps because Kennedy had done an outstanding job of running a major shipyard during the 1914–18 war.)

At Currie's request, I brought him together with my nonpolitical engineering brother, Robert N. Janeway, then of the Chrysler Corporation. Walter Chrysler had become director of the old, pre-merged New York Central and developed an interest in modernizing railroad engineering into line with automotive engineering; he assigned my brother to the project. My brother's collaboration with Currie extended the tradition of head counting begun by Petty to the itemization of railroad equipment—by category, age, capacity, and replacement cost—and to interlocking locomotive, switching, and repair capabilities; the routine data collected by the Interstate Commerce Commission and the Association of American Railroads were inadequate for the management of industrial mobilization, the administration's follow-on experiment from the TNEC. The work my brother did for Currie anticipated the bottlenecks allowed to develop after Pearl Harbor, when the new heavy-duty high-speed locomotives and freight cars backed up, en route from the industrial heartland to Atlantic ports, in front of antiquated bridges and tunnels, vetoed by the White House and the Association of American Railroads as priority candidates for public works projects. Half a century later, these same bottlenecks still loomed as obstacles to the adoption of modern bullet trains.

The political target of Currie's analysis, and of the TNEC's fire, was the insurance industry. Transportation was the principal potential claimant for resources on the investment demand side; the insurance business accounted for the largest factor on the investment supply side. Currie demonstrated that the accumulation of cash reserves by the insurance companies measured the failure of private investment throughout the system. The voluntary idleness of this excess cash guaranteed the enforced idleness of people. This potential political bombshell also threatened the incidental overcharging of policyholders. Corcoran instructed the committee, on Douglas's advice, to hold this consideration in reserve as a deterrent to political criticism.

Currie demonstrated that, as the Depression choked off investment and put the squeeze on consumer incomes, the corporate sector wound up with money to burn, but no customers for its wares, no outlets for its projects, and, finally, no projects it dared to start. His statistical elaboration refuted the classical dictum that the accumulation of savings reinforces the system by flowing back into investment. He proved that the more inefficient the system became, the higher the level of savings rose, but the lower the level of investment fell. The steel industry, for example, used 20 percent of its own production; when it was working at capacity and launching projects for expansion and modernization, it borrowed before it earned. In April 1938, mill operations had been pulled back to only 18 percent of capacity, courting the shortages soon to throw the industry into chaos (a preview of the 1980s, when the steel industry led the pack into scrapping facilities as a strategy to generate cash, to force price increases, and finance the high cost of shrinkage).

Pipeline construction, not a big item in the world of the 1930s, ranked high on the New Deal's transportation investment project list. Public works were in Ickes's bailiwick, and he had built the congressional alliances shrewdly calculated to win money for them; he was emerging as the Washington energy czar. Yet not until the Nazi submarine pack devastated shipping on the Atlantic coast did the push for the Big Inch pipeline clear a right-of-way from Galveston to the inland hills of Moundsville, West Virginia (sheltered from coastal vulnerabilities), and then only after needless and countless shipping losses and casualties.

Huge wartime priority claims for capital investment to modernize heavy power facilities revealed Roosevelt's mismanagement of the economy against his own political interest. If he had developed a recovery program when the economy shut down, preparedness would have come faster and cheaper, saving not only money but lives. In 1935, when the New Dealers won their fight to impose the death sentence on the utility holding companies, they argued that the local operating companies, once freed, would invest to modernize. By 1940, however, when the investment logjam burst, cumulative obsolescence in the power plants was still compounding astronomically. (As late as 1950, in fact, allocations for heavy power construction topped the list of government priorities for rationed industrial nonmilitary use during the Korean War boom.)

The TNEC finessed the twin argument over government spending in general, and defense spending in particular, by escalating the statistical system onto an unrecognizably larger scale. The TNEC hearings tempered Roosevelt's timidity and emboldened him to propose a capital investment program of $4.8 billion, confirming the consensus of the TNEC management that he would ask Congress for just under half the amount my articles projected as the minimum needed to start up the engines of the economy or to prepare it to carry a defense overload. Nevertheless, his request was resisted as too large, despite the stoppage in the economy and the broadened political appeal of combining recovery with preparedness. Even the largest peacetime investment effort imaginable by the political yardsticks of prewar opinion would have failed, because it would have fallen far short of the requirements of that small, stalled economy.

But no sooner had Roosevelt qualified investment programming as safe to consider in public than the first year of the war in Europe resolved the argument. In June 1940, he persuaded Congress to authorize the RFC to begin large-scale financing of investment for the armed services, the key to victory. By the autumn of 1941, Congress appropriated more than $67 billion, with no consideration for the ability of the economy to produce without inflating prices, even with consumer spending subjected to severe rationing and draftees departing in droves from the consumer economy.

Although the main focus of the TNEC was recovery, it dramatized a fallback case for preparedness as a potent job maker. Douglas, meanwhile, had arrived on the Court without abandoning politics in general or the TNEC in particular. Frank was after him every hour on the hour to intervene in the TNEC's disputes, and he did, as the consummate inside political operator he was. A generation before Nixon and Reagan denounced the Warren Court for usurping the policy-making role of the executive and legislative branches, and half a century before Judge Bork's activist crusade to deactivate the courts, judicial activism came naturally to Douglas as an extension of political activism through the exercise of leverage over policy.

Holmes was the first to conceive of the approach, but he was too skeptical, too remote, too lazy, and too busy playing the sage to do anything about it or, for that matter, about anything. Brandeis was the first to work at it, but he was so caught up in his lifelong crusade against Bigness in an era of inescapable bigness that he fell out of step with the times. Douglas suffered from a peculiar judicial quirk of his own, never perceived in any commentary on him: he stood in awe of politicos ("members of the club") able to get elected and reelected. This explained the subordinate role he at first felt comfortable playing on the Court with Hugo Black. It also explained the relish with which he subsequently welcomed Earl Warren, a provincial conservative Republican "hanging" prosecutor and governor, as chief justice, and the fraternal reciprocity he developed with each of these two contemporary giants.

Both Black and Warren, in turn, told me that they held Douglas in intellectual awe because he was an authentic genius with a flair for crystallizing legal quandaries into political policies they could grasp with comfort. Douglas used his relationship with each to dominate the Warren Court, though always from the background. Guiding Warren, he catapulted the Court into the primary policy-initiating force of the American government between the 1950s and the 1970s. This remarkable duo, the cosmopolitan Democratic introvert and the provincial Republican extrovert, revolutionized the most conservative of American institutions and left it armed to resist the raid to reclaim it for the radical Right organized by Reagan in the 1980s.

In the years immediately after Douglas's elevation, when he kept his hand in New Deal politics while establishing himself as an independent power center, Fortas and I spent many hours agonizing over the repercussions of events on the personalities and policies we had been involved with; our three-cornered relationship with Douglas was always the focus. Fortas took the line that Douglas had made his definitive career move when he settled in on the Court. I argued that Douglas had evolved into a revolutionary party of one, though operating within the system.

A more than normally erratic turn of the political wheel ultimately proved Fortas right when FDR, en route to his last hurrah, lost control of the 1944 convention waiting to renominate him for his fourth term, and found Truman shuffled into the succession instead of Douglas, whom he preferred, as his running mate. (FDR was dying by then, and his last running mate was guaranteed a quick takeover of the Oval Office.) Douglas refused Truman's offer in 1948 to step down into the vice presidency and did make the Court his career, as Fortas was convinced he would. Truman's emissaries at the 1948 convention enlisted Fortas and me as allies, even though I was one of those urging Douglas to sit tight on his power base at the Court. But Douglas himself went on to prove me right. He activated the Court as the alternative power center to the White House for the politics of principle, opening it to appeals for counsel and intervention from concerned players, ranging from White House Counsel Clark Clifford (pondering Truman's offer of a justiceship) to Charles Rehnquist (puzzling over how to handle his).

Back in 1941, when Douglas was emerging as the New Deal's heir-apparent, his judicial activism blended with his adroitness in persuading Roosevelt to move forward by flanking ploys at combined objectives, instead of alternately confronting and retreating from them, as FDR had habitually done. Gradually, the fruitless debate over recovery turned into a progress report on preparedness, and theorizing over *whether* the government was to be involved in planning recovery crystallized into the inescapable practicalities of *how* it could best manage preparedness: whether by constructing new plant or converting old plant. Industry pressed for time-consuming construction of elaborate arsenals instead of undertaking to convert commercial facilities onto a war footing. Understandably, business wanted to take advantage of the sudden improvement in consumer spending. Nevertheless, the speedup in war production forced the shutdown of outlets for consumer spending. Meanwhile, therefore, mountainous accumulations of consumer savings built up unfulfillable consumer demand, cushioning the economy against a postwar drop in incomes.

Roosevelt's sorry substitute for an emergency management organization, coheaded by General Motors production wizard William Knudsen, compromised by ordering a token 20 percent cutback in auto production. This token accommodation of emergency requirements to business as usual hurt the booming civilian economy without helping the bottlenecked defense economy.

Labor had lined up with industry in favor of new arsenal construction over conversion; the two supposedly rival blocs wanted to slice up the war watermelon between them. But into the middle of this chaos stepped an unlikely champion of immediate conversion: Reuther, then still an upstart vice president of the UAW. He appealed to me for support in Washington. I told him that Douglas ran the underground network for decision making. When I took him to Douglas, their chemistry clicked. When Douglas took Reuther's case for all-out conversion to Roosevelt, Douglas clinched the argument by asserting, "The business guys don't know their own businesses, but Reuther, who's our guy, knows the score." Veblen had never made it to the Oval Office, but his ghost, lurking in the woodwork, grinned.

Reuther's blitz of pre–Pearl Harbor Washington revealed the changed mood of the New Dealers. They were on their own to find new roles in the war administration. At Reuther's first dinner party in Washington, Lyndon Johnson (with whom I had put Reuther up the night before, dangling dazzling promises of access to the future power in labor, sure, moreover, to control delegates!) was among those on hand at the Fortas's New Deal salon to sniff out the newcomer. When everyone had gone, Fortas said: "He reminds me of how we used to be when we first came here. We're broken to the system. But your Reuther is the revolutionary, still full of enthusiasm, still free to operate, still intent on getting things done instead of just figuring out where he comes out himself, like all of us do now."

Douglas intervened directly with Roosevelt to relieve two other White House headaches on a par with the wrangle over plant conversion. The first involved filling the key job Roosevelt had held during the 1914–18 war, the assistant secretaryship of the Navy. Contrary to Larrabee's assertion that Harry Hopkins brought James F. Forrestal, later the first secretary of defense, to Washington to fill the slot, Douglas and Corcoran had teamed up to do so. Forrestal, as president of Dillon, Read, had been the one Wall Street figure involved in utility holding company finance to line up on Douglas's side of the political crusade over breaking up the holding companies that Wendell Willkie had mounted.

When the war administration collapsed into chaos caused by the escalation of the military clamor for deliveries, and deliveries jammed up, pressure built up on Roosevelt to stop picking business executives as emergency managers. Douglas emerged as a leading candidate for production czar. To evade a draft, he and Corcoran teamed up with Forrestal to find a substitute with credentials too impressive for Roosevelt to reject.

Ferdinand Eberstadt and Forrestal had been junior partners together at Dillon, Read, and Eberstadt already headed the Army-Navy Munitions Board. Judicial activism opened a way to allow statistical realism to allocate resources for victory through the domestic economy. Douglas, jubilant over the status he had won for Reuther in the fight over plant conversion, convinced Roosevelt that the only way to win the war was to expand capacities while stepping up weapons deliveries: an exercise in the application of statistical management to head knocking.

Only a civilian backed by the military could run the new War Production Board: only assured subsistence for the civilian economy could guarantee expansion for the war economy as the decisive weapon it proved to be. None of the civilian misfits Roosevelt had picked succeeded in imposing managerial discipline on the military. Eberstadt, nominated by Forrestal, had what it took. Douglas then sold Eberstadt to Roosevelt, no easy job, because Eberstadt dealt blows, not poker hands. Eberstadt also welcomed Reuther for teaching the industrial establishment an overdue lesson. The two made an unlikely but inseparable and incomparable combination of vision and will.

Douglas was not the only New Dealer who found a second, unofficial role in the war administration. Currie did too. He evolved from the finest statistician of his time to Moscow's clumsiest high-ranking undercover "strategist" in Washington. When Stalin demanded that Roosevelt speed the opening of a second front against Hitler in France, Currie aroused my suspicions. First he started to snipe at Keynes, but not on professional grounds. He jeered that Keynes had become a British nationalist and voiced Stalin's party line: that Churchill and Roosevelt (in that order) were "scheming to prolong the war in order to bleed Russia." Larrabee confirms Currie's unlikely involvement in military allocations. He reports General Marshall being summoned to the Oval Office, "and, in the presence of presidential aide Lauchlin Currie, sprawled in an armchair," being told of a military decision "that had gone against Marshall in favor of Currie." No mention, however, of any ulterior motive on Currie's part!

In our last serious meeting, in 1943, Currie spilled the beans. His bluntness in pushing the party line gave him away. "Everyone knows how close you are to Forrestal," he snapped. "He's the only one who has any influence with Stimson. Why don't you get after him to tell Stimson to open the second front?" (Henry L. Stimson was secretary of war, and by then, Forrestal had moved up to be secretary of the Navy.) Tragically, Currie had the best analytical grasp in Washington of how unprepared America still was to do so, but how quickly she was arming to do so how massively.

"Lauch," I replied, "no one has shown me the cables. I've never discussed that kind of matter with Forrestal, and I'm not about to. If you know what's good for you, you'll go back to practicing economics and stop playing war. Moreover, you yourself made the statistical case we won for you at the White House. You proved that a production buildup to swamp Hitler was smarter and safer than rushing into a big-league Dunkirk."

I told Forrestal the story and tapered off my talks with Currie. Twice, he called late at night in New York and literally begged to sleep on the sofa in our living room: an incongruous request from the economic adviser to the president. Out of pity, I consented, but shied away on grounds of fatigue from any resumption of old-time intimacies, let alone the new argument he had started. He was clearly under cruel nervous strain, complaining more bitterly than ever about his wife and his penury; apparently, his Moscow connection had not relieved it, despite the KGB's well-known preference for controlling its converts by pressing payments on them.

The next news of him that I had was that he had decamped to Colombia; that he was invigorated by a new young wife; and that he was endowing the government there with his economic expertise. Sexual freedom apparently brightened his dour outlook on life. How he converted his Moscow connection to his Bogotá presence, I never knew. Whether with or against his advice, his new client opted against giving top priority to productivity and emerged a generation later as the Saudi Arabia of the dope world. But the Washington authorities never pursued his defection or demanded his extradition. And indeed it would have served no useful purpose. His defection did not undo his contribution. Very few innovations marking the history of the Hot and Cold Wars were more meaningful than Currie's injection of realism into American economic analysis.

Through Currie, I had met Harry Dexter White. His extraordinary technical talents and his imaginative reach won him recognition as the new force needed at the Treasury during the war, when Morgenthau's jealous guardianship of Roosevelt's fiscal change purse was broken. Techniques beyond Currie's grasp of the domestic economic mechanism were needed to structure a postwar international monetary system. White was the one man at the Treasury equipped to meet this challenge to start the world up again, on a dollar standard. White managed the statistical flow of every dollar that left Washington or New York earmarked for a foreign beneficiary. He kept the running tote on the dollars needed to keep the world running. As Washington subsequently discovered, he also kept Moscow tuned in.
    Dollar power called for a new international monetary system allowing flexibility for the currencies of the rejuvenated economies of the freed world by allowing all of them to pivot on the dollar. White saw that the primary purpose of the new system was not to make markets for currencies, but to activate currencies to reactivate markets inside and among the various economies. A prophetic modern dimension was added to Keynesianism in America by this emphasis by White on Washington's indispensable role in managing foreign claims on dollars and ways to use them, though half a century later, Keynes's orthodox American disciples were still quarreling over Washington's domestic role in managing the economy. At that time, Keynes himself, in need of technical support even in his own area of international monetary expertise, was the first to agree with

White's sense of managerial priority, to recognize his gifts, and to rely on them. He showed great generosity toward White as well as candid dependence, despite White's abrasive personality; White was given to storming where Currie merely sulked.

By the end of the war, Keynes was ensconced in Washington as Britain's policy voice, patrolling the pipelines connecting the warring baronies that run the U.S. government with the White House. It is axiomatic that any foreigner enjoying insider status in Washington automatically commands power at home, even a Soviet ambassador, commanding no power base of his own inside the Kremlin. Whitehall did not care about the arguments over America's domestic economic policy in which Keynes's American disciples were embroiling him. But Whitehall did care about the Bretton Woods conference where the International Monetary Fund (IMF) was to be structured. America dominated both the conference and the IMF, yet looked to Keynes to keynote the conference—extraordinary political recognition for a living economist!—and to blueprint the fund. So Keynes used White, and encouraged White to use him, to build their respective positions.

White was too realistic professionally, as was Currie, to have patience with Keynes's stubborn commitment to a holdover compromise with the gold standard; both were voluble opponents of it. Keynes, despite the depth of his differences with White and the violence of their arguments, did press for White's appointment to head the IMF they collaborated in launching. White was rejected on grounds of personality, not security. I never knew what influence, if any, White's rejection may have had on his entanglement with the KGB, or when he drifted or plunged into its clutches.

I had of course observed that Currie and White were close. But this was hardly grounds for suspicion. On his way up the ladder, Currie had accumulated enough professional reason for bitterness against Morgenthau to gravitate toward any opposite number at the Treasury who was also a virtuoso and, therefore, an ally. But after each revealed his defection, I could not guess whether they were coconspirators in the same operation or directed, as I am inclined to believe, by different "controls" against different targets.

Not long after Currie's flight, I was startled to receive a call from White one evening when he was in New York, sounding as desperate as death, asking to see me. I had not had any contact with him for months, though I knew that he had developed reservations similar to mine against the postwar loan to Britain that Keynes was pushing. When I explained that I had guests and could not talk, he hung up abruptly. The next day he committed suicide. I never heard from or about anyone else he tried to contact at the end.

The careers of Currie and White, considered apart from their ignominious endings, marked a new stage in the evolution of statistics into a formative arm of national policy-making. Neither could claim a creative vision, but both were superb technicians who made themselves indispensable to wise and, what is not always the same thing, practical policy engineers. The defections of Currie and White, separate and different though their circumstances were, illustrate the attraction the Soviet apparatchiks have always felt for expertise in economics and the respect they accord to experimenters with the capitalist economic mechanism the Soviets remain unable to fathom. Gorbachev could have made good use of their talents in implementing *glasnost* and *perestroika.*

At that time, the ambitious theory Keynes designed invited quantification of Veblen's pioneering concepts by developing statistical models for the distinctive American economy that Veblen had outlined and Hobson had identified; Currie anticipated Keynes domestically, and White supplemented Keynes internationally. In Americanizing Keynes, both together updated Veblen. They transformed statistics from the tool of advocacy it was in Brandeis's hands and the exercise in observation that Hoover publicized into the indispensable raw materials of policy-making that Petty had envisioned.

11

VOLCKER'S VACCINATION
Measuring Monetary Management

Across three centuries, the realities of economic life have reaffirmed Petty's original thesis that national purpose gives direction to economic expansion, and that political decisions guide economic activity. This same evidence has also nullified the counterthesis Smith developed a century later that political interference stultifies economic growth. The dividends from the relatively modest government investment needed to propel peacetime growth defy calculation. Smith's premise ignores the huge economic paybacks to governments that interfere in the economy when they make wars and win them quickly, as Bismarck did and as Germany demonstrated.

Short of war, the cost-benefit ratio of government investment in economic expansion through harbors, roads, central banks, post offices, schools, hospitals, airfields, and other infrastructure items taken for granted is commanding. Private investment follows government investment in an infrastructure on which investors can build. The auto industry began to mass-produce cars after the public sector began to proliferate roads. (Witness Los Angeles: it was laid out for the explicit purpose of providing a market for the then still nascent auto industry.) The aircraft industry took off after the public sector gave it airports.

Thanks to the mind-boggling expansion of the economy, the age-old argument over free enterprise versus government policy-making has dissolved into academic pillow fighting. The government has resolved the theoretical dispute by evolving into the primary trendsetting, patronage-dispensing, power-brokering force within the economy. Thanks again to the expansion of the economy, the shrinkage of the world has activated governments as the negotiating agents for the economies behind them, exactly as envisioned by Petty.

Under the convergent impact of politics and technology on markets, the basic links tying money to credit, credit to government, government to statistics, and statistics to security—national as well as social—have grown inbred and unbreakable. So are the links connecting the arteries of transportation and communication with modern financial markets—especially in government securities—that depend on the instant international transmission of push-button money in megabillion-dollar lots. The government never operates in a political vacuum when it stimulates the economy to expand. America has allowed the underdeveloped fringes of her domestic economy to expand along with it. But the American government spends a great deal each year that is not necessarily dedicated to stimulating the economy, most conspicuously for defense, social entitlements, and interest. This continuous expansion in cash traffic between the government and the auxiliary systems feeding on the economy has expanded the banks as well. Consequently, each round of expansion in the economy has intensified the interdependence of the economy and the banks on the government to meet, but also to insure, the credit needs that mushroom with the continuous expansion of national outflows: a Herculean statistical responsibility.

In America, Lord Acton's rule that power tends to corrupt applied to the growth of the statistical powers of the government as the postwar boom faltered in the 1960s, as the engines of expansion suffered corrosion by inflation in the 1970s, and as the illusion of expansion eroded into deflation in the 1980s. Because Washington's statistical influence saturated the marketplace, along with the speculative markets (in securities as well as commodities), it dominated media assessment of both. Washington's exercise of its vast statistical responsibilities, originally free of sophisticated scrutiny, became increasingly suspect even to the amateur eye. Nevertheless, the government balked at its obligation to wield this power objectively. Instead, the government grasped at the opportunity to manipulate statistics politically.

Lyndon Johnson, always the brightest and the best at games inside Washington, was the first to take the plunge. He had made his career manipulating everyone of consequence; manipulating statistics was immeasurably easier—they couldn't answer back. His initial try was an instant success. In 1963, when Johnson took over, Kennedy's campaign promises were still unfulfilled. Johnson undertook to fill them all, beginning, however, with the flashy promise to cut income taxes before tackling the spending programs.

The first old cloakroom intimate he hoodwinked was Senator Harry Byrd, Sr., the patriarchal chairman of the Senate Finance Committee, then the bastion of conservatism. LBJ staged an elaborate statistical presentation promising that spending for the fiscal year ahead would be brought down below the magic barrier of $100 billion: Byrd's condition for approving the tax cut Kennedy had promised. Kennedy had failed to cut spending below $120 billion. Just one year later, Johnson had the tax cut, and spending was back up to $134.5 billion. Johnson never stopped grinning as he bragged that he got around not only "old man Byrd" but also the constitutional requirement for initiating tax legislation in the House and the political taboo against initiating tax legislation in the Senate. LBJ, when he was still Senate majority leader, had invoked these twin excuses of constitutional prerogative and fiscal responsibility, in Speaker Sam Rayburn's name, to avoid acting on my suggestion that he move for a tax cut to end the 1958 recession. As he wrote me after we discussed the idea, Rayburn was dead set against "unleashing the wild horses of inflation" by moving for a tax cut with the government running a deficit.

In the mid-1960s, Vietnam gave Johnson his field laboratory for fine-tuned statistical manipulation—in this classic case, manipulation of the functional defense microstatistics that control the visible fiscal macrostatistics. It also provoked my own permanent break with Johnson. He started out insisting that he was not escalating the war, pointing to the absence of escalation in the spending statistics to prove it. He ended up insisting that the war was being won, and now pointed to precisely such an escalation to prove this case as well. In 1966, he was committing $3 billion a month to the war, but admitting only $1 billion—a $24 billion annual discrepancy, then still big money. His statistical cover-up opened new frontiers for the fiscal embezzlement he pioneered, proof that any president can authorize arbitrary uses for diverted funds (as Reagan did in the Irangate escapade and its aftermath), so long as he remains popular and retains plausibility.

Johnson was a past master at the art of channeling appropriations into cash flows. He had his nose in every nook and cranny of the defense procurement system: he had built his congressional power base on his chairmanship of the defense subcommittee of the Senate Appropriations Committee. By the time Vietnam blew up into a national issue of hysterical intensity, senators who policed Pentagon spending and protected Pentagon programming discovered that Johnson was using them to deceive colleagues who were vigilant on the foreign policy side. His "escalation by stealth," as I branded it in my historical study, *The Economics of Crisis,* was rerouting weapons explicitly authorized for other uses to the jungle war. In the end, the conservative southern stalwarts Johnson had trusted as his most malleable allies on the "flag" issue turned him in to the liberal northern and western dissidents who were his most embittered opponents on the foreign policy fraud issue hidden behind the flag.

LBJ's pioneering statistical manipulations were fiscal, inviting Congress to exercise its power to audit them. But Congress is not geared to uncover governmental manipulation of the statistics recording private sector performance, though the importance of this job, a prime statutory responsibility of the government, has grown with the economy. The markets did not suspect the disruptive economic consequences of Johnson's breach of statistical integrity until their own functioning was disrupted. In 1968, the markets suffered a shakeout that contributed to the popular distrust that expansion of the Vietnam War had produced. Johnson read the message as the popular veto of his reelection plans, and it was.

The markets suffered disruption again from a similar executive breach in 1987. At the height of the uproar over Judge Bork's crusade to deactivate judicial activism, and in the face of a crescendo of protest in behalf of the lost cause of fiscal responsibility, Reagan repeated Johnson's defiance of "original Constitutional intent." He committed unauthorized funds without limit to a Persian Gulf naval expeditionary force, bypassing congressional assent. In fact, he was reckless enough to ignore not only the absence of a new defense budget at the end of a fiscal year, and the absence of unearmarked funds for badly needed new weapons, but the absence of any clear continuing count on how many hundreds of billions the defense establishment already owed on past commitments.

With inspired irresponsibility, Reagan also ignored the political power and the administrative professionalism of the congressional adversary he stumbled into confronting: Chairman Sam Nunn of the Senate Armed Services Committee. Less than a year before, honoring Senator Barry Goldwater on the occasion of his retirement, Reagan had responded to the Republican patriarch's emotional draft of Nunn for a bipartisan succession presidency with the wistful hope that "we should make Sam a Republican."

Now Nunn was opposed to the Persian Gulf adventure. When Reagan launched it, Nunn was holding the administration hostage on its ambitious Soviet missile negotiating objectives; Reagan's Persian Gulf adventure tightened Nunn's lock on Reagan's unadministerable defense budget demands. Moreover, Nunn was topping the Eisenhower administration's forgotten bid for "more bang for the buck," calling for "more bang for less bucks," as the balance of power in Western Europe demanded. Reagan's recklessness trapped him in the same "no-win" game of fiscal diversion for military adventure that had undone both Johnson and the markets in 1968. In both cases, unauthorized use diminished the availability of authorized monies.

No one could have linked the 1929 crash to statistical confusion over the federal budget at the turn of that fiscal year. The federal government had no deficits to speak of in 1929 and functioned free of any suspicion of cooking its books. As the Depression spread distress, Hoover lost only his reputation for having a head and a heart; he kept his reputation for counting the dimensions of the disaster over which he presided. By contrast, no one, however simpleminded, could have separated the October 1987 crash from the federal government's mismanagement of its housekeeping and the statistical deception it practiced to cover up its malpractice.

On the broad canvas of history, where economic ideas do battle against political inertia, Petty's aim in experimenting with statistics was benign: simply to monitor the continual interaction between the government and the economy. He undertook head and land counts in England's national interest. Over three centuries later, America had progressed to reliance on statistical cover-ups of government arms commitments in the furtherance of factional political interest. This technique has evolved into a unique American institution. No other free country was subject to statistical sloppiness sufficient to countenance any comparable exercise in routine deception, beginning with deception in defense financing practices.

The distinctive size of America's markets, their elusiveness from hierarchical control, and their susceptibility to media influence have all worked to facilitate routine official exercises in statistical manipulation. Market actions proved responsive to government statements, which in turn proved responsive to market actions, with the media serving as the link. This simple, plausible, and pragmatic sequence has made America's statistical economy—that is, the economy as perceived, defined, and measured by routine statistical announcements—distinctively volatile. It has produced continued divergences between statistics and realities and invited recurrent collisions between susceptible markets and unsuspected developments. In addition, this sequence of manipulation seemed to validate the illusion handed down through the academics that markets functioning on their own embody all wisdom that nobody can fathom alone.

The increasing susceptibility to the official statistical entrapment of America's seemingly free markets explains their increasingly volatile behavior. Statistical power is a prime by-product of political power, and statistical power is market power because statistical power quantifies the power of officialdom to define. As Elizabeth Janeway suggests in her *Improper Behavior,* simplistic exercise of the police power rests on this sophisticated power to explain events by fitting them into familiar patterns. Statistical sleight of hand endows government pronouncements in areas not considered political with a ritualistic aura of magic that markets accept as a form of received religion. Moreover, statistics by definition are subject to continuous adjustment without any embarrassment.

When the Reagan market euphoria took hold in the early 1980s, the market makers were slow to suspect that they were the dupes of political policymakers playing with statistical news releases. Accordingly, the stock market was quick to take recession news from the economy as recovery news for itself. The market makers looked to the government to back bullish bets by accelerating the flow of reassuring statistics. On Black Monday, in October 1987, traumatized portfolio managers could be overheard reassuring one another that "Reagan will do something to prevent anything terrible from happening"—as if it hadn't already. In fact, when he did respond to this desperate appeal, the noises of reassurance he offered were couched as statistics, which electrified the markets into record flashes of recuperative energy, each round, however, was necessarily short-lived. A resurrected Herbert Hoover would surely have been shocked to discover that the publicity he gave to statistics during his tour of duty as secretary of commerce turned out to be more important historically than his impotent and inarticulate presidency.

To take just one example of the incestuous analytical and procedural corruption infecting statistical and market operations, the official oil statistics have been under the steady control of the American Petroleum Institute (API). Yet as late as 1988, *Platt's Week,* a McGraw-Hill publication, ran an article by Patricia Walker pinpointing the API's glaring ignorance of oil inventory flows. *Platt's* grim findings prompted it to warn its oil industry audience, "Building up inventories can be a dangerous game." None of the oil moguls has shown an awareness that he was playing blind-man's buff with the impact of statistics on his own market, despite the lip service all of them paid to "market laws" so long as prices were surging. No market laws can be observed, let alone obeyed, so long as the interplay of inventories on market behavior goes undetected.

Once the oil price runaway of the 1970s was declared unstoppable, so was the inflationary tide, which also gave a free ride to oil reserves in the ground and to inventories in storage. Buying was taken to guarantee profit, and borrowings to buy still more were taken to guarantee even greater profits. Contrary to commercial common sense, the less sold, the meeker customer acceptance of the caper was expected to remain—as if in hypnotized response to some miracle of nature. The resultant buildup of inflated assets also inflated corporate earnings and stock prices faster than cash flows.

For the better part of a decade, the workings of the markets flouted rudimentary market laws. Prices spiraled upward with oil supplies, violating Smith's claim that higher prices buy bargains because they bring out higher supplies that guarantee lower prices. Increasingly, prices often appear to business as objective, external signals to be heeded even when set in motion by self-fulfilling decisions and practices of business itself. As a result, management is habitually confused about whether it is the cart or the horse; that is, when it may be leading its markets or following them. Nevertheless, the statistical economy ignored the vital distinction between transactions supported by current consumption and transactions invited by inventories built in anticipation of future consumption at still higher prices.

Assurances that supply and demand would automatically balance the oil market inhibited any analytical consideration of inventories, the factor relied on to provide the balance; again and again, they turned out to be the source of the imbalance. The confusion over inventories that turned the world upside down misreading the chronic oil glut as an acute oil shortage acted out the absurdity built into classic economic theory. It assumes a theoretically normal state of equilibrium, which is acknowledged never to exist in fact. Instead, the world of economic experience, as distinct from the fantasy of theory, is admittedly jolted into successive states of disequilibrium; fluctuations in inventories determine how fast and in which direction.

The decade-long cycle of boom and bust in oil stands as a classic case of the way that history illustrates how the transactions economy turns on inventorying, and how inventorying consistently crosses up the statistical economy. Veblen had singled out inventories as the key to the performance of the economy half a century earlier, when prices were relatively stable—if only because futures speculation was unknown. Once futures speculation took control of the pulse of the economy, statistical and monetary managers who were trusted to stabilize it persisted in flying blind about shifts in inventories. That put them on a collision course with the oil crash of 1979.

Unnerved but undeterred, the markets spent the decade from the mid-1970s to the mid-1980s bouncing back and forth with each change in government efforts to manage overall price inflation, but without tracing the virus back to the oil price inflation. Accordingly, government scrutiny focused not on the root of the inflationary chain reaction but on its upshot: monetary inflation. Borrowing, not Federal Reserve policy as popularly supposed, is the major cause of monetary inflation; by definition, banks create money with every dollar they lend. The oil inflation exploded into a borrowing spree big enough to run the banking system out of note paper.

From its outset, the oil borrowing boom inflamed business incentives to carry high-cost inventories and, costlier still, to search for new reserves at any price, meanwhile building supertankers as floating storage tanks for the spillover. In the very nature of oil inflation, its price spiral rated any level of inventories as both dirt cheap and inadequate. Consequently, every dollar borrowed to load up on inventories just for tomorrow was seen to justify borrowing fistfuls more to chase after reserves as insurance against the great shortage guaranteed to start next year and to last forever. What ended as outright fraud in the 1970s started with the traditional confusion over statistics. Because the statistics tabulated the oil price inflation as unfolding with the primitive violence of a force of nature, opinion-making sophisticates accepted it as just that, not suspecting the manipulation behind it or the trained ignorance of the analysts evaluating it.

The manipulation supported the propaganda; the ignorance accepted the statistics. In November 1987, *The Wall Street Journal* discovered that the oil deflation had defied the law of economic gravity by increasing the glut. The *Journal* cited the retired president of Shell International Trading, one of the better-managed majors, as complaining, "It's daft. The world's most important industry is flying blind. We don't know whether too much oil is being produced until the market breaks." He might have added that, thanks to continuous dollar devaluation and continuous oil debt inflation, market breaks have come to force contracyclical increases in oil production. The *Journal* went on to quote an economist at Chevron Corporation, one of the worst-managed majors, as opining, "Since [1981 or 1982], about one million barrels a day seem to be consumed but apparently not supplied." This spokesman did not explain the real mystery: despite reports that consumption was increasing with supplies, inventories nevertheless increased as production did. These disclosures, and others, reflected great credit on the integrity, as well as the acuity, of *The Wall Street Journal*'s reportorial corps. Nevertheless, the *Journal* continued to float the routine oil statistical junk without disclaimers, suggesting that its editors and columnists were not among its readers.

Blind borrowing fueled the chain reaction of oil price markups and oil inventory pileups. Intoxicated inventory accumulation, statistically unidentified, happens fast. Distress liquidation, beyond the reach of statistical identification, is slow. So are the consequences of overborrowing to overpay for lemons by any name. The chorus of oil industry opinion assumed that bigger buildups of high-cost oil inventories would support still higher markups for still higher-cost sources of production (notably the far side of Alaska and the bottom of the North Sea).

This statistical shuffling passed muster while production and prices spiraled but turned catastrophic once the price collapse forced an abrupt halt on speculation in inventories. But blind faith in the need for investment to bring still more oil to market withstood the shock. In the face of the second oil price collapse in 1987, from above $20 back below $15 and falling, the momentum of ongoing oil investment sent world reserves surging by 27 percent, by the industry's own admission. As late as 1988, after the price of oil had been beaten down to $8 once and was wobbling under $14, the major oil companies were still increasing their investments to ward off the renewed shortage they saw right around the corner.

Back when oil prices seemed topless, borrowing to buy became à la mode. At $40 a barrel for oil, banks, not suspecting that prices had peaked, rushed to advance $40 a barrel against reserves on the assumption that $80 would provide comfortable loan collateral. But no one charged with statistical responsibility bothered to quantify the oil debt load bulging in the banks. The API computed only the physical traffic in actual production; its remoteness from mere money matters recalled an old joke about a restaurant client asking the time from a waiter standing nearby, whose response was, "That's not my table." If the API limited the scope of its scrutiny to the sources of income and expense, balance sheet considerations remained within the purview of the Federal Reserve Board.

Nevertheless, the Fed ignored the troublesome evidence of pyramiding credit for long-term "oil plays" based on short-term bank debt that no auditor could miss. As I stated in *The New York Times* of February 23, 1988, the Federal Reserve Board, notwithstanding its awe-inspiring reputation for statistical sophistication, neither publishes nor even tabulates the tons of bad oil paper held by American banks as short-term collateral, which has been left uncovered by long-term asset deterioration. More symptomatic still of the analytical disorientation of the statistical authorities, the mandarins at the Fed confessed to complete innocence of the magnitude of the credit pyramid whose inflation they had indulged for a decade and a half. The Fed's backup for busted banks, the Federal Deposit Insurance Corporation, admitted the same surprise when the collapse of major Texas oil banks taxed its deposit-insuring capabilities.

In Egypt, the pyramids stand in a cluster. America's modern-day pyramids also overload the credit structure in a cluster. In Egypt, the pyramids have stood for millennia on top of the desert sands. In America, the credit pyramids on top of the oil sands have sunk underwater. Thanks to the Tax Reform Act of 1986, plus the oil deflation on which it did not reckon, the pyramid of tax-sheltered investment in domestic oil has, too.

The borrowings that financed the takeovers that fired up the stock market explosion have proved equally ill-fated. So have margin borrowings to carry oil stocks; still greater margin borrowings to finance dice throws on oil futures; and corporate borrowings to carry inventories and back ongoing high-cost drilling programs. While the oil inflation was on, and during the long speculation to renew it, the emphasis was on price action. Neither the banks themselves nor their regulators measured the credit bloat needed to pump energy into the oil price speculation. The oil borrowings invited by the oil inflation of the 1970s and forced to support the oil deflation of the 1980s dwarfed even Third World debt. Statistical convenience dictated separate measurement of the two. Again, emphasis on Third World debt distracted attention from its roots in oil debt. Actual oil borrowing practice made the two indistinguishable.

Without exception, Third World oil producers accounted for

the greater part of Third World defaults on bank loans. Also without exception, the major Third World borrowing agencies were government corporations, notably PMEX in Mexico. When one of the elders of the New York banking fraternity pronounced all governments' loans good, his blessing automatically endowed all government oil corporations with impeccable credit ratings, of which they made the most—when oil was top-fishing and again when it was bottom-fishing. Not even the inescapability of ruinous capital write-downs for the banks jolted the Federal Reserve Board into launching a laborious search to disentangle the jumble of Third World bank loans long since gone bad and oil loans that threatened to follow. Instead, the ideologues appointed to administer the rhetoric of economics at the Fed tied up its batteries of computers to search for patterns of market behavior that could not be found in the erratic performance of commodity futures prices on which to base monetary policy.

At the policy-making level, the fight to contain monetary inflation was foredoomed because the warrior in command at the Federal Reserve Board, Chairman Volcker, accepted the underlying oil inflation as inescapable and unstoppable. But the self-deception at the policy level began with sleight of hand at the inventory level. Ostensibly, the inventory factor was recognized in the API tabulations: all statistical releases did routinely classify inventories, but they did so inadequately, as, first, the oil market collapse and, then, its confusion while ostensibly recovering subsequently demonstrated. Admittedly, new supplies pumped into the pipelines of distribution in anticipation of higher prices are always difficult to distinguish from immediate consumption. Weighing the imbalance in the supply-demand equation strains the skill and experience of the forecaster and the manager even when fluctuations are cyclical, but especially when a cartel raid sends a hypnotized market into a buying panic.

Rarely during the exaggerated, compulsive, and prolonged oil spree were questions ever raised that speculation was based on claims of phony shortages—except by me; and I was discounted as an eccentric out of step with the oil inflation whose time supposedly had come to stay. As a result, propaganda for nonexistent statistical shortages converted the propagandists themselves into believers that the shortage was real. The self-deception is built into the market process: again and again, the market manipulators assume they are following independent price actions when in fact they are being led by their own machinations. The API revived the same statistical stunt during the debacle that started in 1986. The institute conjured up the mirage of a gasoline inventory shortage in a commercial vacuum, ignoring the crude oil glut ready to reverse it. Certainly, the official statistical establishment did not question the integrity of the data gathered by the API; nor could the API appropriately dispute the data furnished by its own members. An administration with an effective antitrust operation would have launched an action against the refining subsidiaries of the major oil companies before the phony shortage scheme drowned in oil.

Throughout the 1970s, the markets and the government systematically misread the API data, with no sensitivity to the self-serving role of the oil industry as the supposedly objective source. The statistics issued by the API and blessed by the government enjoyed media acceptance at face value. Consequently, the media disregarded the reports they themselves established of idled super-tankers loaded to the waterline with inventories unstorable on shore, and the markets were left unprepared for the series of shocks that followed the revelation that the oil companies had run out of storage space in 1979. By then, the oil industry had spent eight years blinding itself to the glut with which it continued to swamp its markets. Of all the surprises that tripped them up, none upset traditional calculations more than the discovery that war in the Persian Gulf, the most famous of commercial choke points, actually proved inefficient as the calculated method of price support it was planned to be.

When Reagan took office, he found the oil crisis, the inflation problem, and Volcker waiting for him. Of the three, Volcker proved the easiest for him to handle, despite the aura of statesmanlike independence in which Volcker basked. Thanks to the cues Reagan fed him, Volcker won credit as an inflation fighter, and, under cover of Volcker's image, Reagan won elbow room to inflate the budget deficit. Until Volcker's term ran out, Reagan took cover behind the reputation Volcker built, and Volcker returned the compliment. As long as Volcker kept his mouth shut about the oil inflation underlying the monetary inflation, he stayed out of trouble with Reagan. In fairness to Volcker, the reputation he won was for toughness, not shrewdness, and he may not have known enough to look past the monetary inflation.

When Volcker came to the Fed, he found it under criticism for its failure to stabilize interest rates, though its critics routinely failed to blame the oil inflation for destabilizing them. Volcker recognized Reagan's power over the prevailing winds of opinion. He also saw Professor Milton Friedman, the dean of the monetarist cult, basking in more White House favor than Friedman enjoyed later, after a monetaristic fiasco or two. Meanwhile, however, Friedman was using his White House forum to proclaim the need for a switch in the management of the Federal Reserve from controlling interest rates (and letting the money supply run free) to controlling the money supply (and letting interest rates run free). Volcker was quick to accommodate the Fed to this change in ideological fashion.

Friedman's formula for straitjacketing the economy by establishing an arbitrary and predetermined rate of increase in the money supply acted as a mercy weapon: it was a quick killer, if thanks only to the whirlwind rate at which oil borrowing was expanding the money supply. By mid-1982, credit deflation and interest rate inflation had squeezed the energy out of the economy. This decisive test of the money supply rule as a Fed guideline for managing the economy resulted in two fateful chapters of stock market history. The first, in 1982, trapped Volcker and gave Reagan a free ride, when congressional protest against the slump threatened by Volcker's crackdown on credit triggered the first Reagan bull market explosion. The second, in 1987, left Reagan to bear the brunt at a critical turn of events. Corporate management ignored the telltale deflation of commodity prices and the contraction of the money supply, though both symptoms reflected fatigue in the economy. Nevertheless, management unleashed the frenetic round of product price increases that invited a fatal blow-off in the stock market.

Back in 1982, Congress acted incisively—as the economists would say, contraseasonally—during the normal recess for the dog days of summer. I was scheduled to testify before the Joint Economic Committee on August 2. A few weeks earlier, I had run into my Cornell contemporary and lifelong friend, Congressman Henry Reuss, the committee's chairman, coming up from Washington to New York. In the course of our visit, I spelled out my misgivings over Volcker's willingness to indulge the White House's faith in Friedman's dogmas.

I pointed out that money supply figures vary with bank lending. The restriction then in force was putting a ceiling on bank credit just when the slowdown under way was calling for bank credit expansion. I complained that Volcker and the Fed staff were misreading their data and missing another reversal of trend in the economy precipitated by inventory over accumulation. I emphasized my confidence in how easily the situation would be changed for the better and the safer if the committee instructed Volcker to change foot pedals on the credit engines, substituting the accelerator for the brakes. But I added my judgment that, with Volcker polishing apples at the White House, nothing short of a congressional mandate would produce results—certainly not in time to abort a painful, disruptive recession.

The Joint Economic Committee is an entity endowed with a voice for public policy rather than a vote of legislative intent. Reuss, however, did more than his homework; he did his footwork. By August 2, he forced an unusual confession from Volcker. Reuss expressed the unanimous sense in both houses of Congress that a switch to credit ease was overdue. Volcker's fellow governors and members of the Open Market Committee authorized him to sign a reply that put the Federal Reserve Board on parole to Congress. It promised that "the Federal Reserve Board will obey the law."

The Fed did but was too embarrassed to advertise the fact; the significance of the Fed's compliance was too great for the media to grasp. To compound the prevailing confusion, the president, insulated by his polls, endorsed the Fed's previous policy after the Fed reversed it. The markets had been anticipating a Fed retreat for some months, but they were a full month satisfying themselves that a full reversal had been forced on the Fed. The media discovered it after the markets did; their Wall Street reporters were slow to pick up on the story because their Washington reporters missed it while it was incubating. It was not until September that the bond market had the message that the Federal Reserve Board was opening a two-pronged offensive against the recession that was being blamed on Reagan: the Fed moved forward to lower interest rates and to take the lid off the money supply, then moved backward to steer its course by interest rates again.

The stock market took its lead from the bond market, reversing the traditional pattern codified in Dow theory of running counter to the bond market. This historic divergence anticipated market behavior during the great bull market climax of the 1980s. Meanwhile, the economy took off in the wake of both markets on the simplistic assumption, which proved realistic, that Wall Street knew something it didn't. Wall Street did know that the money supply statistics, and the bank credit statistics that swung with them, were being released from the straitjacket that the monetarist dogma imposed on the stretch in the statistical economy. To paraphrase the poet, that was all the Street needed to know to rocket off the launching pad.

In addition, however, Volcker had left the market makers in possession of a special speculative rocket booster that at first only they knew about: the double standard between the 50 percent margins the Fed itself required on stocks and the 6 percent margins the Commodity Futures Trading Commission (CFTC) allowed on financial futures. Thanks to this glaring discrepancy, futures speculation exerted an irresistible attraction, not picked up in the statistics, for money looking for market action. The contrast was duplicated in the heroic reputation Volcker won for using his monetary powers to take the hard line against inflation throughout the marketplace, and the soft line he was content to take with his margin powers against inflation in the financial markets.

Volcker's inverted priorities ignored the interplay of trends between the economy and the financial markets at the time he took over. The economy was suffering hangover pangs left behind by the decade of oil inflation aggravated by monetization. But the record pool of cash accumulated in productive assets during the decade of inflation was ready to spill over into financial assets. Meanwhile, the financial markets had been depressed by the high interest rates that accompanied the high levels of inventorying responsible for the aggressive price markups of oil and other productive assets.

Therefore, Volcker's choice of regulatory methods chilled the economic marketplace when it was cooling off and ignited the financial markets when injections of excess liquidity were ready to make them explode, as the traders at Salomon Brothers, who controlled the biggest book on the floor, insisted would happen. The markets soon proved the Salomon traders right and Volcker wrong. He and the Fed staff went wrong again relying on their statistics to guide them to their priorities.

With the statistical guideposts Volcker inherited when Carter appointed him to the chairmanship of the Federal Reserve Board went an archaic gerrymandered structure of authority over margins, as irrational as it was unworkable. The board itself had been created to regulate the banks. During the backlash from the 1929 crash, however, when anger was running high against the brokers for luring sheep to the slaughter with irresponsible margins of only 10 percent, the Federal Reserve ran the only regulatory game in Washington. So control over margins, and over the brokers as well, was dumped on the Fed.

By the time the SEC took over policing the brokers in 1933, the issue of regulatory control over brokers' margins was dead, until 1976, when the biggest political "fix" in market history was thrown in Washington at the CFTC when nobody was looking. By a vote of four to one, it decreed that financial futures are not securities, as any dull-witted margin clerk knows they are, but commodities. Logically enough, therefore, the CFTC proceeded to permit trading in financial futures on the same 6 percent basis required for commodities. This nominal basis is reasonable enough for commodities because the businesses that routinely use the commodity futures markets to hedge the prices they pay or charge for future delivery always have their own money invested in inventories on hand and are never exposed to the short squeezes historically responsible for triggering panics in stocks.

In 1982, when Volcker flashed the green light across the front door to the financial markets, the double standard advertised between the 50 percent margin required on stocks by the Fed and the 6 percent offered on futures around the back door to the stock market by the CFTC had the instant effect of a match in a gas tank. For the rest of his term, Volcker officiated over his famous smoke-blowing ritual, mouthing pronunciamentos against inflation and blaming Congress for subsidizing inflation, while the speculative fires burned the investment substance out of the financial markets. With superb timing, he pulled out at the top, in mid-August 1987.

Soon after Alan Greenspan took Volcker's place, I published an article in *The New York Times,* urging him to resist the pressure on him from Wall Street to raise the discount rate and instead to press the CFTC to raise margin rates on financial futures to the same 50 percent basis as the Fed's margin rate on stocks. But Greenspan satisfied conventional Street opinion by leaving the CFTC undisturbed and raising the discount rate. The consequences were disastrous, unmistakable to history and even to him when, in a matter of weeks, the October crash forced him to reverse course in his turn.

Shortly after my article appeared, the CFTC's lone 1976 dissenter, former chairman James M. Stone, wrote to me, congratulating me on it and documenting his own repeated appeals at the time to the Federal Reserve Board, to the Treasury, and to Congress to assert uniform margin control over financial futures in time to avoid the speculative disruption the double standard threatened. Stone subsequently demonstrated his professional mastery of insurance principles in practice by going on to operate a successful entry in the high-risk, highly competitive auto insurance business. After the headlines made by the stock market crash of 1987 confirmed Stone's fears and mine, the media showered blame for the large-scale futures speculation by institutions on the minimargins tolerated by the CFTC. The caper made the 1929 scandal of 10 percent stock margin abuse look tame.

Another bit of history is in order. It centers on the role of the government debt as the carrier of monetary inflation, and as the source of the astronomic growth in the bond market, which of course dwarfs the stock market. In 1929, the government had no debt to speak of, and, therefore, the Treasury bond market was nonexistent as the speculative lever it subsequently became on interest rates and all the financial markets. Consequently, speculation on weekly fluctuations in the money supply was unknown as the presumed key to the combined jumble of market behavior and government operations to stabilize it. Historians and moralists had long since identified lamentations over the money supply as the telltale indicator of inflation run amok, but the money supply game did not take over the money game in the financial markets until the feud between the Keynesians and the monetarists erupted into a national pastime in the media at the crest of the postwar boom in the 1950s. At that point, the economy was reduced to an analytical plaything, like a football kicked up and down by the Fed as fast as the money supply figures blew up recurrent scares over inflation and recession.

In December 1985, William McChesney Martin, Jr., once venerated as the all-powerful, all-knowing chairman of the Federal Reserve Board, gave an unnoticed interview to *The New York Times,*
reminiscing about his time at the summit of power and knowledge
between 1951 and 1970. He recalled telling himself:

> My gracious, here I am the new chairman of the Fed and
> I'm doing my best—I'm not the brightest fellow in the world but
> I'm working hard on this—and I haven't the faintest idea of how
> you figure the money supply. Yet everybody thinks I have it at
> my fingertips.
>
> They don't really know what the money supply is now,
> even today, they print some figures—I'm not trying to make fun
> of it—but a lot of it is almost just superstitious.

The *Times*'s profile of Martin's rise, encapsulating forty tumultuous years, pictured him as a bureaucratic Horatio Alger who
blazed his way to the top from a lowly start as a bank examiner. In
fact, Bill Douglas snatched him out of the obscurity of regional
"white shoe" stock brokering in St. Louis and thrust him into the
presidency of the New York Stock Exchange to solidify the transfer
of control over Wall Street from the Morgan power structure to the
SEC. Martin fit Douglas's preference for Main Street Ivy League
types to man the power stations Douglas brought under his control.
Martin's father-in-law had served as Coolidge's secretary of war and
endowed the Davis Cup. By the time Truman came in, Martin was
in place to serve as the Wall Street ornament for Truman's Missouri
machine.

Martin's confession made clear that no one responsible for the
money supply statistics, during his regime or subsequently, knew
what they meant or even how they were concocted. Only a few aged
history buffs were still aware of the thread Martin provided between
the New Deal's statistical breakthroughs tracking the transactions
economy and the fuzz of complications in which the managers of the
statistical economy shrouded their calculations. Soon after Martin's
revelation, the Fed deemphasized the money supply as its measuring
rod for regulating bank credit and influencing short-term interest
rates. It went back to answering emergency alarms at endangered
banks and squinting at interest rates for guidance on the direction

in which its statistical guesswork suggested that the economy might
be lurching.

With disillusionment over money supply statistics spreading in 1985,
a new statistical story was needed to keep the markets at a boil and
the market makers off the backs of the market regulators. Accordingly, the Commerce Department, Hoover's old bailiwick, moved
into the breach to advertise the standard statistical concoction that
was the flimsiest and most familiar in the government's entire statistical arsenal: the leading indicators.

    Not all the leading indicators are equal. The Dow Jones Indus-
trial Average, supposedly still representative of the performance of
the stock market as a whole, is given top weight; the money supply
comes second. In simpler times, marked by the absence of monetary
inflation and the presence of investment values, both the stock market and the money supply did serve as bona fide leading indicators.
But once Washington developed the nerve and the know-how to
manipulate perceptions of the money supply, and, therefore, the
behavior of the stock market, its statistical managers had a new
handle on the leading indicators. They used this leverage to give the
financial markets another powerful upward shove. Thanks to this
statistical exercise, Wall Street's reintoxication early in 1986 painted
a deceptively rosy tint onto the persistent lackluster surface of the
economy.

    The statistical managers exploited the interacting buoyancy of
the money supply and the stock market to bull the index of leading
indicators, and they kept the game going by reversing the batting
order between its two top indicators. Once the institutional market
makers were bedazzled by the leading indicators, they bought the
Dow stocks blind; the money supply followed automatically in response to the injection of liquidity into the banks. The Fed finished
by unleashing a Frankenstein monster: once it put the financial
markets into the business of mass-producing liquidity on a scale that
dwarfed its own operations, it put itself out of business controlling
liquidity.

    Ironically, the Wall Street gunslingers, in their institutionalized
ignorance, ran wild during the years of the blow-off, looking to the
Fed for signs that it might be clamping down on the money supply,
not realizing that Wall Street was outgunning the Fed as the source

of new money. Meanwhile, Volcker was winning kudos for his daunt-
less assault on inflation while the Fed was inviting the financial
markets to inflate, in defiance of the deflation in the economy whose
working the markets were supposed to mirror. The oldest swindles
leave their victims feeling most comfortable. Both Barnum and Ponzi
could have made conventional careers on the bureaucratic escalator
and ridden it up to top status in executive suites during the euphoric
Reagan era.

The statistical bureaucracy operates a full arsenal of weapons to train
on the complacent media and the susceptible markets. A familiar
blunderbuss on the analytical rifle range is the series on employment
and unemployment. The labor statistics and the banking statistics
run neck-and-neck as market influences. The markets set their sights
monthly by the creative countdowns from the Bureau of Labor
Statistics (BLS). In January 1986, the stock market and the American
economy were behaving like partners in an open marriage, each
going a separate way—the stock market heading over the moon and
the economy slipping under water. In the first week of February, the
BLS released its version of the employment and unemployment figures for January. The headlines proclaimed employment up from
December, as likely a phenomenon as waterskiing up Niagara Falls.
Just as December is the strongest employment month of the year, so
January is the weakest: the postholiday drop-off usually approxi-
mates 40 percent.

At the outset of 1986, the market-making activists in the futures
pits had little patience for such functional facts. They did not question the gem dropped into the news stream by Commissioner of
Labor Statistics Janet Norwood, a valued Carter administration
holdover. She attributed 30 percent of the employment gain claimed
for January to 123,000 new hirings by retailers. This flagrant disregard of the basic pattern of seasonal performance was refuted the
very next day, when the January retail sales results showed retailers
suffering a more serious drop from December than in a good year.
Early in 1988, the BLS reasserted the same claim. Against a background of stagnant-to-slumping retail sales, it found 161,000 jobs
added to the retail work force in January and, in February, another
118,000 retail jobs, on top of an increase of 25,000 in wholesaling. In
the arcane language of statistical adjustment, the tabulation of fewer

jobs lost than expected, from month to month, is translated to mean actual job gains.

By the spring of 1988, however, retail employment collapsed, with a powerful assist from Robert Campeau, the flashy Canadian empire builder. Campeau borrowed himself into a trap in the course of winning a takeover fight for the Federated Department Stores chain, a faded blue chip. The stock market had misread these high jinks as good news for the retailing sector of the economy. But as the old newspaper business saying goes, the "merger submerged" large segments of the retailing work force. The layoffs exposed the flimsiness of the BLS's reporting technique.

Monthly seasonal adjustments offer lures, then snap like traps. January is a big "down" month; February is a moderate "up" month. Therefore, euphoric seasonal adjustments for January, even when the statistical managers do not go so far as to claim improvements over December, invite downward adjustments in months earmarked for great expectations. Predictably, the hype whipped up over the January 1986 results took just one month to dissipate.

Nevertheless, the standard statistical device of seasonal adjustment gulled the market makers into accepting the optical illusion in the labor market headlines. The stock market took its customary quick look at the vigorous statistical results and imagined that it saw the real world of the transactions economy taking off too. The sheltered world of Wall Street responded by speculating before it searched for any confirmations.

The BLS extended its orbit of creative counting to its reporting of growth in the work force. No single statistic excited more enthusiasm on the political stump or on the market screen, encapsulating as it did the role of Reaganomics as the pep pill getting the American economy moving again. The BLS adopted four statistical conjuring devices to transform the shrinkage in the economy to expansion. The first was the inclusion of the armed forces in the work force, with no sensitivity to the tendency of recruits to enlist and remain in uniform when opportunities for skilled private employment turn shaky. The second was the elimination of those unemployed for more than six months, on the blithe assumption that they were no longer looking for work in an expanding economy assumed to be looking for them. In 1986, Albert Sindlinger, whose demographic samplings had proved reliable, counted at least 8 million such nonpersons subsisting in the shadows of the statistical economy. The third was the inclusion of part-time, low-paid "temps" as fully employed, replacing laid-off members of the permanent work force whose benefits stopped along with their paychecks. The fourth was the fiction of still counting people as fully employed after they were cut back to part-time schedules or fired altogether but given severance pay.

The Keynesian approach had encouraged the expectation that short and manageable injections of debt into a depressed economy would reflate it; Friedman's fears had reinforced this hope with warnings that the Keynesian remedy would result in overkill, igniting reflation into inflation. Now, however, continual and unmanageable injections of debt into an economy presumed to be expanding on its own were having a drag effect. But statistical deception perpetuated complacency about the economy and intensified speculation in the stock market, diverting attention from a troublesome new phenomenon: the coexistence of inflation in the federal budget and stagnation in the economy. Nevertheless, by 1985, the markets jumped to the conclusion that Reagan had forced a profligate Congress "to pass Gramm-Rudman to bring spending under control." But the political perception of Gramm-Rudman was the exact opposite of the political reality behind it. Congress palmed it off on Reagan to clear its reputation for fiscal responsibility. Reagan welcomed the ceiling Gramm-Rudman put on social spending, but had no intention of respecting the ceiling it put on military spending.

By March 1986, the economy was rife with layoffs. Admitted unemployment was back above 7 percent. Undaunted, the BLS pointed to 5.9 million people cut back to part-time jobs, all paid less than subsistence incomes after withholding, as evidence of a nominal rise in seasonally adjusted employment. By April 1988, this employed corps on subsistence had grown to 15 million people, no less than 12.5 percent of the work force; Norwood was explaining that these workers "do not want full-time jobs." She made no mention of the pressures on these people to limit their work time in order to care for their young, old, or infirm, lacking community support.

Instead, the BLS smuggled this corps of workers on subsistence into the labor force, and cited their "jobs" as bona fide evidence of the dynamism claimed for the economy. Adding the approximately 5.5 percent admittedly unemployed and recognizing the economic reality that part-timers are effectively unemployed by any realistic standard of subsistence, Norwood's own statement conceded an unemployment rate of 18 percent! All the way to full-fledged European stagnation, and far from the land of purported plenty. This before the "nonpersons" were eliminated from the work force because disemployed for more than six months.

Three telltale divergences, each not only confirming but aggravating the others, signaled deep-rooted distress in the American economy between 1986 and 1988. The simplistic euphoria trumpeted in the statistics distracted the markets from all three, despite the normal sensitivity of the markets to divergences. The first unfolded in the glaring and stubborn contrast, counted by the BLS, between dynamism claimed in the employment figures and sluggishness undeniable in the retailing figures (the latter punched in receipts and, therefore, immeasurably more reliable than the former). The second invited detection in the eerie parallel between the dynamism bubbling over on the supply side of America's industrial and commercial markets, buoyed as they were by the simultaneous upsurge in domestic production and import value, and the suspicious lags in actual absorption of manufactured end products: evidence abounded of consumer preference for services over goods. The third advertised this same parallel in the dizzying spiral of producer prices—nickel quadrupling, copper doubling, for examples—but retail prices (especially of imports) slashed to the bone and home as well as auto prices—the big-ticket items—plunging.

The resulting miscalculation revealed not only the naïveté of the supply-side credo; it came as a rude reminder that clichés about supply and demand ignore the third side of the eternal triangle that haunts supply and demand: inventories, alternately strengthening and overburdening the supply side, that bewitches markets. Not only romances are bemused and bewildered by eternal triangles. A more subtle and sophisticated reminder followed of Veblen's forgotten warning that the key to statistical realism lies in realism about inventories. In the autumn of 1988, revelations of unprecedented inventory distress, accentuated by interest rate increases in the face of earned income decreases, updated the practicality of his prophecy.

Shortly after the Labor Department admitted disappointment over the February 1986 unemployment figures, *The New York Times* published a telling article by a dean of economic academia, Geoffrey

Moore, director of the Center for International Business Cycle Research at Columbia and, therefore, the titular successor of Wesley Mitchell and Arthur Burns. Moore preceded Norwood as commissioner of labor statistics; she had worked for him at the BLS. He offered a number of overdue practical suggestions for improving the leading indicators, especially in the computation of the critical ratio of inventories to sales. But he also cautioned against taking the February drop in employment too seriously and assured the country that it could "look forward to stronger growth in 1986 than in 1985."

Moore's solemn assurance was overruled the very next day when a front page headline in the *Times* proclaimed: GLOOMY DATA MAKING ECONOMISTS UNCERTAIN ON OUTLOOK FOR GROWTH. Before the month was over, the barometric automobile business registered a strong dissent from expressions of complacence that were reminiscent of 1929, and the stock market took a preliminary tumble, discounted at the time as a false alarm. The stock market trusted the statistics to protect it against bouts of vertigo on its fatal run to the top.

Presumably, Moore meant growth in the statistical economy and, presumably, he rated the growth recorded in 1985 as impressive. True, his shopping list of overdue improvements in the leading indicators did put realistic emphasis on modernizing the foreign trade statistics to account for the powerful negative leverage exerted by import deflation. Inconsistently, however, his assessment ignored the shrinkage in the transactions economy, directly attributable to the harsh impact of this same import deflation. Moore's call for more realistic handling of inventory and foreign trade credit factors conceded the shortcomings in the February leading indicators that he claimed to find so encouraging.

Moore cited the rise in nonfarm employment as a plus. But the raw numbers, unmentioned, recorded another minus, confirmed by jumps in unemployment shared by states as diverse as California, Illinois, Texas, and New York. Moreover, the rise in nonfarm employment ignored the disturbing cumulative evidence of lower weekly earnings, fewer hours worked, and fewer help wanted ads— all reliable workhorse leading indicators. Nevertheless, by the time the headlines and the claims publicized by the statistical managers and their academic allies could be scrutinized by the naked eye, the markets had long since taken the bait of the statistical propaganda and charged ahead toward their October 1987 rendezvous with reality. Moore later stepped up his criticism of the leading indicators. In May 1987, *The New York Times* reported him at swordpoint on a variety of technical counts with the bureaucratic keepers of the leading indicators.

The reforms Moore recommended showed that he knew better than to echo the alibi offered by Norwood. Yet her official disclaimer dismissed half the February disappointment as due to farm layoffs forced by bad weather in California, and so did Moore's academic disclaimer—proof, if any were needed, of the united front cementing the bureaucratic-academic complex. Like their counterparts in the military-industrial complex, they take turns passing through a revolving door. In the discussion Sindlinger and I had with Norwood and then Secretary of Labor Bill Brock, Norwood's explicit reason for vetoing our recommendation that the seasonal adjustment be dropped was that "the academics like it." The weather alibi, glaring in its unprofessionalism, sneaked by unnoticed. A basic purpose of the seasonal adjustment is to exclude changes in the weather as causes of change in business activity.

Leads and lags prevail in politics as well as in markets. Thanks to the statistical illusion that debt growth was fueling inflationary excesses, opinion of all shades was beguiled into the belief that cutting the budget deficit would cure all inflationary worries. In fact, the trouble went deeper. The official budget deficit admitted to the public—and to Congress—was considerably smaller than the total cash deficit the Treasury was scrambling to finance. Opinion tracks the official bookkeeping deficit but is blind to the large cash deficit. Consequently, any debt ceiling imposed on the admitted deficit was foredoomed to be a sham, because it could never reach the supplemental deficit hidden by numbers games played outside the budget. Here are samples of tributary deficit flows from the Treasury into the economy, lost in the statistical shuffle.

Budgeted borrowings.   The shadow deficit reduced budgeting to an anachronism. Reflecting this reality, the U.S. government began to base its permanent operations on a continuing resolution that authorized temporary increases in the debt limit. Congress refused to trust any administration—least of all one headed by a popular president—with extensions as long as even a year. Its members responded to each successive showing of presidential need with a show of reluctance. Congress greeted each deadline as a usurer does: by exacting tribute. The White House paid for the right to borrow more by obligating itself to spend more on pet projects pushed by congressional members with the votes to authorize or block a higher debt ceiling. The White House exhorted the country to support Reagan's crusade to press Congress to impose fiscal discipline on itself. But Congress sold the White House the right to practice fiscal indiscipline and collected ransom from the Treasury for doing so.

Just the same, budgeting under the continuing resolution remained as dependent on the statistical economy as bona fide budgeting had. Guesstimates on the size, frequency, and maturities of Treasury borrowings proved no more reliable than the wishful forecasts for the statistical economy that stimulated the exaggerated responses of the markets. Each of these guessing games represented raffle tickets on the eventual performance of the transactions economy.

Unbudgeted refunds. A bias toward expansion was built into the statistical economy as well as into the budgeting process. Manipulations in both blinded the Treasury's policymakers to the surprise of reversals, as well as the discovery that, in 1986, expansion was easier to talk about than to bank on. In addition, the policymakers ignored the automatic drain of refunds on the Treasury, which accelerate in the wake of setbacks. Slumps penalize the Treasury for overselling optimistic forecasts and overcharging taxpayers who buy them. Refunds, thanks to the Keynesian pay-as-you-go system incorporated into the tax return forms, are the quickest-turning dollars in the economy: the fiscal equivalent of litmus paper in testing the trend of the economy. No budgetary discipline can prevent a refund hemorrhage in a disappointing year. Any rise in refunds refutes claims of expansion. Refunds serve as the pivotal indicator—though not recognized among the official leading indicators—measuring the continuing interaction of government operations and business decisions.

As Minsky has demonstrated, high-velocity instability, precipitating one financial crisis after another, has superseded the traditional corrective rhythm of the business cycle. The decision to commodify the dollar, turning its exchange rate over to the tender mercies of the futures pits, inescapably commodified all other traditionally stable values. Volatility guaranteed that losses would rise as the velocity of trading turnover accelerated; losses put players on the speculative equivalent of welfare, feeding them with entitlements to refunds. The banking structure has suffered, and so has confidence in Fed monetary management (already called into question by Wall Street's rise as a rival pumping center for excess liquidity into the financial economy, cloning ready money for takeovers at any price).

Refunds complicate fiscal management, which is scarcely simple to begin with, as well as analytical observation of the Washington money managers. Refunds also invite cross-ups between the Treasury and the Federal Reserve Board during dips. By the time political pressures persuade the Fed to ease clamps on the economy, the automatic refund valves at the Treasury have already anticipated the need for first aid. Keynes meant refunds to serve as stabilizers. But instead, refunds tend to intensify fiscal instability, especially when the trend happens to be turning.

Gramm-Rudman could not legislate a limit on spending in the name of conservatism so long as refunds were forcing Treasury disbursements mandated whenever income returns fall short of the expectations projected in tax returns. Nevertheless, in 1986, refunds rose under cover of market euphoria; traditional barometric stock market performance would have called for refunds to slacken in a strong market year. During just the January–February period hailed by the academic-bureaucratic complex as the takeoff into an auspicious year, refunds rose 60 percent, from $3 billion in 1985 to $5 billion. During fiscal 1986, therefore, refunds worked as an automatic fiscal offset to the budget austerity mandated by Gramm-Rudman. But despite the first forced cuts in federal spending, scheduled to begin on Tax Day 1986, this telltale jump in refunds went unnoticed, although they exerted powerful leverage on the bottom line of the budget.

Cumulative refundings.   To listen to the lamentations about the budget, the annual deficit is the problem. The deficit is by no means the most troublesome part of the Treasury's borrowing problem; refundings are. As they matured, refundings of old borrowings (not to be confused with refunds) built up to troublesome proportions by 1986. The Reagan administration spent its first three years promising that it would bring down the need for new money in the next five, when its original borrowings fell due. Accordingly, between 1981 and 1986, the bulk of the budgeted debt load surge was concentrated in three- to five-year maturities.

Consequently, beginning in 1985, annual refundings have been running on an ascending scale at a trillion-dollar clip. Alongside them, the $100–$200 billion deficit rate looks like petty cash. The torrent of refundings has dwarfed the annual totals in the budget. The markets have absorbed them, and the media have reported them. Yet their magnitude and their frequency have combined to dramatize the dilemma confronting the Treasury: whether to protect the market from peppering by refundings, or, whether, in the hope of keeping interest rates down, to risk rushes of refundings. Only the market professionals, congenitally illiterate, have taken alarm. The Treasury has congratulated itself for staying ahead of the game.

The bond market does not know or care about the difference between new borrowings and refundings of old debt. It travels in search of buyers, with its eyes on the total debt load up for bid at any yield at any time: not only refundings, but investor holdings offered for resale. Yet when rates are rising, the markets come under strain and worry more about old debt in need of bids for refunding (as well as for repurchase) than about new debt. In 1985–86, the bond market fed on the belief that its appetite to invest for income was endless, despite the flood of new Treasury offerings. The wonders the sham worked, while it did, to bring interest rates down and to put bond prices up made the exercise profitable—not, however, for bond buyers investing in income, but for price players gambling on quick turnovers, and of course the Treasury, which merely paid more interest.

The 1985–86 market euphoria responsible for lowering interest rates on lowered economic expectations fueled hopes for a meaningful cut in new borrowings. Not only did this assumption ignore the cumulative inflation of the 1981–86 refunding load; it shrugged off the step-up in liquidations of bond holdings by Americans needing money or chasing stocks, as well as by foreigners running out of the dollar or into spendable cash. Sheikhs filing for welfare as the result of the oil collapse turned sellers of Treasuries from 1983 on. Altogether, the jump in new annual federal borrowings accounted for only a fraction of the overload on the U.S. debt markets. The statisti-

242 AMERICA'S YARDSTICKS

cal fog obscured the magnitude of Treasury demand for money to manage refundings and redemptions.

By the end of the fiscal year following the September 1985 dollar devaluation, the Federal Reserve system had eaten no less than $221 billion in unredeemed securities bought back from the market. A year later, however, despite the most spectacular bond market boom in history, it had swallowed still another $252 billion. Half a trillion dollars of government securities bought back from a market supposedly dedicated to buying government securities ain't hay, especially when the sole designated buyer happens to be the Federal Reserve authorities, supposedly dedicated to independence of the government doing the selling and, more than incidentally, to respectful following of the workings of the market. This gargantuan official exercise in market support belied the belief, shared by foreign central banks and their various domestic satellites, that the market had an unlimited appetite for government securities.

Contrary to misgivings in the media, fears that the Treasury might see its new offerings crowded out of the bond market proved false alarms, like speculation over a chain reaction of default on Third World debt. Government buybacks continuously reliquified the markets. The Treasury could always count on the Fed to keep credit conditions easy enough to accommodate its borrowing needs. But fear of drops in market prices precipitated by unexpected rate increases haunted the Treasury as a continuing danger. The more the government bought back from the market, the more its total borrowing needs rose. Whether the Treasury or the Fed put up the money, the result was the same. Unbudgeted government outlays were confusing concealed government spending with authentic third-party purchases of new government paper to be held for investment.

Till tapping. Till tapping by the Treasury was one of the official malpractices responsible for the compounding divergences between rosy statistical appearances and grim fiscal realities that proved insupportable by the spring of 1986. After years of dipping into the Social Security and Unemployment Benefit Trust funds, the Treasury, in a stormy confrontation with Congress over the debt limit in December 1985, was forced to disclose its habitual use of them as its private petty cash box. Reagan's remarkable popularity provided cover for the details. During the histrionics of that fixed fight, the

Treasury admitted an emergency dip of no less than $13.5 billion from the trust fund float.

But the market enthusiasm generated by the crusade to balance the budget blinded market opinion to the significance of revelations such as this. The amnesia endemic to an increasingly illiterate, media-conditioned society immunized Secretary of the Treasury Baker from scrutiny when, during the preliminaries to the presidential election of 1988, he looked out from the TV screen and assured the country with solemnity that diversion of Social Security trust funds from investment in government securities is forbidden "by law."

Cash emergencies. The Treasury's budget emergencies are an old story; so are the contrived freezes on its cash disbursements that go with life under a temporary debt ceiling. But the cash emergency the Treasury engineered when its devaluation boomeranged started a new story. The Treasury had originated the big, unbudgeted, unanticipated multibillion-dollar cash hemorrhages until the Pentagon joined it in 1987. The Treasury engineered its own cash distress. Between the Treasury's reluctant discovery in November 1986 that the dollar, instead of floating in devalued stability, was collapsing, and July 1988, amidst admitted chaos in the currency markets, it dissipated $13.7 billion of unbudgeted cash on frantic, futile stabs at buying restabilization for the dollar by playing the currency futures markets. Market turbulence in 1988 forced the Treasury to continue throwing new cash by the billionful into emergency efforts to restabilize the dollar. Emergency borrowings by the federal bank deposit insuring agencies overnight raised over $25 billion; and, on the side, without so much as a wink of approval from Congress, handed out kings' ransoms in the form of tax benefits to well-heeled benefactors of busted banks.

Defense spending. No one—not even the secretary of defense—can ever know how much the Pentagon will actually spend in any budgeted period, or how fast any spending speedup may force still more to be spent. The procurement process calls for bids on a scale larger than can be spent in any single year—up to $50 billion or more at a clip. Defense contracts are always extendable, and, in quixotic pursuit of perfection, subject to change. With appropriations, authorizations, disbursements (or recurrent holdbacks), and "supplementals" for "improvements," the Pentagon is always on a quadruple schedule too intricate for the markets, or even for most members of Congress, to track.

In fairness to the Pentagon, the White House ordered the unexpected cash drain on the armed services sent to the Persian Gulf in 1987. To Congress, the administration admitted running up a cash cost for the armada of only $16–$20 million a month. But this preliminary audit excluded arrangements with Kuwait for furnishing fuel, as well as the burden on defense overhead, the cost of forces diversion, and accruals for casualties suffered plus repairs needed. It also ignored the intangible costs of insuring an expeditionary force wide open to attack because navigating in waters controlled by an identified enemy, without the rudimentary protection of air cover.

As if defense budgeting weren't complicated enough, military pensions and health benefits are continually compounding—again, no one knows how fast or how much. By 1984, the Treasury and the Pentagon between them had tolerated a $546 billion bloat of unfunded liabilities to military personnel and their families, untouchable by any debt limit or constitutional amendment. At that point, however, contrary to the myth exploited during Reagan's television opportunities, Congress imposed a salutory regime of fiscal responsibility. To fund this open-ended accrual, it obligated the Treasury to dedicate a set-aside, not included in the defense budget, of $10 billion a year for sixty years. Since then, this item has been on a pay-as-you-go basis: a welcome innovation.

The U.S. Treasury can never run out of money, though Congress is adept at exercising its constitutional power to suspend the Treasury's ability to create new money as needed. But the Pentagon has long since gone insolvent, and it has improvised a variety of methods for operating without plans to recoup its solvency. In fact, the operating hole the Pentagon has dug itself into has deepened, even though the Treasury has managed the government's overall insolvency, as it always can. But no matter how much more the Treasury may borrow, it has not increased borrowings enough to enable the Pentagon to catch up with its monumental accruals of unfunded obligations.

In 1981, the Joint Chiefs of Staff furnished the White House with an unpublicized finding of the then-unfunded total of the Pentagon's accrued obligations: $750 billion (in the face of the $925 billion legal debt limit at the outset of that fiscal year). The size of the amount explained the secrecy surrounding it. To ensure continued secrecy, the Joint Chiefs stopped counting, and the White House restrained its zeal for balancing the budget by refraining from requests to resume the tally. Reagan spent eight years popularizing his insistence that defense spending was not spending, but rather the measure of military strength. Meanwhile, back in the real world, his administration's default on the management of the statistical flow of defense commitments, relative to Pentagon use and acquisition of defense assets, ended by paralyzing the military.

Well before Reagan rushed into the trap he had set for himself in the Persian Gulf, the Pentagon had lost its capability for starting new programs along with its capacity for funding programs under way. All it could do was juggle the immense cash flows given it by the year, to pay Paul what it borrowed from Peter. It had no money left to invest in the new weaponry urgently needed to replace its vast accumulation of obsolescent weaponry. Like the private sector sinking in debt around it, the Pentagon was reduced to leasing what it could no longer spend cash to buy.

Four programmatic sources of claims swamped the Pentagon: ongoing programs with escalating bills, new programs to overcome the aging of those unfinished programs, current needs for supplies and repairs, and contingency funds for emergencies. The day after the first obsolete American helicopter crashed in the Persian Gulf in July 1987, Senator Nunn told me that the Senate Armed Services Committee "had been after the Pentagon to build minesweepers for seven years"; modern minesweepers would have avoided dependence on the ancient helicopters that America had flown into her rendezvous with disaster in Vietnam. Insolvencies and managerial controls don't go together, least of all when accelerating changes in military technology accentuate the insolvency. Inside each service and among the services, all semblance of calculation between the sinews of war and the weapons of combat—or prevention of combat—was lost.

One statistic reduced the gargantuan Pentagon operation to a systematic, unaccountable, mindless drive for "planned inferiority" against not only Russia, but Iran and even Grenada, let alone Nicaragua. No Democrat pinned this label on Reagan's impersonations of John Wayne. William L. Dickinson, the ranking Republican member of the House Armed Services Committee, did (and his state, Alabama, had the biggest single stake in helicopter operations!). By the mid-1980s, the development period for any new weapon, not merely a helicopter obsolete on activation, had stretched out to fifteen years; death throes for a dud had stretched to five years. The resultant debacle recalled a more than normally cynical response of Forrestal's to the question "How many people work at the Pentagon?" "About half," he snapped.

The statistical snarl that paralyzed the Pentagon left Forrestal looking like a star-gazing mystic. My own sardonic approach called for selling the air rights over the Pentagon to a gang of Iranian arms dealers and lend-leasing out to them the option to double-deck the Pentagon to provide office space for its successive new crops of under- and assistant secretaries and military brass acting as clerks. Of all the government's extrabudgetary malpractices, the mind-boggling games played with statistics at the Pentagon were the most self-defeating. They have remained the main obstacle to getting the budgeting process off to a new start.

Managing crises begins with assessing them. Omission runs outright deception a close second in the statistical manipulation sweepstakes, and the government managers practiced both forms. As the markets looked to the government for reliable information on which to base their behavior, the government intensified market vulnerability to statistical manipulations and omissions. The government's default on its managerial responsibility began with its mismanagement of statistical responsibility. This fatal flaw infected all follow-on governmental defaults—right down to the loss of control over the sinister procedures involved in arms sales to Iran.

America's Distinctive Liabilities

12

REAGAN'S ROOSEVELTIANISM
America's Illusions About Economics

America's many difficulties fall into two categories: first, mere problems overdue for managerial action, no end of them outrageous, all of them disturbing, but none of them beyond the reach of managerial remedy; and second, liabilities. Design and procurement of military aircraft typify her urgent problems, overdue for handling; America can save billions by harnessing technological advances to upgrade performance, without risk of losing military muscle. But if action is needed to solve America's problems, inertia explains the besetting liability responsible for her default on the challenge confronting her sense of urgency and stamina, her patience, and her ingenuity. Inertia accounts for the comfort she takes from the clichés that justify the bargaining advantages her foreign adversaries squeeze from her.

Action—even helter-skelter—can create the appearance of coping with problems, but only a strategy can direct the actions aimed at shrinking accumulated liabilities. Problems arise from habitual activity or neglect and are brought to a head by events; liabilities measure states of mind stimulating self-defeating moves and inhibiting advantageous ones. People can activate technology to solve problems, but liability control calls for them to rethink their assumptions. While accrued aircraft obsolescence is merely a problem, however costly and dangerous, the economic preconceptions that hold America's politicians and her public hostage are a liability limiting managerial ability to identify problems, let alone to mobilize technology and human potential to cope with them.

The new economists of industrialism preached the secularization of faith in miracles with remarkable success; Marx was their most avid pupil. Just when Marx thundered in 1848 that "religion is the opium of the people," the emerging electorate was broadening the teachings of the Enlightenment. History has enjoyed many laughs at Marx's expense since, but he played a great joke of his own on history. In his capacity as the Attila of capitalism, he politicized economics—more precisely, he repoliticized economics, giving due credit to Petty for having done so before modern industrial capitalism was born. Economics has since returned the compliment by theologizing politics. In fact, economics has established itself as the opium of the politicians. In America, the hold over politicians exerted by the preconceptions of economics has emerged as a major liability interfering with their capacity to identify and cope with America's complex variety of operating problems.

But this pervasive liability is relatively new. America's assets easily outweigh it. They are tangible and ample enough to supply and support programs of action. By contrast, the liability with which economists saddle politicians by propagating institutionalized dogma is intangible. The mind-set structured by the premises and the methods of establishment economic thinking suppresses the pragmatic functioning of politics; and pragmatism, after all, is the acid test for politicians.

Time is the critical consideration when encrusted intellectual liabilities obstruct the activation of viable assets. Drives to avoid and/or reverse depressions lose effectiveness with the loss of time. America's assets are so rich and so diversified that they free her from any risk of exhaustion in short sprints: she is equipped to renew her resources in an endurance contest, as she demonstrated in the 1939–45 war, which she finished with more energy on tap than when she was drawn into it. But preconceptions drilled by prestigious professionals, which economists are, into the reflexes of pushy performers, which politicians are, jell quickly and dissolve slowly. Erasing the policy implants economists have placed in the minds of politicians would not be a worry if time were not.

America's markets and managements will rush to buy any gimmick economists sell to politicians. The distinctive American asset of big financial markets has spawned a distinctly troublesome contingent liability, which hinges on a naïve American susceptibility to medicine men sporting economists' lingo. America is the only country blessed with public markets broad enough for professional operators to affect, and also too broad for the government to calm down once Washington provokes or permits hysteria to stampede Wall Street into panic, whether to buy or to sell. As its October 1987 meltdown demonstrated, no foreign market is insulated from heat radiated from New York.

Today's economists are armed with more data—some incomparably better, some unbelievably worse—than their predecessors ever dreamed could be quantified. But their elaborate equations have not ensured them against qualitative miscalculations built into their methods. The analytical difficulties of American economists stem from the theoretical frames, or lack of them, around their data. The numbers punched out on their computers to quantify the behavior of the economy do not fit the catchphrases of their premises. Nevertheless, American politicians defer moves to satisfy constituent pressure until they can win a consensus of economic approval. Politicians regularly come under fire heated up by clichés charging the subordination of readily identifiable moral principle to Election Day expedience. In reality, they have subjected themselves to the tyranny of arbitrary principles asserted by economists. Suicidal political advocacy holds a fatal attraction for economists.

To take just one example, Professor Ravi Batra, of Southern Methodist University, who established himself with *The Great Depression of 1990*— a runaway best-seller topping sex, scandal, and spy-catching—as the economist of the moment with the most popular appeal, revealed his political innocence in November 1987 by declaring his preference for Mondale as Reagan's successor. The public flocked to buy Batra's book with no concern for his glaring political disorientation. To take another example, at the climax of the 1987 deficit crisis, Reagan and his Democratic congressional opposition waited for a bipartisan consensus demanding budget reduction, then agreed to disagree on how to provide it. Reagan insisted on token spending cuts, and the Democrats on token tax increases, but the consensus across political lines settled on the budget deficit reduction solution beforehand. Under their very noses, meanwhile, import deficits and economic drag were combining to inflate the budget deficit more than any political negotiation could deflate it. Bipartisan agreement among economists failed to detect the budgetary deterioration from either source, let alone the political backlash from both.

Three hundred years after Petty extended the clinical approach of medicine to society, economics remains hidebound in voodoo land while medicine has evolved into a pacemaker of scientific advance. George Bush's celebrated sneer at Reagan on the 1980 campaign trail for practicing "voodoo economics" hit its mark, but the Reagan brand was by no means unique. All schools of economic thought have persisted in preaching varieties of voodoo. The guessing game economists play is neither a disciplined science nor a clinical art. The confusion goes back to the education they receive. They are trained to think of themselves as scientists and to regard the history of ideas as excess baggage. Few, if any, have heard of Petty, though all have been conditioned to accept Smith's dicta as gospel.

But a century after Petty, as Chapter 9 showed, Smith led economics onto a protracted detour away from data gathering in the real world of conflict calling for pragmatic compromise and toward the dogma of lawgiving in a behavioral vacuum. Two centuries after Petty, Marx colored economics with the lure of revelationary history and revolutionary politics as a swap of substance for slogans. By the test of performance, Petty in his obscurity has worn better than either Smith or Marx in their prominence. The complexities of economic life in modern political society have forced practicing economists to mix measurement with moralizing and to substitute it for evaluation. The trick is to combine sophistication of analysis with practicality of judgment, and shrewdness of quantification with wisdom in assessment. Economics enjoyed an impressive paternity but matured under auspices of dubious integrity, with disastrous results for many of its clients, especially its political clients, who need realism from their economic advisers most and get it least.

Back to Petty for guidance on the rudiments of practicality in approach and method: Behind the razzle-dazzle of computerization, the practitioners of medicine and of economics still depend on the parallel data-gathering procedures he enumerated literally by hand. Yet any deserved comparison between the needs of the two professions accentuates the contrasts between their methods. Physicians have come to base their procedures on hypotheses adjusted to results and to check their diagnostic surmises and prescribed remedies systematically against clinical experience. With a few honorable exceptions, economists in America have not. Instead, they hedge their bets by echoing their opposite numbers across party lines and cover their combined confusion by bullying their respective political clients. If physicians, however experienced, treat patients solely by eye and ear and by aping each other, they risk malpractice suits. Clinical discipline calls for diagnosing symptoms by ordering objective tests trusted to produce quantified results reducing the margin for human error to a minimum. The consensus among economists compounds the propensity to human error.

Economists are notorious for failing to achieve reliable results, although they too have computerized their approach. Most national economies are structurally similar, as people are, but America is a special structural case: not just because she is bigger, and not even because her economy is subject to market gyrations involving masses of people, but essentially because she is dependent on doing well at home. When she does, she makes prosperity her major export. Nevertheless, her economic physicians persist in prescribing for her dislocations and disturbances as if she were Denmark or Mozambique or, increasingly, Japan, whose system cannot serve "special case" America as a reliable model.

The parallel between physicians and economists that history suggests now extends to decision making based on data gathering in other fields, beginning with politics. The new breed of politico draws on data banks that substitute backyard hearsay systematized to reflect yesterday's impressions for seasoned experience and judgment anticipating tomorrow's performances. Accordingly, polls dictate campaign decisions. Any American election campaign, including many foredoomed to defeat, starts out with the candidates demonstrating their ability to raise the money to pay for polls, which reciprocate by rating the money-raising potential of the candidates. Polltakers double in tea leaves as policy advisers. Old-time wheeler-dealers have been phased out and replaced by printout filers. The exorbitant cost of polls, and of advertising their results, inflames the incentive to buy them. It is axiomatic that any campaign that can be financed can be launched. Pollsters and economists collaborate to give safe-conduct passes to their political clients as worthy investments for likely contributors.

Like executives and politicians, present-day commanders also rely on economists. The military profession has been automated and subjected to interdisciplinary indoctrination. Commanders operating without paying attention to calculations from engineering and economics are throwbacks to spear-carriers and daredevils on horseback. Military confrontations have come to be managed with the same dependence on computerized data as surgical procedures, political polling, and market trading. Methods, weapons, and operations have been mechanized, their capabilities predetermined. War games simulate battles based on the quantified strengths and weaknesses of "the enemy." Military personnel fight according to the data they recite from the handbooks that accompany the equipment they tend. They are trained to trim their tactics to the weapons tailor-made to support their missions. The automation of war has transformed "bear arms" from a functional figure of speech to a phrase reducing heroics to mechanics. Arch-individualists can no longer play the hero. Warriors in modern dress have been cut down to size.

Politicians operate under scrutiny to put faith in their economic advisers as well as in their pollsters and military chiefs. The military hierarchy in turn consults its own economic advisers for directing its weapons designers. Time was when politicians and soldiers operated by intuition. No longer. Johnson was the last of the breed to follow his nose when he roamed the political jungle, Patton the last on the battlefield. Both came to untimely career ends.

Rulers with an instinct for survival long ago learned to rely on their commanders and assign the cost of covering their wars to their treasurers. (Elizabeth I stood out as the one exception. She was poor and predatory enough to convert her navy into a piratical profit center for her treasurers to tap.) In modern times, politicians have continued the tricky tradition of trusting treasurers. Functionally, their military advisers have stood in the line of command, and their economic advisers have been downgraded into gray eminences— originally, charged with responsibility for minting hard coin; subsequently, with finding takers for sovereign paper; finally, with learning how the system works in order to take in cash fast enough to find takers for paper even faster.

Gradually, all three interrelated professions—politicos, warriors, and economists—have found themselves functioning subject to engineers, who affect the balance but, like economists, have no direct power. Instead, they derive their influence through their responsibility to recommend designs, as economists do policies. The applied arts have played their own game of musical chairs over time: economics began as a branch of medicine, and medicine has evolved into a branch of engineering, which is also a composite of sciences and an on-site activist exercise. Complications in this four-way traffic between engineers and their three powerful fraternities of clients have been compounding ever since Petty supplemented his exercises as a medico and a politico with excursions into shipbuilding and pioneering economics.

Petty recognized these interconnections one hundred years before the steam engine triggered the industrial revolution, which, in turn, revolutionized war and subordinated the economy—at peace and, even more, in war—to technology. Politicians, by definition knowledgeable about people, necessarily play catch-up ball with technology. The presumption is that economists will factor the dynamics of technology into their calculations on market equilibrium. Instead, they have been diverted by the numbers game. Consequently, economists have remained as technologically ignorant—especially about defense—as the captains of industry Veblen castigated, reason enough for economic projections to go wrong. As Veblen observed, economists have clung to technological ignorance as a symbol of power exactly as the captains of industry before them did, and as the conglomerators after them have done.

The firepower deployed by the military in any decade embodies the blueprints drawn by engineers of previous decades and is funded by cost and capability guesstimates furnished by economists during the planning period. Keynes's observation that the ghosts of anonymous economists dictate the rantings of lunatics in power goes double for the faceless, nameless engineers collaborating with economists to dictate to the military brass that comes afterward. Wars that politicians blunder into are decided beforehand by the capabilities that engineers and economists combine to put into place and to get financed in time for use in combat; the French never knew what hit them in 1939. Usually, engineers and economists have been let out to pasture by the time their work is adopted (though the 1914–18 war saw the unveiling of the tank, the submarine, and poison gas along with air warfare, and the 1939–45 war sped research on the atom bomb in which the shooting culminated).

Political leaders whirl their economic counselors and military chiefs on a continual and uncertain waltz in uneven time, unaware of the ongoing impact of the engineers—including health engineers, that is, physicians—on the structure of finance, or on the economy in its alternating cyclical roles as the vehicle of peacetime prosperity and the mechanism of wartime destruction. Politicians, economists, and the military came to regard their engineering auxiliaries as necessary but manageable evils and to discount value judgments from them: witness the long resistance of all three fraternities to warnings from engineers of the need for an overhaul of nuclear arms programming. The nuclear engineers started the outcry in the 1940s and were all but outlawed. By the 1980s, politicians of all stripes began to echo their call. But economists shrank from their part in the calculation of costs compounding with the conspicuous overproduction of nuclear arms and the accrued underproduction of conventional weaponry.

People mistrust what they cannot understand, especially in a media society where the media seem baffled by political and military misreadings of prevailing drifts. But media opinion is conditioned at one and the same time to delight in revelations that economists have gone wrong again, and to exude confidence over their next consensus. Before America's financial crisis came to a head in the late 1980s, repeated basic failures to measure the impact of economic performance on political expectations and military calculations shrouded these ingredients of policy-making in mystery bound to be demoralizing—not merely for political and military leaders, but for confidence in the security entrusted to their joint leadership.

Voltaire's throwaway line—that if God didn't exist, we'd have had to invent Him—applies to the indispensable role economists play at the summit of power. Just as politicos, commanders, and economists rely on the designs of engineers, so do politicians, the military, and opinion makers rely on the latest projections of economists. Only economists provide a continual head count of youthful entries into the work force, as well as of aging exits; adjustments are needed to account for immigrants, moonlighters, older women, single mothers, retirees returning to the work force, and temps, as well as overall counts of the electorate. Economics begins with demographics. Moreover, America, as the only diversified superpower, is responsible for funding both her social and national security. Economic forecasts are essential for each. Unfortunately, economists have fallen into a statistical rut, continuing to count heads but neglecting to distinguish the quality of demographics behind productivity. Even more, the premises they publicized for discussion underwrote the dubious conclusions their political clients bought for action.

America had herself to thank for the troubles that overtook her in the 1970s. More precisely, she was misled by her own preconceived ideas about national policy-making institutionalized by the economic texts. She was still gullible to the hoax that she was short of resources, not just of oil in the 1970s but, even more, as the 1980s arrived, of productivity. Not until the slumps of the 1980s in the working economy started to pinch did America begin to suspect that clichés about the statistical economy that her leaders swallowed and spouted were the source of the policy indigestion she suffered. Ongoing interchanges between her politicians and her economists spread it. Each developed an insatiable appetite for the other's sales pitches. The more general the judgments were, the more cheerful they tended to be. Conversely, the more specific observations became about economic performance in particular regions and functions, the greater the concern observers were willing to admit. By 1987, the condition of the economy was lagging disturbingly behind the statistics trumpeting its progress.

American politicians seem to regard one another with reasonable humor and cynicism. All of them campaign under pressure to top each other's bids. They run scared of letting their opponents get either too close or too far away from their own positions. But politicians of the Right and Left alike are less prudent in their responses to the economic advisers of their rivals. Again and again, candidates and officeholders rush to buy policy insurance from economic stargazers on the other side in order to bolster recommendations from their own gurus.

Reagan provided the most effective example of an instinct for appropriating the objectives of his political opponents. In 1986, he pronounced himself a convert to the crusade for tax reform that a generation of Democratic economists had urged to ban "tax breaks" for special interests. When he did, he demonstrated his genius for stealing Democratic thunder. Under cover of accepting this chant from the Democratic economic brain trusters, he hornswoggled their congressional clients into going along with a cut of over 25 percent in the corporate tax rate.

He had already pulled off his first coup. As a crusader for fiscal responsibility, Reagan presided over deficits larger than any Democratic president would have dared apologize for, let alone fight to feed, and larger than Nixon had covered up when he declared "we are all Keynesians now." The shock effect of Reagan's deficits converted the most liberal Democratic spokesmen into crusaders for a balanced budget. Dukakis went even further in his presidential campaign, promising to pay down debt with projected surpluses in the Social Security trust funds (though—small detail!—belonging to benificiaries in the work force, not the government). Keynes himself had advocated a vigorous return to surpluses in response to any reduction of unemployment below levels of distress. He was explicit in limiting his recommendation to resort to deficits only in periods of unacceptably high unemployment. He would have recoiled in horror from acceptance of them amidst claims of prosperity.

Economists are likely to take one another more seriously than politicians take their coconspirators across the party divide. In America particularly, economists take comfort in each other's thinking across party lines; their agreements about principles override their differences of opinion. They view the workings of the economy from the same starting point, even though they end up putting emphasis on different remedies for its malfunctions. As a result, their unanimity on theoretical matters contributes to their bipartisan clout with politicians. Practitioners of the ancient art of placating crowds are conditioned nowadays never to utter, let alone move, without consulting their own economic medicine men and then taking added comfort when authoritative voices agree across the ideological fence.

The broader the analytical consensus among economists, the firmer the authority each faction exercises over the assumptions of its political clients in support of its remedial proposals. The bipartisan unanimity with which economists accepted the phony oil shortage of the 1970s provided a striking illustration of how their wrongheadedness grew with their prestige. Thurow furnished another with his "discovery," as a liberal Democrat, of the conservative belief in wage cuts to help cure trade deficits.

Gibbon blamed the fall of Rome on bishops within and barbarians without its gates. America can blame the jeopardy she has compounded on the homogenized thinking her economists wholesale to her politicians, who retail it to their customers. The weight of the consensus that economists command has aborted efforts to scrutinize the theoretical foundations of the thinking they share or the practicality of their bipartisan proposals.

Witness the stubborn grip two interrelated articles of economic faith hold over the economists, and hence, the politicians, of all persuasions and both parties. Free trade is the first; America's supposedly sound growth from a manufacturing to a service economy is the second. The latter, an illusion, conditioned economists and their public to welcome the disastrous consequences of the first, a shibboleth. Claims of overall statistical growth, overriding admitted industrial shrinkage, supported the impression that depression for producers could create prosperity for consumers. Common sense anticipated the spread of shrinkage from the country's manufacturing industries to its service industries that surfaced in 1985. But it came as a complete surprise to the economic establishment. Nevertheless, even this fiasco failed to discredit the fraternity with its political clients.

The practitioners of the "dismal science" do best when they live up to their reputations, viewing with alarm any possible changes. But a funny thing happened to them on their way to the crash of 1987. They changed their image from undertakers to cheerleaders. When they did, they showed no historical sensitivity to the derision their profession brought on itself in 1929. Fisher, then the chief mandarin of the academic establishment, was not alone in hailing the storm warnings of 1929 as the birth of a "New Era." In 1986, even Professor

John Kenneth Galbraith, of Harvard, the dean of iconoclasts, joined in the renewed chorus of complacence.

Galbraith broadcast his confidence in the safeguards left over from the New Deal—particularly, unemployment benefits and farm support payments—to hold any 1986 reversal within tolerable limits. But systematic statistical understatement of the need for supplemental unemployment benefits, though serious, distracted attention from strains more urgent. So did the inadequacy of farm supports to prevent a brush with an agricultural trade deficit, a setback repeating the long-drawn-out nightmare of the pre-Roosevelt decade that first bankrupted America's farm economy, then brought farm banks down wholesale on top of it, preparatory to wrecking the city banks as well.

Nevertheless, in the 1980s, economists fastened on blinders when they discounted farm troubles because farm income had shrunk to a nominal fraction of the GNP; they mistook the bustle in the hub of rural America for the minuscule weighting assigned to farm production, a gross statistical distortion. Back in the 1920s, when the farm factor approximated a full third of economic activity, the GNP yardstick was still undreamed of as the consensus substitute for observation of the component parts of the national economic mechanism at work. Economists, like metaphysicians before them, trusted the abstractions in which they believed to illuminate the world in which they lived. Because the farm crisis of the 1980s was not measurable in meaningful GNP terms, the economic metaphysicians dismissed it as a problem for the working economy, as if the working economy could work without the farm economy or without the rural society enveloping it. They ignored two fatal symptoms: the tendency of a banking crisis to spread from the farm economy, as it did in 1929; and the vulnerability of the world economy, in whose viability they believed, to a deep depression in commodity prices and, consequently, to a crisis of bank debt default. American farm prices are invariably the long-term bellwethers of both disturbances.

At least the members of the economic fraternity had taken cover behind a plausible alibi for their professional embarrassment in 1929. Fisher and his peers were traveling in the very best company when they missed the message in the stock market top: their political and business clients, who were responsible for running the world as it was then, did too. In those days, the stock market did not revel in a world of its own, as it did in the orgy building up to the 1987 collapse. It functioned as an integral part of the workaday world: so much so that Clarence Dow devised the Dow theory at the turn of the century as a handy, commonsensical method of forecasting the business trend outside Wall Street, not the speculative atmosphere inside.

Up to 1929, the stock market still played the authoritative business forecasting role; in 1929, the American economy was strong, stable, and, certainly, competitive everywhere, except on the farms. Accordingly, the economic opinion makers blamed the stock market for selling a vibrant economy short, instead of blaming themselves for failing to sell the market short. They agreed with the Hoover administration that the sustained performance of the economy through most of 1930, until the bank trouble started to spread, guaranteed a quick and powerful recovery for the market. Hoover believed it, and he was still admired. Mellon, recognized as the power behind the throne, insisted that the market collapse would work out as merely a speculative aberration (a phrase echoed in October 1987) punctuating continued prosperity, and he was trusted to know.

The October 1987 market crash popularized comparisons with 1929. But the contrast between the strength of the American economy before the 1929 crash and its weakness before the 1987 crash was eerie. Although the 1985–87 market bubble had given comfort to political gullibles, market opportunists, and managerial Babbitts as the measure of self-energizing expansion, the euphoria it inspired covered up a slump spreading from region to region and sector to sector. Nevertheless, the chorus of "blue chip" economic forecasters spent most of 1986 anticipating a 1987 looking better still, and a 1988 guaranteed by the magical powers attributed to Washington in a presidential year. They took comfort in 1986–87 from the simultaneous surge in the stock and bond markets, betting their reputations that Washington would add momentum to it and put a safety net under it as well. Economists hypnotized themselves with the same confusion that they spread among their clients for operational use, among politicians for policy use, and throughout the media for use on a public that, however, was growing increasingly suspicious of the discrepancy between professional pretensions and pocketbook performance.

The economists covered their confusion with three prevailing assumptions. The first accepted Reagan's aims as uncheckable by any threat, even when the clock began to run out on his term. The second accepted the stock market as unstoppable by any scare until after it crashed. The third accepted foreign products as unchallengeable within America's domestic markets by any reactivated American effort.

The third assumption commanded the most stubborn acceptance because it hinged not on what the president did nor on how the stock market performed but on what economists had led people—not just politicians and publicists—to believe; settled beliefs are resistant to contrary evidence. Economists are always content with leading indicators ripe for discarding, and never on the alert for replacements called for by changed conditions. In 1986, they missed the new leading indicator that would have quantified the boomerang effect of dollar devaluation—that is, if they had alerted the government to start tracking it. Yet Professor Moore himself had identified it, although he had isolated it in the foreign trade sector of the economy: the cash-to-credit import ratio. In his revealing *New York Times* article emphasizing the obsolescence of U.S. foreign trade statistics, Moore noted the persistent official failure to distinguish between imports paid for in the time-honored way, with cash, and cargoes brought in on the cuff (tariff charges included), carried either by the exporting producers or by banks, their own or American.

Politicians did not awaken to the difference between cash and paper in payment for imports until after the damage was done in 1986. Economists were blinded by the statistical lag, which counted all import transactions equal. In the interim between devaluation and admission of its failure, therefore, the economists persuaded the politicians that dealing with Japan called for price competition in supposedly free markets, not political negotiation over Japanese credit subsidies. Though economists grudgingly conceded the need for political negotiation over Japanese subsidies, they could not bring themselves to believe that the Japanese were resorting to any such tricks, further flawing competition that was already imperfect.

In September 1985 the unanimity of approval that greeted the devaluation of the dollar demonstrated the failure of the economics profession to put sophistication of analysis and practicality of judgment at the disposal of its political clients. The subsequent reaffirmation of bipartisan confidence in the devaluation expedient, despite its continuing failure to work, revealed the susceptibility of business opinion and market behavior to any backward-looking advice from economists that forward-looking politicians continued to accept.

In that fateful month, nevertheless, America was driven to negotiation as the last hope of retrieving marketplace losses: negotiation not across the spectrum of the marketplace for products but selectively over the market for her currency. She was anxious to avoid admitting the need to abandon the fetish of free trade, but she was desperate to hammer out a deal for the products money is supposed to enjoy the freedom to buy. The devaluation decided on by the five cooperative governments (the United States, England, West Germany, France, and Japan) was welcomed with cheers by the world economic establishment. There's nothing like temporary relief from impending disaster to whip up stock market enthusiasm. Dollar devaluation did it instantaneously in 1985.

Weaker countries had accepted devaluations forced on them. But the lament a popular torch singer once chanted—"I wasn't pushed, I fell"—applied to America in 1986, when her own government decided to devalue her dollar assets and, along with them, her distinctive geopolitical asset: active and continuous control over the bargaining power of the government of the United States. While standing tall and talking tough, the Reagan administration decided to weaken the dollar. Once it did, it put the government out of the business of negotiating on behalf of dollar power and subjected it to the vagaries of the market on the innocent assumption that its workings were free.

In 1982, when Congress saved Federal Reserve Board Chairman Volcker from the folly of his start-up experiment with the monetarist formula for tight money, the congressional watchdogs over the economy were content to get action without making any noise. Consequently, the markets were slow to catch fire. But, in 1985, Secretary Baker's advertised exercise in persuasion of the other big four financial powers rocketed the markets into instant orbit beyond the reach of any gravitational tug. The economists had drilled their devaluation dogma into all their clients—in politics, in business, and among market makers. None of them doubted that the move would work. Their confidence was bolstered by the spectacle of the four other governments cooperating in time to defuse the danger that Washington might start a currency war by starting down the devaluation road alone. The only question raised by consensus thinking was how long the benefits would take to materialize.

Neither politicians nor managements understood what went wrong with the assurances the economists were providing. Corporate managements and their lobbyists, however, were quick to grasp the specific opportunities offered to exploit the free-trade shibboleth. They scrapped domestic facilities and accepted subsidies to start up Third World ventures—often in partnership with their foreign industrial competitors—from which to ship identical products into the domestic market at little or no capital cost, less proceeds from their domestic scrapping. In addition, the U.S. Treasury offered a cash tax refund to owners of scrapped plants on the hunt for foreign tax havens, who could salvage cash as they booked losses. As icing on the cake, this thriving traffic in the export of plants judged obsolete inside America eluded the reach of tax reform.

The result of relocating the productivity of Toledo from Ohio back to Spain was to catapult manufactured imports from backward economies into big-business status, a further explanation of the failure of dollar devaluation to hold back imports and the failure of economists to notice the reversal. By 1977, Japan and Germany were no longer the main targets of import complaints; Korea, Taiwan, Singapore, and Brazil topped the list. The anomaly of tiny Taiwan rolling in $120 billion (and mushrooming) of free cash, which had been accumulated by breaking the bank in the United States, finally raised suspicions that free trade might be working as a cover for subsidized leeching. Later, when avant-gardists in the Reagan administration complained, Singapore retorted that over half of the offending dump originated with runaway subsidiaries of American corporations—this in the year of the bipartisan Tax Reform Act, which lowered the corporate tax rate from 46 to 30 percent!

The calculation that devaluation would work encouraged the Fortune 500 to anticipate a double advantage in its pricing from this new Third World strategy. Corporations relocating plants into the Third World would benefit from minimal capital and labor costs. Moreover, the dollar export standard prevails in the Third World countries of their choice. Accordingly, these corporations could count on enjoying the best of both worlds: cashing in on cost advantages outside America, while nevertheless quoting the products they

brought back into America in dollars rather than in priced-up foreign currencies. Their economic advisers assured them that the domestic price level would inflate enough to absorb their subsidized
products from plants in Third World economies with rich profit
margins, free from the domestic profit squeeze.

Simultaneously, industry projected steep price increases for
near-capacity production from remaining domestic plants. Altogether, industry bought the devaluation scenario with every anticipation that the transplant to Third World production plants would
yield a two-pronged (domestic plus import) boom from shrinkage,
and that the blend of domestic price increases and imported price
cuts would build a price structure reaching all the way to the promised land. During the golden Reagan years, freedom from fear of
antitrust scrutiny invited price fixing to blunt the import price competition that was promised in the traditional texts as an ongoing
domestic dividend from free trade.

Enter the stock market. Its time-honored role is to anticipate,
and so it did, or so it thought. In fact, the market makers spent three
full years celebrating the success of dollar devaluation before the
suspicion spread inside the financial markets that its failure might be
triggering a full-fledged debt deflation and market bust. By that time,
however, all the institutions were locked into positions too heavy to
be lightened without wrecking the market. Once the market makers
fell for the devaluation illusion popularized by the economists and
the establishment of institutional investors and corporate managers
tagged along, who could blame susceptible politicians for scrambling
onto the bandwagon? As long as the clichés of free trade and currency devaluation continued to dominate economic sermons, political hallelujahs followed.

In Wall Street, the market makers relied on their economists to
keep their customers, institutional and private alike, coming on the
buy side of the market so that the sell side could remain open for
insiders leaving. Long before they started paying for computers and
economists, the old-timers in the Street were given to bragging that
they knew "the price of every stock and the value of none." Their
suspicions about the fate awaiting the surging bull were aroused
when their yuppie protégés started to believe in the values behind the
merchandise being hustled, declaring American stocks underpriced

because cheaper than Japanese stocks. As the senior partners of the big firms listened to this drivel, they watched private investors who had been the market's backbone liquidate positions held for a generation, and they themselves joined the silent minority heading for the exit gate. The simple calculation that good stocks paying double-digit yields at cost were paying 2 percent or less at market was turning investors into sellers opting to collect 8 percent and more for taking no chances. The same senior partners tolerated economists, whom they were given to saying were "good for customers," and accumulated cash as they waited for the crash.

Along the country's Main Streets, moreover, customers who had been trusted to provide the institutions with a safety net for the distribution of blocs were airing suspicions, more angry than uneasy, that projections from economists were better ignored than accepted. Reagan's skilled pollsters picked up the static in their routine interviewing just when Wall Street ran wild between 1985 and 1987. They reported a strident outburst of sneers and snorts against miracles promised by politicians who fell for panaceas dreamed up by economists. But Reagan's pollsters were only mechanics paid to hunt for complaints, not to consider how Reagan and his economists might cope with the failure of the economic schemes he sponsored.

As the stock market exploded into its blow-off and slipped its moorings in the economy, Reagan's polls found solid families of believers in the conservative credo, whom Nixon had saluted for wearing "good Republican cloth coats," blaming Reagan's economists for their own frustration with his efforts, yet still crediting him for managing so well in spite of the advice he was getting. This distinction was gratifying to Reagan, but the message signified a deferral of danger, not an exemption, and a warning that time was running out on his charmed political life. He had only one image to lose.

This finding was not the proprietary Republican problem it may have seemed. As candidates eyed 1988, they were on notice to handle the pet theories of fashionable economists with care. Crank economic cure-alls launched under prestigious academic sponsorship had wrecked three Democratic presidential campaigns. McGovern and Mondale ran on proposals, hatched by academic economists, that pinched sensitive tax nerves in middle-income brackets normally inclined toward Democratic tickets. Carter was his own prophet of the drummed-up hysteria to cut oil and gas consumption, a nostrum universally favored by academic economists and resented by voters. By 1985, Reagan's polls, assessed against this background of boomerangs thrown by Democratic presidential candidates, called for cooling the indiscriminate evangelical fervor Reagan had displayed all along for his economists and, instead, experimenting with a disclaimer. Their sponsorship of devaluation won them a reprieve, thanks to the deep-seated belief in it as a cure-all and the market enthusiasm it whipped up.

By the time the voodoo cult of Reaganomics came a cropper and the dollar devaluation bubble burst, the awe in which presidents and their challengers held economists had accumulated a long tradition. Hoover started it half a century before Reagan put his charisma behind it. He used his years as secretary of commerce to publicize the double duty his engineering credentials did for him as an economist. As president, he exploited the expertise he had claimed as an economist to warn that the Depression would deepen if he did anything to check it. When nevertheless he did, and it did, Roosevelt's response was not to decry economists, but to break Hoover's franchise on the copyright. Accordingly, FDR affirmed his respect for professional economists as opposed to self-styled pragmatists from the front office of the Commerce Department, who were conspicuous by their absence from the regular interagency bull sessions of the economic advisers to his chief political lieutenants. He relied on the young academic disciples of Veblen whom he brought into the power structure to keep his political appointees on their toes and fighting among themselves about how to advise and protect him.

Roosevelt accumulated economists during his long stay in office as fast as he accumulated problems. He made sure that they were of every liberal persuasion, and he even kept a few housebroken conservatives for insurance—not that his glibness was up to reeling off the differences among them. But once he disposed of Hoover, FDR felt the need to keep the bankers—along with their economic credo and the sins it invited against political prudence—on display as the villains responsible for the idiocy he had driven out of the White House. This strategy led him to promote the new breed of economic and legal academics whom he presented as knowledgeable yet untainted by greed and, therefore, as he invited the customers to infer, qualified to repair the damage done by the Depression.

Reagan made the most of this tradition. He duplicated Roosevelt when he arrived at the White House attended by a well-publicized entourage of economists sporting all visible conservative colors; he just looked to the Right instead of the Left for his applause. Like Roosevelt, he sang the praises of them all and soft-pedaled their factional differences. What passed for Reaganomics was never a distinctive unified product, but always a hash held together and flavored by Reagan's own brand of intoxicating sauce and his persuasiveness in coloring the contents for media consumption. More important, the patronage he divided among them carried princely endowments, worth immeasurably more to each of them professionally after serving in Washington than while on tap there.

The appearance of academic economic authority was more important than any cohesive substance. Even conventional conservative advocates of tax increases in good times as antidotes for deficits were allowed a place, if not a hearing. Most of the noise was made by apostles of Reaganomics in its supposedly pure form of "supply-side economics," especially Professor Arthur Laffer, of the University of Southern California, who first formulated the credo, and Congressman Jack Kemp, of Buffalo, New York, his most ardent popularizer. Yet when Laffer put his patter to the test with Republican voters in the California senatorial primary in 1986, and Kemp did at the same time in presidential primaries, both were swamped. When the time came for Vice President George Bush to exercise his prerogative as Reagan's heir and pick a running mate, he passed over Kemp, the economic ideologue, and draped the mantle on a purely political designee trusted to recite on the public relations of standing tall for defense.

One signal from his pollsters was all Reagan needed to stop trying to sell his economists. Instead, he concentrated on selling his personality while he let his economists unsell themselves. In an unheralded reversal, counter to the trend of the times, he wound up taking the laissez-faire attitude toward his economic partisans that he promised but failed to take toward the economy. When he won his reelection mandate in 1984, he scandalized the hierarchs of managements at the policy level by his promise to ignore his economists and, instead, to trust the advice of his pollsters to stand pat on his astronomic deficits.

At the operating level, however, his followers were relieved by his rejection of Mondale's call, which was dictated by orthodox economics, for a curative tax increase. But when advice came from his pollsters as early as 1985, he induced withdrawal symptoms among the captains of industry and the moguls of finance by snorting that America was suffering from an overdose of economic counsel. The moment Reagan went public with this diagnosis, some private sector presidents promptly took the ax to their forecasting staffs. Corporate economists have been suffering unemployment ever since. The gag about them started to make the rounds: "About half of what they know is wrong, and the trouble is that you can't tell which half."

Reagan's stroke of genius inspired him to threaten to punish his own economists by blaming them for his failures. This trial balloon was as good a political idea as could be aired without being implemented: an eloquent bluff, full of sound and fury, revealing everything. Reagan was winning credit on every side for single-handedly converting hope into a measurable source of economic energy. He split the credit for performance beyond expectations among employers motivated to reach for more, working people conditioned to settle for less, consumers invited to bargain for more, and government committed to do less. Still, he needed a scapegoat for unmet expectations galore. Reagan himself intoned the death sentence against his own economic high priest swiftly and safely, as Henry II did not dare when he put out his famous contract on Thomas à Becket.

The target of Reagan's wrath was the incumbent chairman of his Council of Economic Advisers, Professor Martin Feldstein of Harvard, the personification of classical conservative intellectual propriety in an academic economist committed to the free-market principles laid down by St. Adam. When the White House cleared Feldstein for the appointment, it did not know the difference between a proper conservative, which Feldstein was in his abhorrence of deficits, and the wild radicals spouting Reaganomics who felt comfortable indulging deficits. Feldstein's use of his position to air alarm over Reagan's deficits provoked Reagan's decision to purge him. To show that he meant what he said, Reagan threatened to scuttle all the government's economists. He went so far as to say that his first second-term budget would call for the abolition of the President's Council. It was a case of the wish metamorphosing into the urge to kill, but the inhibition aborting the urge. The White House substituted a monetarist nonentity for Feldstein, then accepted the drone's resignation and finally persuaded him to stay by conferring cabinet status on him: the political equivalent of revoking his professional franchise.

But economists operate under a compulsion to mouth clichés even when elevated to the level on which their clients weigh their recommendations and arrive at decisions. Feldstein himself, duly chastened in exile at Harvard, put the seal of professional approval on Reagan's hovering presence over the economic scene. The occasion was the 1986 Janeway Lecture, dedicated to Schumpeter, that Feldstein gave at Princeton's Woodrow Wilson School; lecturers are invited to apply their experiences in the public sector to the economic theories with which they work. Feldstein was candid in recognizing Reagan's difficulty in grasping the intricacies of economic policy discussion but emphasized Reagan's flair, as a decision maker, "for getting it right." Becket's ghost gave no such posthumous pardon to the royal source of his downfall. Feldstein's loyalty oath came as both a reassurance and a warning to the political establishment: a reassurance that modern economists are easier to handle than medieval priests were, but a warning that servility is not very useful.

Reagan dropped Feldstein as the designated Republican custodian of economic principles, but that didn't stop Dukakis from picking up Professor Lawrence Summers of Harvard two years later to serve as the designated Democratic custodian of exactly the same economic principles. Soon after that, dissident Democratic political operatives and assorted ideologues were jeering at Summers as the "Democratic Feldstein." *The New York Times* scrutinized Summers and his Republican counterpart in Bush's shadow, Professor Michael Boskin of Stanford, and found them "made from the same cloth." *The Wall Street Journal,* assessing the preferences of the absolutist advocates of free markets undisturbed by government intervention, gave Boskin ideologically acceptable grades, which put the professional seal of good housekeeping on Summers—to the echo of expletives from Reagan and Bush charging Dukakis with unflagging dedication to the "liberal" credo.

The question for political incumbents is not whether, but how, to use economists. Their projections are as necessary as they are risky but are only as safe as their premises prove practical. For better or worse, no president can collaborate productively with the leaders of Congress on decisions binding their successors, and the successors to their successors, without projections on the record from their respective economic advisers. All presidential decisions boil down to calculations of funding: who will put up how much for how long. Responsibility for counting—and finding—the bucks to back decisions stops with the economists, even though they do not participate in the power to spend. Just as surely as water cannot rise above its level, neither national policy nor the technical performance of politicians, commanders, physicians, and engineers in fulfilling it can rise above the quality of the supporting economic analysis or the unstated economic premises. The 1987 crash confirmed the case that the institutionalized economic wisdom fed to America was wanting in theoretical relevance and practical reliability.

Tragedy is never wanting in its comic accompaniments. On the eve of the October 1987 crash, *The New York Times* reported, on September 1, that the Union of Radical Political Economists, "which represents the nation's Marxist economists," admitted to a new sense of confusion and caution, conveyed by the success of Reagan's dollar diplomacy on the domestic market front. When the young academics of radical bent threw in the sponge and accepted the bull market with awe, echoing their establishment seniors, no bet was surer than that the crash really was around the corner and traveling on a fast track.

13

EINSTEIN'S EPISTLE
America's Role
in Nucleonics

All twelve chapters up to this point have focused, successively, on America's distinctive characteristics, dominated by her distinctive assets, all tangible, visible in her geography and sociology, and functioning in her economic mechanism and financial structure. She suffers from underutilization of all her assets, especially in her foreign dealings; their deterioration remains unchecked. By contrast with her underutilized assets, the one distinctive domestic liability that burdens her is intangible; as Chapter 12 showed, it is institutionalized in her ideas about economics.

This chapter focuses on the special burden America carries for nuclear statesmanship that remained with her when she lost her monopoly over nuclear power. Once it was diluted into a head start, however, the risk of nuclear disaster emerged as a contingent liability shared by every country in the world observing the nuclear race and the two principal participants in it. During the Truman years, America herself recognized her special responsibility for limiting this liability; under the farsighted leadership of David Lilienthal and Bernard Baruch, she took constructive steps to anticipate and avoid the world's subsequent exposure to nuclear anarchy. But she did not recognize her special responsibility to understand Russia's history and problems, especially the historical interplay between Russia's pre-Revolutionary technological evolution and the Kremlin's political ideological regression.

No suggestion of being "soft on Russia" is implied in acknowledging this understanding as a unilateral American responsibility. No one can realistically expect Russia to understand America; no other country does, not even England, which shares a language. But America's multiplicity of distinctive capabilities includes a potential to assess the tugs that are straining Russia's purposes and preferences in opposing directions. America's failure to meet this analytical responsibility has jeopardized her opportunity—providentially, without dissipating it—to develop an approach to Russia calculated to switch nuclear rivalry into cooperation. Mere cutbacks in levels of Russian and American nuclear weapons, which leave enough in place to trigger a holocaust, will not turn Russia and America into effective policemen (especially against terrorists bearing nuclear arms) nor end the arms race. Until America gives top priority to this twin effort, she will find no freedom from her deficit burdens or for her recovery priorities (itemized in Section IV). Nuclear disarmament has been conceived in a vacuum, insulated from the need for increases in conventional weaponry to offset Soviet superiority, as well as from the opportunity to turn U.S.-Soviet trade into the pivot for reversing the worldwide drift into depression. The dividends spread across the world economic landscape by U.S.-Soviet nuclear cooperation would seal political bonds between the nuclear superpowers to the point of inviting economic cooperation extending into space: the economic frontier waiting to be opened for our congested world in the twenty-first century.

Since Japan's rise as an economic superpower, American academic economic wisdom has downgraded the overriding priority of trustworthy nuclear agreement with Russia as the indispensable means to stabilize and patrol nuclear peace. Instead, the bipartisan establishment view has supported a defensive counsel of desperation, urging America's corporations to concentrate on regaining market competitiveness by copying Japan's corporate methods. America's inferiority complex about her managerial failure has been earned. But her underlying confusion of purpose as a country is at fault, not Japan's superiority in buying the lead position in any particular product market and accepting America's open invitation to use dollars to do so. This confusion of purpose feeds America's fear that deficit pressures will increase no matter what she does or doesn't do.

But deficits never sprout in a policy vacuum. They are the consequence of policy failures. America's road back to policy success begins not with more introspection over failure but with a broad look outward toward success: approaching the 1990s resolved not to limit her inventiveness by trying to copy Japan's economic version of a corporate state, which is programmed and subsidized by the government. Such a changed stance, from failure conceded to achievement consolidated, would eliminate Japan as America's mentor to copy and identify Russia, first as America's subject to study, then as her pupil to teach, and finally as her market to tap. A by-product of this achievement would demonstrate American methods that work, as America makes them work again. No limited partnership could be more fruitful. Therefore, an understanding of the history of ideas involved in America's nuclear confrontation with Russia emerges as vital to a joint effort to minimize missilry and to harness it in a hospitable environment where scientific creativity, technological skill, and political vision merge to produce economic health while respecting institutional differences.

America can extend this opportunity by converting Russia into a fellow traveler behind her, up the inclined plane to a permanent plateau of worldwide prosperity. A prime benefit, condition, and talking point would be modernization for Russia, free of the traditional blight of cyclical setbacks and financial disturbances built into capitalist economies and based on bona fide, verifiable arms reduction agreements. The opportunity hinges on America's understanding of the road Russia has traveled into the nuclear age. Unfortunately, the policy inertia that indulged the escalation of America's government debt in defiance of budget discipline, at the very time when cutbacks of nuclear missiles appeared on Reagan's final agenda, distracted her from the pressing need to bring this domestic liability under control and to tackle her equally distinctive international nuclear responsibility. Her ability to overcome her suscepti-bility to the wares of her economists, and to lessen all-around nuclear vulnerability, is hampered by her ignorance of the ideological cross-currents whirling about the Kremlin.

Only America shares with Russia the challenges of space, of nucleonics, and of conventional military power present on a massive multifront scale. Consequently, only America shares with Russia the burden of supporting all three at once. Only a vibrant economy, skillfully managed, can provide the wherewithal to carry the load. America has done so in the past. Russia has come under great pressure to adapt America's experience of economic mobilization to the workings of whatever bastardized compound of capitalism and socialism, of militarism and modernization, may be hammered out within the Kremlin. The Kremlin's dealings with Washington may well determine Russia's chances of success at modernization—and everyone's chances of peace. Gorbachev's failure would escalate risks beyond control because it would abandon Russia to anarchy and yield the Kremlin's decision-making initiatives to the militarist politicians.

Churchill's famous remark about Russia being a riddle wrapped in a mystery inside an enigma exaggerates the intelligibility gap Russia presents to America. Four bridges of understanding lead across it. America can cross the first hand in hand with Russia, as its people rediscover Russia's past, with emphasis on its creative heritage in science, which matches the glories of its literature and art. Accordingly, this chapter will undertake a recapitulation of scientific development in Russia and of the indefensible ideological suppres-sion of this legacy at Lenin's instigation, which the Kremlin's politi-cal intelligentsia have covered up. America can also cross the second bridge together with Russia, as it rejects the ideological imprison-ment that the country suffered under Stalin and reverses the dog-matic myopia inflicted on it by Lenin's ignorance. America is over-due to recognize that Russia itself is struggling to build the third bridge: to bring its technologically backward civilian economy that doesn't work onto the same plane as its technologically advanced military economy that does work.

Of the four bridges under construction, the first three already have clear throughways awaiting American exploration. But Gorbachev has rushed to cross the fourth without a blueprint, let alone a supporting structure. Aganbegyan's 1987 book on *perestroika* reveals an abysmal ignorance of the intricate rudiments of the socialized capitalism the Kremlin has been rushing to graft onto the junk heap of a Soviet economy the Gorbachev regime has inherited. Specifically, Gorbachev's experiment in enlightenment proposed to adopt the pricing mechanism of a market economy under the misimpression that it could function by direct price fixing between large enterprises in behalf of consumers at large. It also proposed to open its own marketing process to wholesalers under the misimpression that they could function without credit.

Such ignorance leaves Russia free to follow America from the frying pan into the fire without fear of sharing the heat America has brought on herself by her inertia. America will move into the 1990s at her peril—and the world's—if she does not cross all four of these bridges that can connect American policy-making with Soviet transitional pains. Insight from such bridging into Russia's cumulative imbalances and ignorance can strengthen America's hand in bargaining, as well as her leadership in managing nuclear risk and in restructuring the limits of the world economy.

The way for America to insure herself against nuclear danger is to insure her military superpower antagonist as well, along with all the others: whether innocents caught in between or toughs itching to pull their own nuclear triggers. Only America commands the distinctive combination of military, economic, financial, and health technology to provide such insurance and keep it in force. Just another familiar rhetorical exercise in the reformulation of American idealism will not. Neither will the equally familiar American expedient of dollar subsidies, financed with borrowings, given away for no tangible benefit in return. This was the offer Gorbachev put on the table as Reagan was leaving office. Reagan had no better alternative to show him. Gorbachev was on trial for his political life to bring in the money. His adversaries inside the Kremlin had good reason to press him to do as well with Washington as with countries possessing less nuisance value.

Russia itself lacks the developed economic resources needed to undertake any lead toward nuclear peace. In 1987, Gorbachev's offer to reduce the European nuclear arsenal was an approach out of one of Aesop's fables; Aesop's moral is to beware of bargains ensuring entrapment. Gorbachev lured Reagan into conceding Russia's over-whelming superiority in conventional weaponry. Gorbachev's ploy dramatized Russia's powers of political initiative, which stem from its muscle as the military superpower. Even such technological virtu-osity as Russia commands is specialized within military limits.

Just as the business of America was business in the days when America was doing well, so the business of Russia is war, whether Russia ends up making war or not. Just as drift is standard operating procedure for democracy, so preparedness is routine for a garrison state and, by default, Russia has been preparing to threaten and/or conduct war professionally. Moreover, the *nomenklatura* (the desig-nation is part epithet, part brag) have made a particularly good thing of their position in the military segment of Russia's domestic struc-ture, and an even better thing out of their stake in Russia's major export growth industry, the arms business, which is more stable as well as more lucrative than the oil-export business. In fact, the worse the oil business gets, the better the arms business gets. Soviet strate-gists are single-minded adherents of the twin rules of victory in combat: concentration of striking power and control of the initiative. Gorbachev proved it by downgrading nuclear missile parity in order to capitalize on conventional arms superiority and to position Russia to cash in on American economic aid. He demonstrated to his die-hard opposition that he could bluff more out of America by concilia-tion than his predecessors had by confrontation. America never tried bargaining.

In Russia's mind-set, however, the immediate source of nuclear paranoia is not America, surprising though this may seem to Ameri-cans. Moscow is increasingly nervous about the number of militant Moslem entries into the nuclear club. Some are simply terrorists, some are governments, and some are both. Further advances in nuclear technology will give greater mobility to terrorists, as nuclear bombs are compressed to briefcase size. By a fateful geopolitical coincidence, the most aggressive nuclear muscle flexers poach within Russia's sphere of influence. Moscow is alert to the reality that these potential nuclear aggressors pose first-strike threats to Russia, as America does not. Their striking power, after all, is not interconti-nental. More elusive still, none of them presents a centralized target for preventive counterstrikes in force by conventional weaponry. Witness Afghanistan.

When Reagan permitted Gorbachev to seize the nuclear initiative, he voiced no awareness of the universal dependence—including Russia's—on America to lead the world in her two interrelated roles as the world's only diversified nuclear and economic superpower. Although America, from the Kissinger period through the Reagan years, was blind to the enormous leverage this distinctive dual role put into her grasp, Russia did recognize it, beginning with America's economic leverage over Russia itself. In the hands of a nuclear superpower, economic leverage always carries a decisive premium. Nuclear leverage over an economic superpower is usable only for a destructive purpose, and then only once. But economic leverage always is readily and continuously usable constructively, on terms controlled by the superpower with the economic leverage. Specifically, the scientific community (its Soviet members included) knew that Russia lacked the capability to utilize nascent space technology to build and operate a command/communications center from which to patrol terrorism on earth. But no one could doubt that Soviet collaboration with an American lead would ensure a gigantic step toward internal security for Russia and toward joint U.S.-Soviet policing against terrorists. As early as 1981, Senator Nunn had urged President Reagan to take such an educated initiative, to no avail; although George Bush, early in his presidential campaign, did float a trial balloon "expressing interest" in Nunn's call to limit SDIs to defenses against accidents and terrorism, a feasible goal for the 1990s.

But the Reagan White House preferred paranoia matching Stalin's Kremlin in assessing America's nuclear exposure to Russia. Toward America's creditors, however, the Reagan administration retreated to a supine acceptance of economic dependence. America allowed them to call the tune on the terms of their lending through their manipulation of the currency markets, although none of them enjoyed military superpower status and none of them would have enjoyed creditor leverage over America if they had borne their fair share of the cost of her protective transoceanic burdens. Both miscalculations by Reagan resulted from his reversal of priorities: negotiating on weaponry instead of with resources.

Only a superpower with a full and active economic arsenal has any chance of turning a peacekeeping initiative into a bona fide negotiation to tie nuclear peace and world prosperity into one knot. Russia and Japan don't fill the bill: Russia for lack of modern economic muscle, Japan for lack of modern military muscle, as we saw in Chapter 6. Washington's first overseas stop en route to a solution in search of a military truce and a policing partnership is bound to be Moscow. Japan's status as an economic and financial superpower has won it no chips at the nuclear table, not even a seat. Moreover, the spread of affluence within Japan has sharpened new incentives among its people in all age brackets to avoid any danger of another Hiroshima or another firebombing such as General Curtis LeMay unleashed on Tokyo beforehand, much less even indirect responsibility for ongoing complicity in nuclear missilry; its young people have been developing a flair for life and its pleasures and seem intent on resisting any effort to turn the clock back to the era of militarization. All the other political powers and weaknesses were quick to copy Japan's stance of nuclear aloofness from the Cold War, as if the hot blast over Hiroshima had charred them as well.

In May 1986, the Chernobyl disaster revealed Russia's unpreparedness to deal with even the peacetime fallout from a power plant failure. More than incidentally, the damage Russia suffered, ranging from casualties to chaos to epidemic to mere embarrassment, nursed suspicions that Gorbachev's impressive public relations breakthroughs were packaged as staples for export consumption. Russia's commitment to *glasnost* started out warily selective, and the Chernobyl episode kept it that way. Moscow took a full three days to confirm foreign findings of the nuclear contamination, after first denying it; Gorbachev himself took eighteen days to offer a lame explanation for the delay. Despite *glasnost,* the Kremlin's military hierarchs, not their political spokesmen, still appear to dictate Russia's domestic priorities, as well as policing political appointees and the latitude they enjoy. The 1987 case of Yeltsin, whom Gorbachev had sponsored as head of the Moscow party apparatus, illustrated the point; he was not shot, merely stripped of his power, and with Gorbachev's acquiescence: an advance, but hardly a breakthrough.

All Europe woke up traumatized on the morning after the spillover at Chernobyl. The shock reflected fear of entrapment in an uncontrollable drift of nuclear waste and of being sucked into the chaotic maelstrom of incompetence (traditionally accompanied by brutality) inside Russia. Chernobyl flashed the word across Europe: "The Russians messed up, and sheep from Romania to Wales are dying."

As Chernobyl made clear, nuclear power has made Russia vulnerable as well as formidable. No outbreak of conventional warfare in Europe could be contained within a military front. Western Europe's commercial commitment to nuclear power has evolved into an effective defense, unplanned at the time, against conventional attack. From the Atlantic to the Rhine, Europe is studded with nuclear power plants. Any shooting would trigger a series of Chernobyls, and the prevailing winds in Europe move eastward. The Chernobyl disaster forced the Kremlin for once to admit not only error but weakness, specifically, at the time, need of medical backup as well as nuclear engineering on a scale only America could furnish. Soon afterward, the Kremlin made its admission of weakness official when Gorbachev made his historic offer to Reagan to cut nuclear arms in Europe, signifying, on the most suspicious of interpretations, Russia's need for time to recuperate from the disaster. The ever-shrewd elders of Beijing suspected that the disaster of Chernobyl, not any change of heart, forced the retreat to *perestroika*: witness the jailing of the managers (a reversion to a forgotten technique of Stalin's).

Yet America did not understand the enormous long-term political opportunities opened to her by this revelation of Russia's economic weakness. She rushed volunteer aid to Chernobyl without realizing the significance of the emergency call she was answering or the distress she was curing: that Russia was as helpless in the face of nuclear disaster as Ethiopia was in the face of famine. America was the superpower relied on to solve environmental and humanitarian problems everywhere, as well as to supply the money needed to buy technology to cope with overseas disasters. Never mind that America herself depends on borrowings—increasingly from abroad—for the aid she lavishes around the world. She is still the source from whom aid is sought and generosity is expected.

The popular American attitude toward the nuclear revolution remains ambiguous. The stressful, inventive work at Los Alamos was and still is a source of pride—a paradigm of American "can-do," although many of the doers were not American-born. The nuclear burnout at Hiroshima and Nagasaki demilitarized Japan, but in a manner that presented uneasy humanity with a vision of horror and the continuing fear of a repeat visitation. But the nuclear revolution did not happen by itself. It was the progeny of a cultural revolution that mated new concepts of philosophical speculation with new concepts of scientific method, which, in turn, have spawned uses that are both productive and humane.

As usual in scientific advance, thinking preceded technology by decades—in this case, by over half a century. America and Russia have dominated the nuclear scene in the immediate follow-on phase of technological development and political negotiation. But other countries with rich cultural traditions—Austria, Hungary, England, and Germany, to name a few—played memorable parts in the initial philosophical and scientific inquiry.

The complex structure of American life has invited her institutions of learning to overspecialize in professional disciplines and to disregard an interdisciplinary cultural approach to understanding and managing the world. The pressing problem of how to translate the history of the nuclear revolution into the politics of nuclear disarmament and the economics of reciprocity has been left to ignorant goodwill. In addition, America has been overcome with enough guilt for unleashing the nuclear bomb to inhibit candor about the nuclear record. Americans are as overdue to understand their part in the historical drama as they are overdue to understand Russia's. Meanwhile, the Soviet educational system suffers from the opposite fault: an extreme of rigidity. Everything developed under the regime of the Romanovs was outlawed as reactionary; everything Lenin attacked was outlawed as unworkable.

Nevertheless, both the American and Soviet educational systems, however alien their spirit and their methods, sprang from a common root planted deep in Europe's cultural soil. All European intellectual authority, respected in Russia and in America, has flowed from Aristotle, who brought Plato's airy speculations down to earth. He dealt in experimentation and he fortified his findings by classifying them. Modern breakthroughs that supplanted his teachings did not discredit his reputation as the embodiment of common sense (though both the metaphysics he formulated and the sciences he described are now historical memorabilia). Yet Aristotle's logic stood the test of time pedagogically long after Europe's laboratories superseded it functionally. His deductive method survived the Baconian empirical revolution in the seventeenth century, by contrast with Galen's Roman medicine, of which Harvey disposed.

Europe's structure of knowledge and basis for scientific creativity pivoted on the Aristotelian view of the world drilled into children in the classroom. Aristotle segregated the categories of thought into compartments, beginning with space and time, separate and independent, though interrelated. His first canon stipulated that no body could occupy more than one point in space at one point in time. The experience of events sustained the rule, as the old-time Brooklyn Dodgers learned the hard way when they managed to put two men on second base at the same time one fine afternoon: a philosophical student of baseball, if present, would have appreciated the force of Aristotelian observation. The companion Aristotelian tenet held that stretches of time are needed to traverse stretches of space. Thanks to the nuclear revolution, however, significant stretches of space can be crossed in insignificant—almost instantaneous—stretches of time: a change in quantity that has produced functional changes in social and cultural qualities of life, as well as in techniques of killing.

The nuclear revolution was conceived in virtue. Although its questions had been addressed earlier by many pioneers, in its genealogy the popular counterpart of Adam has been Albert Einstein, universally revered as a symbol of saintliness in science. He was already living in exile as a senior seer at the Institute for Advanced Studies in Princeton (which Lauchlin Currie, in his salad days at the White House, chose to sneer at as the "Institute for Advanced Salaries") when FDR's war administration undertook to discover and build a road toward practical and grim uses of nuclear science as a basis for weaponry. Explorers such as Enrico Fermi of Italy, Leo Szilard of Hungary, and many other admiring apprentices in the scientific revolution that Einstein's researches had initiated were part of America's unique crash effort to unleash the forces of atomic fission.

But their mandate to disregard cost and nonpresidential authority was subject to one limitation: strict instructions to exclude Einstein from the work. His scientific protégés in America came to feel that they bore the guilt for engineering America's fall from scientific purity into technological sin. The top secret team assembled at Los Alamos knew, of course, that they were in a race against Hitler's team and felt deeply the need to prevent and preclude a success that would ensure a Nazi victory. It was Hitler's Germany that started the race to militarize nuclear power. But a sense of guilt over the demonic violence that they had let loose welled up after they won the race. The Nazis had taken a wrong turn early on when they turned to the use of heavy water as raw material, when the American Manhattan Project had concentrated on uranium.

Although Einstein was indeed an innocent in politics, his name was political magic. When, at the instigation of Szilard, he wrote Roosevelt a letter in 1939, its effect was profound, though delayed. In it, he warned FDR that Hitler might have already gained a head start in developing atomic weapons. As matters turned out, Einstein was alerting Roosevelt to a future risk not visible, at the time, to the naked political eye, but unmistakable to trained scientific scrutiny. The secrecy FDR imposed on the Einstein letter suited his campaign strategy of standing pat on America's unpreparedness while Hitler was running wild in Europe, and Japan in China. The master politician knew what he was doing when he kept Einstein's letter under cover. But the alert it flashed gave him documentation that he used as future justification when he set up the Manhattan Project at Los Alamos, without compromising his visible inaction.

Einstein published his first paper on radiation as early as 1904. In 1905, his flair for experimental improvisation led him to select radiation salts for his original probes to demonstrate the interchangeability of mass and energy. An early by-product of his work showed how the conversion of mass into energy involves splitting atoms and unleashing explosive effects. By 1939, when Hitler's marshals on the scientific front were still groping for the right questions, Einstein had the answer. He knew how the atmosphere could be bombarded with overwhelming blasts of energy released by split atoms. No one involved in Einstein's early warning to catch up with Hitler could guess that America would ultimately use the nuclear weapon she originally aimed against Germany to devastate Japan, and only after Japan had been beaten to its knees by conventional weapons, the only ones Japan had: proof of the high-speed erraticism inherent in the nuclear war game. No player knows how to be prepared—or against whom.

Einstein's coactivators included two of the most hyperactive and articulate members of the budding fraternity of atomic scientists (as nuclear physicists called themselves in the early days of the nuclear revolution). Szilard was one; Edward Teller, another Hungarian, was the other. After Hiroshima, the start of the Cold War split the two apart as violently as their collaboration had split the atom, Szilard advocating nuclear disarmament and Teller nuclear armament. But back in 1939, still allied in revulsion against Hitler, they led a push for the initiative that Einstein took with Roosevelt. The *Dictionary of Scientific Biography,* edited under the auspices of the American Council of Learned Societies, explains why Szilard, Teller, and their cohorts chose Einstein as their intermediary with Roosevelt, and why FDR, never moving in a straight line, welcomed the alert. But, the *Dictionary* is careful to add, "Einstein neither participated in nor knew anything about these efforts" at Los Alamos. The nuclear activists shied away from involving him.

The Struggle for Survival, my examination of how America won the 1939–45 war as a war of production before she got into the fighting, classifies 1939 as "the lost year." FDR spent it spreading uncertainty about his willingness to engineer a third-term draft for himself, while alternately denying plans to prepare for war and promising to remedy America's unpreparedness. Einstein's letter, if made public in 1939, would have let loose as much explosiveness politically as the bomb later did militarily. In that year of ambiguous but irritable jockeying for position, FDR neither advertised Einstein's letter nor let it drop through cracks in the White House floor well known as conduits to the press. The *Dictionary* credits this letter of Einstein's to Roosevelt with initiating the American effort that eventually produced the nuclear reactor and the fission bomb after the fighting was over and the war won, when ironically the time arrived to prepare for peace.

Pacifism, of course, was in the air during the period between the wars. Where Mahatma Gandhi attracted the attention of the world as its most colorful and controversial activist, his nonviolent methods seemed eccentric, and the aggressive nationalist cause he espoused won few adherents in the West. By contrast, Einstein emerged as the world's most knowledgeable pacifist, and its most disinterested one as well. Misty-eyed, fiddle in hand, his unimpeachable attributes typecast him as the symbol of peace, living testimony to Roosevelt's determination to preserve it. The need for military secrecy muted the alarm sounded by the original nuclear warriors. Their coconspirator in the White House, however, welcomed their undercover prod. He loved schemes, especially those that challenged his ingenuity to do what he was itching to do anyway.

Altogether, Einstein's letter served the double purpose of arming the president with his scientific knowledge but also protecting Roosevelt with his pacifist mantle. No president receiving such a solemn warning from such an eminent figure, who personified scientific wisdom and the hope of peace, could have ignored it. Nor could FDR reveal his response. Einstein's identity as a German Jew in exile from the German scientific establishment in its Nazi captivity gave weight to his words of warning. His letter provided FDR with all the support he would later need.

Urgencies dictated by the political calendar prevented FDR, posing as a harassed second-term lame duck, under a cross fire for moving both too fast and too slow, from initiating a preparedness program so vast, so uncertain of success, and so sure of disapproval by significant segments of the electorate, and therefore resisted by Congress. Meanwhile, the outbreak of the 1939–45 war, which Roosevelt took as forcing him to run for an unprecedented third term, concentrated his attention on the 1940 election campaign, during which he paraded all his skill and experience as indispensable to safeguard the supposition of American neutrality. It called for him to balance his statesmanship against Hitler's threat and to inspire confidence that he could contain nazism without blundering into war.

Having considered America's approach to transforming creative conceptualization into results of unheard-of violence, let us pick up the postwar trail taken toward the exploitation of nuclear science by the military superpower. How did the drama of preparation for nuclear war unfold on the stage of Stalin's Russia to jar the peace that America had restored? History reproduced art. Science and politics were the vehicles connecting separate chapters of the past with the future. The irony is that Stalin's Russia incurred ineradicable hostility in America for stealing the bomb from America, only to discover afterward that the knowledge responsible for its development originated in Russia before the Revolution.

Professor Walter McDougall, of the University of California at Berkeley, has produced a classic account affording insights into the split between the researches of the pre-Bolshevik era in which nucleonics originated, and the Bolshevik dogma outlawing the application of nucleonics. In *The Heavens and the Earth,* he shows how "modern rocketry and social revolution grew up together in tsarist Russia." His findings are vital to an understanding of the Soviet regime's eventual success as the military superpower in the twentieth century, as well as to its consistent failures to overcome economic weakness.

Nevertheless, his focus is limited to the careers of Dmitri Mendeleyev and Konstantin Tsiolkovsky—the stars in the first act of Russia's nucleonic drama—and the scientific leadership they provided for Russia and the world before the Bolsheviks organized for business. Mendeleyev's discoveries are spread across the encyclopedias. He activated the chain reaction of modern scientific thinking in Russia with his codification of the atomic weight of elements in the late 1860s; he discovered some of them himself, notably the rare earths, and was apparently the first scientist of stature to work with uranium, the nuclear raw material. But although science was creative in czarist Russia, technology was sterile: nothing productive came of it inside Russia. Only violence did, as the product of social ferment that accompanied the scientific ferment.

A profile of the Russian past the Soviets inherited is encapsulated in Mendeleyev's career. Mendeleyev's progressive mother, herself a budding industrialist, arranged for the youngest of her brood of fourteen to be educated in a hotbed of democratic dissent. Despite the suspicion this attracted to him, his genius earned him his rise to the academic summit at the St. Petersburg Technological Institute. Though the czar was its patron, it stirred up a ferment of social unrest as well as scientific inquiry: Russia's indigenous schizophrenia at work! As McDougall writes, the institute inspired "an almost sacramental reverence for chemical mixtures." One of these special brews demonstrated its power by doing a devastating demolition job on the czar himself. McDougall introduces his probe with the gruesome tidbit that the czar's "lower body was blown inside out." Understandably, his son reigned in terror of the same ceremony. It was he who ordered the execution of an obscure provincial student activist named Ulanov, whose younger brother turned up in history as V. I. Lenin, remorselessly dedicated to revenge on the Romanovs.

Meanwhile, Mendeleyev was hewing out the principle known to Soviet history as Bukharinism: industrialization of a peasant society through the modernization of agriculture, rather than by crash industrialization. Mendeleyev bought an estate and turned it into a model farm factory, pioneering the use of fertilizers. As early as 1863, he pounced on the discovery of oil at Baku as offering, along with coal, "the sole means of saving Russia." The naval ministry commissioned him to conduct "a large-scale . . . high priority . . . project on smokeless powder," of obvious strategic importance in protecting troops against detection. Other ministries trusted him to show them how to develop metallurgical production in the Arctic, as well as in Siberia. The last phase of his career involved him in the study of radioactivity.

On top of all this, Mendeleyev jeopardized his senior status in the czarist academic hierarchy by coming forward as a leader of the student protests of the day. Yet in spite of the line of descent the Bolsheviks claimed from the student movement, Lenin and his colleagues remained ignorant of Mendeleyev and his work. Their ignorance condemned Russia to a serious handicap in the nuclear sweepstakes.

Mendeleyev was a world figure, honored everywhere. But Tsiolkovsky, left deaf by a childhood illness, was self-educated and forced, as he wrote, "to work alone for many years under unfavorable conditions and not even to see the possibility for hope of assistance." Though he never did benefit from official patronage, Tsiolkovsky blueprinted the advanced technology for exploring the unlimited reaches of space over half a century before the hardware materialized. As McDougall writes, "Two decades before the Wright Brothers demonstrated heavier-than-air flight within the atmosphere, Tsiolkovsky imagined reactive flight outside of it. By 1903 he had published the mechanics of orbital mechanics and designed a rocket powered by the (precocious) combination of liquid oxygen and liquid hydrogen."

At the age of sixteen, he proclaimed a war of liberation from the force of gravity. Within ten years, he solved the theoretical problems of interplanetary flight. As his *Dictionary* biography explains, he established the interdependence "between the velocity of a rocket at a given moment, the velocity of the expulsion of gas particles from the nozzle of the engine, the mass of the rocket, and the mass of the expended explosive material"—all before aeronautical engineers were trained to develop rockets.

From the vantage point of embattled America in its war to control costs in 1986—especially military costs—one of Tsiolkovsky's more pedestrian suggestions had the most contemporary thrust. It called for using "twin screw propellers rotating in opposite directions." In mid-July 1986, *The New York Times* reported Boeing, McDonnell Douglas, General Electric, Delta Air Lines, and other leaders of American commercial aviation racing to catch up with this modest proposal, which increases payloads by conserving fuel in flight. The military application would be decisive, thanks to the direct payoff from fuel savings in combat performance.

Lenin dominated the second act of Russia's nucleonic drama, before America knew the show was on; he was blind to his own role in it. As Marx's self-proclaimed apostle and heir, Lenin operated under a compulsion to lay claim to expertise in every branch of knowledge his mentor had mastered. Marx was himself a mature genius in the library, even though he turned into an obstreperous child when dealing with people. But Lenin vulgarized concepts as if they were epithets to hurl at debating opponents, a fault that offset his political genius in manipulating masses and governments. Lenin's ignorance of even a smattering of data on the marvelous Russian contribution to the European store of scientific knowledge, which provided the background to nuclear technology, lends irony to the Cold War and laughs the last remnants of intellectual dignity out of Leninist dogma.

In 1909, Lenin had directed an entire volume of vilification, *Materialism and Empirio-Criticism,* against an unconventional pioneer in the concepts and techniques that opened the way for the nuclear revolution. He was a gifted Viennese academic named Ernst Mach (for whom supersonic speeds are named). Lenin, in his Zürich pamphleteering phase, launched a fierce know-nothing attack on Mach, provoked by Mach's reputation as Europe's preeminent philosopher of science. Americans may minimize the importance of

Lenin's blunder in targeting Mach as a scientific charlatan, but the extent of the embarrassment Lenin's ignorance made for his heirs cannot be minimized by anyone with a stake in understanding Russia's commitment to ideological rigidity.

Soviet Russia's sense of achievement in catching up with America's postwar nuclear virtuosity was marred by hard evidence that the Soviet scientific and educational establishment had freed itself from the cult of Lenin's infallibility. Plainly and painfully, the deans of Soviet scientific education faced the responsibility of documenting to their peers and their pupils how Lenin had gone wrong and the supposedly hapless Mach, a butt of ridicule in standard Soviet texts, had been proved right. The scandal in the intellectual history of the Communist Party paralleled its criminal guilt over Bukharin. No doubt the scope of this intellectual scandal of Lenin's lay beyond Gorbachev's understanding; though his equally charismatic wife took her advanced degree in philosophy as taught according to Lenin. The enormity of the intellectual cover-up of the Mach issue loomed as a decisive test for *glasnost.*

The revelation threatened to explode a scandal inside Russia immeasurably more disruptive than admission of Stalin's judicial murder of Bukharin. That was merely a crime, by Soviet standards; Lenin's lunacy over dogma was a full-fledged political blunder, undermining the authority of the credo and its supports. Americans do not understand that the Soviet system relies on ideology to pave the way for force to ensure conformity. To find Lenin wrong, even posthumously, in any argument he started on ideological principles is to undermine the entire Soviet system that stands as his monument. Even at the height of Stalin's insane abuse of power, he justified everything he did by invoking Lenin's authority.

Before Einstein's rise to prominence early in the century, Mach had achieved no fewer than five enduring breakthroughs that paved the way for Einstein's. His investigation of supersonic speeds was the first. As early as 1887, he laid out the principles and parameters of ballistics for artillery use. Mach armed the decrepit and decadent Hapsburg Empire with the prototype of the ballistic gunnery of the future. It was an exercise in the theater of the absurd. The Royal Austrian Naval Base at Fiume, on the Adriatic, was an antiquated mechanical toy, and though the Hapsburg generals knew enough to ignore the admirals, they did not know enough to seize on this discovery by the dean of the Viennese academic establishment. But their idea of war never progressed through the 1914–18 war beyond rounding up millions of young peasants from the empire's ramshackle structure of feudal baronies and sending them out to fester in pest-ridden trenches, exposed to rifle shots, bayonet thrusts, and epidemic at close quarters.

Indeed, although the 1914–18 war unveiled many lethal innovations—the submarine, the airplane, the tank, Big Bertha, poison gas, the machine gun—Mach's innovations found no place in its annals. In all the reams written about the war, including the reminiscences of Winston Churchill, a personification of technical ignorance on parade, generals and admirals were still believed to be fighting wars with armies and navies, rather than with the weapons their engineers furnished. Mach demonstrated ballistic missilry more than half a century and one murderous war before Hitler's generals took it up. Once again, the military mind asserts its propensity to lag at least one war behind scientific advances.

Mach's second momentous achievement was an extension of his theory of ballistics. Science benefited from this exploration, but the military prompted it. Having related the speed of a projectile to the speed of sound, he proceeded to identify and measure the ratio of the speed of a flying object to the speed of sound: hence the modern scientific staple of Mach numbers and the flying start they gave modern aerodynamics—not to mention Einstein, who undertook to establish the speed of light. When the sixteen-year-old Einstein began to ponder the significance of the constant speed at which light moves, a new exploration of the universe and its laws was under way. The theory of relativity and the nuclear revolution followed. Einstein gave generous credit to Mach as the principal source of his youthful inspiration. But notwithstanding his devotion to pacifism, he knew that Mach's pioneering insights into the structure and dynamics of the universe were tainted by their potential for destruction—and that his own were too.

Einstein was not the only admirer of Mach troubled by the flaw of war in his work. In America, William James was too; in fact, Mach and James formed their own mutual admiration society; "genius of all trades . . . an absolute simplicity of manners and winningness of smile when his face lights up," was James's description of Mach. James's anxiety over the military application of Mach's work explains why his call for "the moral equivalent of war" was prophetic and practical, not merely idealistic, as popularly supposed. Mach practiced "radical empiricism," as he showed by dedicating one of his books to James, who defined his new, distinctively American philosophy of pragmatism as just that. James found philosophy preoccupied with fixed ideas and preconceived principles and left it alerted to how things and people function. Mach found no experimental basis for Newton's concept of absolute space, time, and motion. His heretical move to question it illustrated James's novel concept of radical empiricism, though Mach was at pains to assert, "I am not a philosopher, but only a scientist . . . above all, there is no Machist philosophy." Not all original thinkers are skilled and intuitive investigators and balanced human beings as well. Mach's humility befitted the mentor of William James and Albert Einstein, not to mention the target of V. I. Lenin.

In a monumental paper on the origins and scope of relativity theory, Einstein used the term the *Mach principle* to explain Mach's third achievement: that the inertia of any isolated body reflects the tugs on it from all other bodies in the universe. Mach had developed this proposition to refute Newton's theory that bodies maintain independent motion in space. Mach formulated a strange new term that defied common sense by confusing the straightforward Aristotelian categories of thought. He spoke of "space-time" and established it as standard in scientific usage and training. According to the Mach principle, the distribution of matter and energy posits the existence of space-time as a measuring rod to track the speed of bodies propelled across the curvature of space. Einstein applied the Mach principle to demonstrate that the motion of bodies is relative to one another across space-time. In Einstein's lifetime, the trauma of "blitz minutes" brought home the meaning of space-time with a brutality that no lucidity of expression could match.

Mach's fourth achievement was educational. In German-speaking Europe, learning was institutionalized under the orders of drillmasters in mufti. Beginning in 1887, the year of his breakthrough in ballistics, Mach established himself as the source of definitive textbook wisdom on physics. All classroom presentations and laboratory practice followed his precepts. The achievements of German science and technology, until Hitler nazified German education, stand as a monument to his teachings.

His fifth achievement was unplanned. Mach won immortality in Leninist hell by attracting Lenin's fire as the unlikely symbol of subjectivism in philosophy. Lenin's vituperation derived from a profound misreading of Mach's writing. If words could singe the flesh, Lenin would have outdone Torquemada and the Inquisition in disposing of Mach—but not for the good of Mach's soul, and with no sensitivity or respect for Mach's "pure intellectual genius," as James put it.

Mach did indeed investigate sound and (in connection with ballistic weaponry) light, but not as a tribute to philosophical subjectivism, the bête noir of Communist thinking. He did so as a study in mechanics. Lenin, suspicious as a revolutionary needs to be and ignorant as any political genius can be, jumped to the conclusion that Mach was bogged down in the verbal quandaries of the seventeenth-century philosophers, who questioned how people could know intangibles, like colors and sounds, that they saw and heard. In fact, nothing could be more objective than Mach's explorations of scientific principles. Lenin's thinking was mired in the era of rifles; Mach anticipated the era of missiles. If Lenin's vision had been broadened by a glimmering of either Russia's heritage of scientific achievement under the czars or the scientific revolution launched by his own contemporaries, he would have known enough to salute Mach as a mentor and embrace him as an ally. Instead, Lenin cut off his own political nose to spite his ideological face.

In *Materialism and Empirio-Criticism,* he charged that "the philosophy of Mach, the scientist, is to science, what the kiss of Judas is to Christ"—a classic case of the Devil quoting Scripture, this time with righteous indignation. Lenin's successors continued to parade behind Lenin's slogans but learned to operate by Mach's methods. In demonstrating their mastery of Communist politics in Russia, they advertised his blindness to the creative intellectual ramifications of the revolutionary scientific ferment inside pre-Revolutionary Russia. The milestone he ignored was established in 1897, ten years after Mach demonstrated the feasibility of ballistic gunnery and ten years before Lenin vented his contempt on Mach. That was the year when Tsiolkovsky applied Mach's theory of mass turning into energy in space-time to the specifics of interplanetary rocketry.

If we convert the scientific concept of space-time into a convenient and apt metaphor for social and economic usage, we are invited to connect ideas that have hitherto been taken for granted separately. Making the connection forces us to think of the consequences that the passage of time factors into policy and decision making. When efforts to conquer space, whether for purposes of commerce or of policing, involve America in decades of costly development and experimentation, the metaphor helps to avoid succumbing to visions, mirages, and slogans, like SDI.

Dealing with such intractable realities as space and time can be therapeutic. American society has always found a place both for dreamers and for practical tinkerers. When America puts both modes of living, working, and thinking together, she sometimes finds they don't mix. But sometimes the interaction of dreaming and doing can produce constructive revolutions. A look at the strengths, as well as the weaknesses, of some of the unconventional thinkers whose lives and ideas have been noted here invites an attempt at innovative, cross-disciplinary combinations that mimic the space-time marriage.

Technology is unforgiving in what it can or can't do, but scientific ideas often flourish in strange soil. Lenin's denial of the possibility of progress under the czars took its toll. Simultaneously, Veblen rated the Russian Revolution as sure to survive because Russia was immune to mechanized intervention by the West. Indeed, Veblen, in his perennial skepticism, showed more perceptive faith in the revolution from its outset than Lenin, who expressed genuine anxiety that it would not survive. Then, Veblen identified vestigial feudalism in Germany and Japan, nurturing militarist roots in both countries after the 1914–18 war, as the carrier of the war danger that materialized in the 1939–45 war.

True, Veblen did not live to see how America's victory on both fronts of the 1939–45 war culminated in her blind jump into the Cold War. Nor could he have anticipated the evolution of Bolshevik Russia from the Revolution at bay, repelling invaders, on through its subsequent stages of military development: ally of Nazi Germany, target of Nazi invasion, captor of Germany's arsenal, and military superpower occupying Eastern Europe, dominating Western Europe, threatening China, checking Japan, and confronting America, while reeling into a replay of czarist chaos. Nevertheless, the metamorphosis of Soviet power from invincible on the defensive at home to imperial threat on its outposts bears an eerie resemblance to the combination of backwardness and ultramodern efficiency that Veblen had identified as sinister in imperial Germany and imperial Japan. Soviet power was inherently unstable; it offered American power a standing invitation to determine its direction.

By the mid-1980s, the terms of the Cold War had fleshed out the skeleton of the new American economy Veblen had sketched as a legacy of the 1914–18 war. The nuclear climax of the 1939–45 war committed the American economy to continuous investment for automated war in time of peace, as laid out in my *Economics of Crisis*. The distinctive functioning of the military cycle, from investment to activation, from modification in development to maintenance in use, split the traditional analytical atom that joined economic space and financial time. Only the Mach-Einstein metric of space-time could keep track of Veblen's distinctive model for America of continuous compulsive investment in instant obsolescence, requiring long time lags for incubation but producing split-second intercontinental transit from earth across space. Only America has developed an economy gaited to run on the routine application of the concept of "relativity" to the financing of industry, in the absence of war but in standby readiness for war, dictated by the nuclear jeopardy every country has come to share with all the powers.

Einstein was entirely insulated from the economic or financial consequences of his work. His inquisitive mind was never taxed with computations about the investment requirements proliferating from the technological activities—military and commercial—stemming from the revolution in physics he engineered. Though Einstein, whose career overlapped Veblen's, had no knowledge of Veblen's work, Veblen was too clear for comfort about the pressure the scientific revolution was putting on the economy to change its ways, beginning with its ways of creating demand, generating capital, putting capital to work, and rating its results.

The concept of space-time also invites application to the new cycle of continuous investment in advanced technology. Keynes had been ignorant of Veblen's work and unaware of the applicability of Einstein's to economics—at least in America—when he viewed demand levels as fundamental and in need of cyclical stimulus to offset historically cyclical dropoffs in investment. Technology dictates a high continuous level of investment demand, regardless of the level of business activity.

This blind spot of Keynes's revealed the central gap in his approach that limited his enormous contribution: the insulation of his monetary thinking from the impact of the technological revolution that dominated the two great wars of his lifetime and transformed economic behavior, social relationships, and financial mechanisms under his own eyes. The economic impact of the Cold War—particularly, of the nuclear arms race it produced—has distorted the simplistic structure of the traditional short-term business cycle beyond recognition and divorced rates of investment from rates of return or use.

Yet Veblen's achievement turned into a tragedy. He discovered the engineers, but they never discovered him. The connection between his concepts and their functions was invaluable, but America failed to make it. Consequently, myopia on the American side of this intellectual rift proved as distractive, and destructive, as the myopia Lenin displayed on the Russian side. Just as Lenin, in his obsession with fixed ideological premises—such as "the workers"—mistook Mach for foe instead of friend, so the American consensus, in its obsession with "business confidence," mistook the engineers, and the scientists allied with them, for clerks instead of creators.

To hindsight, the subsequent conceptual lapses on the American side of the Cold War reveal tunnel vision at least as narrow as the routine orthodoxy perpetuated on the Russian side. America has subjected herself to a higher standard of scrutiny by the very fact of her achievements and her higher level of self-evaluation, as well as the expectations they spread among influential segments of her population. The failure of Russian opinion to develop a clear view of America did not excuse the failure of American opinion—beginning with policy-making American opinion—to take advantage of its greater advantages to evaluate crosscurrents in Russia realistically and to identify opportunities to strengthen American power by helping Russia to grapple with its weaknesses. Once tried, the magnitude of Russia's deficiencies, and the fit they make with the diversity of

America's strengths, hold out a promise of realistic negotiation for mutual benefit.

Major historical coincidences have a way of converging into causes, unnoticed at the time but undetected by posterity at its peril. One such momentous coincidence was a three-way collision in 1919: the American economy collapsed after its 1914–18 war boom, Veblen published *The Engineers and the Price System,* and the Bolshevik Revolution hunkered down and defied Western intervention. Subsequently, the American economy bounced back into the unsustainable prosperity of the interwar decade. When it collapsed again, the New Deal institutionalized the principal reforms Veblen had advocated, although it supported the existing price system and ignored the potential decision-making role of the engineers. The Bolshevik Revolution staggered from one failure to another on its way to consolidating its success as the military superpower. Between the 1920s and the 1990s, though America and Russia talked and worried a great deal about each other, America failed to grasp the underlying clash between ideology and technology in Russian life and thought that tied the admittedly backward regime of the czars to the supposedly progressive regime of the Soviets. Russia's sloganized thinking, with its glib generalities about capitalism, advertised its inability to recognize the distinctive complexities of the American system.

But back in 1919, Lenin, for all his intellectual limitations, was guided by his matchless instinct for the political jugular to seize on Veblen as the link to America he felt the need to seek. Though Lenin was the blindest of bigots in matters of ideology, he saw in Veblen a future for America that America could not see for herself, even though he failed to sense the vision Veblen also had for the future of the Revolution in Russia.

Soon after the jerry-built peace treaty was signed at Versailles, Veblen followed up with *The Nature of Peace.* Such topics were judged out of bounds for a conventional economist, as were his observations on the chaotic influences of war on orderly business-cycle sequences, to say nothing of his explorations of the intangibles of sociology. The challenge confronting America in the 1990s is to update and work to combine Veblen's vision with Russia's drive to align its history in the construction of a workable future. The engineers, the price system, and the nature of peace, viewed from an interdisciplinary perspective, hold the key to a future that could profit both countries. Since 1940, the world has awaited a sequel to these two milestones in analysis by Veblen, titled *The Engineers and the Investment Cycle*; the engineers carry the concepts of scientists from the laboratory, through the financing process, to the production floor. This book is offered as a belated and modest substitute.

Russia has unwittingly developed a case history for America to ponder, illustrating how much America can learn from taking a hard new look at Russian history. The Kremlin has structured its success as the world's military superpower on a foundation of failures in economic and political policy-making. America can turn her two liabilities—the distinctive economic one in her head, and the nuclear one she shares with every other country—into assets: first, by changing her idea of herself from economics student to Yankee trader; second, by changing her idea of Russia from evil empire to her most eager customer. Once American society sees how to manage these liabilities accruing from the past, her body politic would be freed to manage the problems confronting the present.

This elementary question is nearly as familiar to the naked eye and the untrained mind as the proposition that $2 + 2 = 4$. In America's case, $2 + 2 + 2 + 2 = 8$ because America allowed herself to be dragged into the crisis of the late 1980s juggling eight out-of-hand problems at once. They invite solutions in pairs, and that's how they will be addressed in the chapters that follow. Her eight policy quandaries can be reduced to a rudimentary equation of arithmetic: $-2 \times -2 = +4$. Though we cannot manage this formula on our fingers, we know that it works. Multiplying negatives produces pluses—not only in the abstractions of arithmetic but in the realities of political arithmetic as well.

The problems of food exports and oil imports make a twosome. So do the budget deficit and the trade deficit. Enriching Social Security entitlements and fighting the flight of people, capital, and cash into the underground economy make a third pair. Superficially, the last two are unconnected. But each drains and diverts money from the Treasury—one legally, the other illegally—and each invites reversal by the same legislative action. Defense, too, is overdue for policy pairing with social programs: the Cold War and the welfare state have contrived to pair them as contenders for funding in the legislative process.

Problems loom as negatives on the national balance sheet. All four pairs of negatives can be converted into national assets, by applying solutions in pairs. All eight share a characteristic: the cash flow to each has attracted a formidable political constituency for each, as largesse always does—unfortunately, funding has spread insecurity along with its benefits. Arithmetically, each problem in each pair adds to the trouble made by the other, but, subjected to the seemingly magical equation that combines them, they can together multiply America's assets, as the next section shows.

Failing this packaging approach, "borrow as you go" will remain the norm for all of them, but none of them will win authority to borrow enough. Over a fifty-year span, which Kondratieff and Schumpeter have characterized as the "long cycle," the comparison was eerie between the dilemma that alternately challenged and frustrated the Roosevelt revolution in 1938–39 and the dilemma that confronted America at the end of the Reagan years in 1988–89. In the earlier case, as we have seen, Currie's Americanized version of Keynes's general theory demonstrated the scale on which government financing could have reactivated the American economy. But the president who authorized the demonstration needed no counseling from any economist to recognize that, realistic though the abstract arithmetic calculation might have been as a calculation, no follow-through try at the political arithmetic needed to convert the numbers into an operation could succeed. Congress can never be expected to vote appropriations faster than events force their need.

So far the comparison holds between the Roosevelt and the Reagan years, but no further. In Roosevelt's time, the borrowing could have solved the problems paralyzing the economy, but it was the war that did. By Reagan's time, not only the borrowing had exhausted its stimulative power, war had too. Consequently, another, still grimmer comparison loomed before America in her transition from the Reagan years: with Gorbachev. In 1988, his own battle cry in behalf of *glasnost* and *perestroika* reminded his troops and enemies alike that there was no retreat for Russia to seek, no alternative except to get on with his program. But here, too, the comparison stopped. Gorbachev did not have the foggiest idea what going forward meant or where it would lead. America does.

The French saying about treason—that it is a matter of timing—applies as well to solutions. Some of the solutions proposed here are not new but are presented here in sharpened form. My *Prescriptions for Prosperity* (1983) advocated the barter of food for oil and also urged the export of Treasury zero-coupon bonds as a device to earn credits for defense services and, at the same time, minimize interest payments on the federal debt. Bartering food for oil became desperately needed by 1987. So did the zero borrowing alternative to defense giveaways and interest ransom. The other solutions paired in the chapters that follow, though less familiar, answer urgent needs too. The country has needed time to try out the formula for failure called for by the economic catchwords in vogue and to find that they work—as formulas for failure.

By 1988, America had exhausted her various bromides. Drift was the danger. Politicians able to produce policies that work have no trouble packaging them into politics that pay off—especially when irresponsible substitutes for policy have brought hard times and awakened awareness that new remedies are the order of the day. For Americans, their businesses, and their livelihoods, the quick and effective way to begin is for their government to update the old-time Yankee trader in its bargaining with foreign governments. The program I present in the chapters that follow would prove not only politically acceptable but politically popular. It would also work, as nothing politically popular is expected to do. None of these solutions would cost any money or call for any new taxes. On the contrary, some would produce dramatic savings and spread continuing benefits. Nominal modification of the machinery of the system, engineered quickly, would enable Americans to keep what they still have, get back what they have already lost—both substantial—and raise their goals—potentially limitless—for achievement.

Distinctive
Solutions
for America—
and the World

14

EXON'S EFFORT

America's Offer of Food for Oil

The oil-import flood put the American oil patch out of business, and the crop-export failure did the same to the grain belt. The 1988 drought was a temporary blight, but an avoidable one. As early as 1958, after the Soviets scored their first in space with Sputnik, Senate Majority Leader Lyndon Johnson, as he then was, called hearings to explore the implications of the space race. Edward Teller supplied a sensational aside that snapped LBJ out of one of his customary snoozes in the chair. Under prodding from assistant counsel Cyrus Vance, later secretary of state, Teller explained how simple devices operated from space could precipitate precipitation in drought-ridden and desert areas.

The oil-import and crop-export disasters together contributed massively to America's ruinous, demoralizing trade deficit. Together, they have undermined both sides of the flow of goods essential to America, inflating oil imports and deflating crop exports. Only a correspondingly massive American strategy, aimed simultaneously at both these causes of her trade deficit, can roll it back.

The conventional argument against political action to cure the

U.S. trade deficit is that it would necessarily turn America protectionist, shut down world trade, and bring on a replay of the Depression. Reaganomics added its characteristically polemical revisionist view of history. It indicted American protectionism for having precipitated just that chain reaction in 1929. Reagan's spokesmen were ungenerous to the extreme of denying credit to the derided Keynes for having been right about the false boom of the 1920s, and for the right reason, even before Senator Smoot or Congressman Hawley wrought their tariff damage.

In any case, the Smoot-Hawley Tariff levied *only* a 10 percent surcharge on 1930 duties, and only on one-third of the items imported; petty cash alongside the nominal markup imposed on imports by the collapse of the dollar resulting from the Baker devaluation of 1985. Professor Barry Eichengreen, of the University of California at Berkeley, makes the telling point that the Fordney-McCumber Tariff, effective at the outset of the boom decade of the 1920s, raised duties nearly as much as the Smoot-Hawley Tariff subsequently did at the onset of the Depression decade of the 1930s. Nevertheless, America's ability and willingness to power prosperity with her capital exports made the decisive difference in trend between the earlier boom decade and the subsequent Depression decade. Higher tariff rates proved no obstacle to world prosperity in the 1920s. By the 1930s, however, America had stopped exporting capital, and had started to press for repayment of past investments. Reaganites revising tariff history also revised Depression history, blaming the calamity on the Smoot-Hawley tariff as if they never heard of Hitler or of the insupportable structure of war debts and reparations that he pulled down on his climb to power.

In targeting Smoot-Hawley, the Reaganomists blithely publicized the easy premise that a guiding protectionist philosophy had inspired the United States. But in scrutinizing its legislative history, Eichengreen raises a more practical question, which has assumed steadily increasing urgency since 1930. He notes that the 1920s opened up the legislative process to invasion by lobbyists, which, in turn, subordinated policy aims to logrolling compromises. Though he fails to blame the Prohibition lobbyists for opening the floodgates to the lobbying scourge, he documents a plausible case for classifying the Smoot-Hawley package as a formless collection of compromises rather than an assertion of protectionist political direction.

At any rate, the various opponents of American government action to cure her trade deficit had the better part of the argument in the late 1980s. As a result, not only America but world trade suffered as if America had already fulfilled the gloomiest protectionist prophecies of doom. Predictably, such action failed, and not merely for the familiar reason cited: protectionism provokes retaliation. The functional reason is that conventional American protectionist action forced on powers fearing market saturation has always relied on tariffs to raise import prices, as America did in 1929. In the 1980s, however, price subsidies by governments of the countries importing into America became the source of the problem. Tariffs invite the governments putting up subsidies to increase them as fast as America increases her tariffs. Dollar devaluation, and the Eurodollar slot machine open to all governments wanting more dollars, have made tariffs a no-win game for America. Thanks to the vicious circle of devaluation and subsidies, raising the cost of imports no longer can force reciprocity. Only managing the volume of imports as a negotiating lever can.

By the simple expedient of switching import surveillance from cost to volume, America can adopt a quick and constructive method of taking incisive action to cure her trade crisis. She will reverse the shrinkage in world economic traffic the moment she does. The way for her to start is to forget the price game as it is played in world trade today. She cannot hope to hold her own at it, let alone win it. She will do better to rethink the whole structure of her trade and to rely on the cargoes of food she has to sell as a fair exchange for the cargoes of fuel that the oil producers want to sell to her. Ironically, her need to buy oil—popularly mistaken as a source of strategic weakness—arms her with a decisive bargaining weapon. In addition, wielding it will enable her to use her need to sell food as a means of raising the quality of nutrition around the world, and to do so without raising the cost of living.

The distinctive structure of the American economy invites such a reciprocal, interrelated solution to both its food and fuel problems on a world scale, not just a domestic one. But only the American government can manage such a miracle with a single stroke. Mention of still another program of government action invariably provokes shudders over the cost, let alone over the possibility of bureaucratic bog-down in failure. Such knee-jerk reactions are shortsighted. Ambitious solutions do indeed loom as scary, but all the familiar failed expedients have spread desperation.

Yet the consequences of failure to cope with America's food and fuel problems have put her in overdue need of a new policy broad enough to manage both sides of the commodity crisis. Though the two make a natural export-import fit, Washington has dealt with them separately because it is not conditioned to deal for reciprocity. Contrary to the impression created by ideological opposition to remedial government action against the trade deficit, the international commodity crisis is accelerating the cancer eroding the Third World. Ironically, however, opponents of action branded as protectionist make noises protective of Third World victims of distress. Contrary to the impression created by self-inflating government subsidies, and contrary to anachronistic American taboos against government management of markets, policy intervention into commodity markets need not be costly; it can be profitable. A precondition of effectiveness, however, is recognition that foreign governments already run these markets. The only agency large enough to take effective action is the American government itself. Besides, it's the only entity with a legal mandate to try.

In basic budget and trade terms, the twin solution proposed here calls for a law authorizing one commonsensical expedient: to switch the basis for oil imported from dollars paid out to food swapped. Farm subsidies are routinely listed as one of the items on the spending side of the budget that are fixed charges; only token cuts are mentionable in serious political discussion. From 1985 on, relief from the cost of farm relief has awaited the success of devaluation to revive American farm export markets ready to welcome American crops in volume. At the worst of the spring 1988 drought, with Reagan railing against "liberal profligates in Congress," his own administration remained staunch in its support of subsidies paid to take large tracts of land out of production: self-defeating, self-perpetuating welfare.

If markets were efficient, or even sensible in their adjustments to obvious commercial incentives, their workings would long since have brought America's thirst for imported oil into balance with the world's hunger for her foodstuffs. When the world's stability and her own security are at stake, America cannot afford to wait for market players to discover what political players can grasp as common sense. The way for America to free her markets to work efficiently is to barter food for oil. This is also the way for her to fulfill the primary purpose of national policy, as envisioned for any country by Petty; that is, to harness the political arithmetic she manages in support of the market arithmetic by which she hopes to prosper. Adding what's good for American agriculture to what's good for American oil would guarantee the oil bloc in Congress against a veto on principle by the congressional delegations from the oil-buying states; the oil votes in Congress plus the farm votes add up to an unbeatable combination when food-oil barter is negotiated in order to protect food and oil consumers as well as producers.

Whenever crises get out of hand, the government is sure to enjoy high visibility directing traffic on the edge of the nearest cliff. Washington has been mismanaging the country's farming and oil troubles for years. It has also been the official nursing home of last resort for all the banks broken or jeopardized by the interlocking farm and oil crises. The more the government has put up to help the banks, the more help they have needed.

So the question of keeping the government out of any crisis is academic. By the time a crisis has intensified from chronic to acute, the government is sure to be throwing new money at it faster than devising new ways to solve it. In 1988, it began to throw unlegislated tax shelters as well as money at buyers willing to take over busted Texas banks and S & Ls. Moreover, the depression threat traveling with America's twin deficits has made the sick sectors of her economy sicker, and it wrecked the sector that seemed healthiest: the stock market, the artery that animates the entire financial mechanism. The solution offered here calls for the adoption of procedures by existing administrative arms of the U.S. government, and is guaranteed to save substantial expenses authorized routinely and to bring in new revenues—not only public, but private as well—without raising any taxes or hopes of psychological manipulation of the economy by the White House, as the poll management behind Reaganomics did. Clearance through the political process would be required, but without any need for institutional changes, administrative additions, or budgetary burdens.

If an oil-import quota were adopted without a food-export tie-in, the benefits to the domestic oil industry would be limited to

American producers. But the oil states are also farm and ranch states. Some of them are blessed with mines vulnerable to shutdowns, and all of them have been ravaged by farm, real estate, and bank failures. Accordingly, no mere oil turnaround would take the oil states off the sick list. Any solution to the oil problem needs to be broad enough to turn around other sick segments of the economy.

America's annual oil and gasoline import bill from 1983–87 averaged just under $47 billion a year; it fell under $50 billion with the price of oil, but the rising tide of high-priced gasoline imports buoyed it. Her net average annual food surplus came to the better part of $20 billion. This means that America's food earnings failed to pay for even half of her oil bill. A move to stop paying cash oil bills but to continue collecting on cash food sales would endow America with something like $65 billion-plus for her trade account. The positive flow from this barter, produced by eliminating America's oil-import bill, could then work to restore her farm-export income. Oil barter would free American consumers from the bogey of a boom whipping up an "orgy of waste" and threatening a return to the nightmare of gas lines.

The quick way to take American agriculture off welfare, and to keep it off, is to start selling the privilege to import oil into the United States and to require payment at U.S. ports of entry in a new currency: American food products, monetized for export use only. The thrust of the Food-for-Oil legislation I support would be to recirculate within the United States all the dollars now dissipated around the world to pay the country's annual oil bill. But no protectionism is implied or risked, nor are any pipe dreams about free trade indulged. The more oil America took in from abroad on a bartered basis, the more food she would sell abroad.

A welcome by-product of the food-for-oil swap system would resolve the no-win argument over whether to pay farmers to take land out of production or to pay them for planning still more overproduction. Once cash income from domestic oil sales rewarded farmers for producing more rather than less, they would operate under a new incentive to produce as much as they could. The brief shortage scare stirred up by the 1988 drought stands as a reminder that my proposed food-oil barter system would welcome more farm production, not as a burden, but as one form of insurance that would earn income in the process of avoiding risk.

An American food-oil barter would shower two immediate benefits on the entire economy. Both would be massive and continuous. The elimination of America's oil-import bill would be the first; America would collect on food shipped out to pay for the oil taken in. The second would cut another $20-odd billion from the top of the budget deficit by substituting taxable oil income from food exports for government farm benefit payments (including land taken out of production); neither farmers nor oil producers are big taxpayers. This double spur to farm activity would bring benefits to rural America impossible to count in advance or to undervalue as a force in American economic society. For decades, agriculture has been discounted as a shrinking component of the revenue base. The legislation I support would mandate the routine swap of American food products for imported oil and would expand the tax base of American agriculture again, and not as a burden to be resented. No one ever goes broke paying taxes on windfall cash income—or objects to its being recorded.

In the summer of 1986, America went so far as to offer Russia an export subsidy to move some of her huge wheat surplus. When Russia refused, Washington lowered its own offer, negotiating against itself. Cutting a price and still finding no takers is an unmistakable symptom of a damaged market. Government subsidies, called loans, routinely move mountains of unmarketable crops every year into government storage bins, reducing free-market doctrine to wasteful academic make-believe.

The arrangement I propose could be described as giving hard-hit farmers an insured claim to strike oil already in barrels, and to sell it for cash up front, free from wildcatting risks beforehand and from storage or transportation costs afterward. Producing food that is guaranteed an export market as payment for imported oil would turn farmers into middlemen between the foreign governments bringing in oil and the domestic oil companies lined up to buy it. Such barter legislation would divert import payments from foreign governments to American exporting food producers (in some cases, individual farmers or farming corporations; in others farm cooperatives; in still others, food processors). The government would be relieved of the need to dole out nonproductive payments to farmers and, instead, would receive revenues from oil companies documenting commercial payments of taxable earned income to farmers.

Not only would the new food-for-oil swap system free farmers from their age-old exposure to low crop prices and high debt burdens: the lure of steady oil payments would reverse the long, dispiriting trend away from agriculture as a way of life. Moreover, thanks to the parallel but seemingly unrelated drift away from urban concentration and the growing attractions of farther-out suburbia for younger people, the broadening blur between suburban and rural America would offer a variety of incentives to younger nonurban families. The trend of the times has induced or forced them to split their lives between holding nonfarm jobs or living on productive farms. A food-for-oil swap would permit a life-style that offered the assurance of cash income from farms lived on to supplement take-home pay from other jobs paying cash.

Bad statistics have been institutionalized for so long that no one can come within a country mile of guesstimating the hard core of work-force membership in households routinely commuting between nonfarm occupations and rural homes supporting subsistence farming. Certainly, this pattern of two-job existence supports increased auto sales, but in the absence of incentives for steady farm income, this changing trend in work-force distribution has not been examined for other useful side effects.

If the food-oil hat trick is so easy for America to pull off, how come no administration has tried? Answers are obvious. The Middle East did not take over center stage in Washington until America had put her distraction with Vietnam behind her. By then, Nixon had turned America's chronic crop surplus troubles acute by blundering into his decision to embargo soybean exports. Carter, when his turn came, put all his zealotry into his blind belief in the reality of the oil shortage. The Reagan people shrugged off the twin problems as reserved for market forces to decide. During the long years while America was waiting for a market solution, she was not ready to devise one that would work because it was manageable with her own resources.

No other major oil-exporting country is also a food exporter, though Canada would count as an exception if the Canadian oil producers were not captive suppliers to the U.S. market. America is the only industrial power that is also a major oil producer, accounting for more than 60 percent of her enormous needs. Henry Kissinger, in his day, had dismissed economics as "for bookkeepers." The shrunken and glutted world of the late 1980s made one axiom self-evident to the most literal minded of ledger keepers: The combination under one flag, in one well-populated continental haven, of food-export need and oil-import leverage is a marriage of resources made in market heaven. This alliance needs only a bit of political management, free of any blackmail or shakedown by either foreign oil suppliers or foreign food customers, to run profitably on both sides of the barter and collect political dividends from the intangibles of power. A reversal of recent form!

The media have focused on the shock effect of the total trade deficit. They have ignored the potential for a solution offered by its two weakest performing components, both "swing" items. Between 1983 and 1987, America's cumulative trade deficit surged to $678 billion. Her oil-import bill alone accounted for $305 billion, no less than 45 percent. The debacle stemming from the bankruptcy of what passed for America's food-export policy during the same high point of the Reagan presidency cut America's food-export surplus by more than half, from its 1981 peak of $26.6 billion to a 1983–87 average of just over $12 billion, and a 1986 low of $5.4 billion; another $74 billion in the wrong direction. Adding the two brings the combined half-decade damage from the oil drain and the food-export failure of the 1980s to close to $370 billion (allowing for a $14 billion average annual dropoff from peak), or an average of $74 billion a year. No pair of entries in America's foreign trade ledger come close to making a comparable splash.

While the Federal Reserve Board was tearing passions to tatters fighting inflation, America was suffering from both deflationary trends: on the farm-export side, from the entire loss of her traditional export income; on the oil-import side, from the devastation spread throughout her rich domestic oil economy, interlocking as it does in each of its regional bases with her distressed farm economy and her overdone real estate boom. The renewed oil collapse of 1985 caught all these key segments of her economy overborrowed together, as if in anticipation of the Hallelujah Chorus soon to be sung in the stock market. The climactic celebration in Wall Street did not start until the damage had been done at the base of the economy, until agriculture, oil, and real estate had collapsed onto her undercapitalized banking system, and until the consumer economy had sagged into stagnation, intensifying the pressure on countries importing into America to slash retail prices in America still further. The market crash of 1987 coincided with chaos in the world oil economy, which fuels every continent, and paralysis creeping into the American real estate market, to which every community in the country pulsates.

Any time an emergency calls for innovative remedies, reliance on historic precedents structured into American political arithmetic is the quickest and the most reassuring way to proceed. In the absence of presidential theatricals—of banks to close and to reopen, of gold to seize, of war mobilizations to proclaim, of assassinations to mourn—commodity economics offers the safest legislative precedents and the most solid congressional majorities; coalition politics operates across party and regional lines through combinations of commodities, as with crops and oil (with real estate values and bank bailouts thrown in for good measure).

Two formidable precedents of American political arithmetic are on the side of a two-tier price structure solution for America's oil problem. The first is an oil precedent; the second, a food precedent. Before OPEC tried to subject the world to its demented version of the law of supply and demand in a never-never land of oil shortages that could never be, America charged $4 for domestic crude and paid $2 for imported crude. Then, however, production controls were needed to protect the Treasury from the cost of price supports. The structure of America's domestic oil industry anticipated OPEC, but worked, as OPEC did not. Ironically, the independent oilmen ran it. The major international oil companies merely complied. Texas, the power base of the independents, produces 25 percent of America's oil.

In 1931, when the price broke all the way to 10 cents a barrel, the governor called out the National Guard to shut down the wells, and the Texas Railroad Commission was empowered to regulate oil production in the state. The hallowed cause of conservation was invoked to insure the battered economy of Texas against this devastating price deflation. The commission started by cutting production back to seven days a month. As late as 1961–62, after many ups and downs, it still kept production cut back to only 8–9 days a month. It made its decrees for *prorationing* (production rationing) stick in the homeland of free enterprise. Because the majors recognized the political clout of the Texas independents and were content to let production controls work in Texas, production in the new areas being opened up remained free. When OPEC's chance came, its strategists entered the field in entire ignorance of the price-fixing and production rationing history to which the oil industry owed its original prosperity.

The rising generation of American oilmen dedicated to free-market principles shared this ignorance. Governor William F. Clements of Texas, who started as a hard hat and evolved into an oil tycoon as well as deputy secretary of defense under Reagan, personified their thinking. He did, at least until the American oil depression of 1987 converted him into a born-again believer in market stabilization, this time by support controls—despite his blind faith in the power of OPEC to control the market it had debauched. Clements's pragmatic, if not intellectual, conversion serves as a reminder that studying history is cheaper than learning its lessons the hard way.

On the food-export side of America's food-oil trade equation, the strength of the dollar all during the worst of the OPEC siege of the 1970s did not price American agriculture out of its export markets. America's farms had always earned a heavy portion of the country's export income. America's average annual net export "take" from competitive agricultural products held comfortably above $25 billion during the half decade from 1980 through 1985 and peaked above $30 billion. By May 1986, however, this comfortable cushion was entirely gone. For the first time, the trade balance of the American farm economy followed the overall American trade balance into the red. For 1987, American agriculture turned in a hair-thin trade surplus, negating the giddy claims made for export gains from dollar devaluation. The bare figures revealed a performance failure more scandalous, in light of America's incomparably superior agricultural system, than Russia's crop failures.

Unrealistically, American agriculture remained structured into dependence on cash exports and, therefore, under pressure to be priced competitively. Around the world, government subsidies for production, controls on imports, and bartering for exports have steadily deprived American farmers: first, of their market margins and, then, of their markets. Washington had been resigned to protective moves of this kind by the world's traditional bulk crop exporters—Canada, Argentina, and Australia—but the decision by the industrial powers of Europe to follow suit undercut America's participation in the world's demoralized crop markets. The Third World debt crisis completed the damage; America had been buying crop exports with the "loans" she lavished on the defaulters of the 1980s. When the "loans" dried up, so did the exports.

In the 1970s, my first call to mobilize American agripower as a weapon to counter petropower met with the smug response that the concept of using food as a weapon is immoral. At that point, the chance of depression seemed a mirage and America was thought to be a fertile and friendly giant, happy to share her resources with all comers. But the danger of depression transformed food bartered for another basic commodity in oversupply into an expedient that is not merely moral, but a blessing for both partners in the deal.

The second precedent to be found as a guideline for oil is sugar: like oil, sick enough to be incurably political. In fact, fear of shortage has never distracted attention from the crying need for continuous control of sugar production. As a result, America's sugar producers have enjoyed a decisive advantage over her oil producers: at least they know they are weak. They inject politics into their business dealings for the same reason the Russians do: because of their need to offset weakness in the marketplace with the strength they bring to the political table.

By contrast, America's oil producers have never balanced their swaggering style with the lobbying that reflects their weakness and is admitted in their need for protective legislation. America's sugar producers, importers, and processors have combined to structure a system that works; foreign producers are allowed limited access to the American market and have learned to live with it and, considering the alternative, to be grateful for what share of the market they get.

America's sugar industry operates under a quota system that protects her domestic producers of cane and beet alike. It even guarantees an extra allowance for freight to the higher-cost beet producers. The country consumes 7.75 million metric tons of sugar annually, with American producers allotted no less than 7 million tons. Foreign producers make do with just under 10 percent—slim pickings indeed. This production schedule supports a price structure set at 18 cents inside America, and fluctuating around 8 cents in the world market and the U.S. futures markets (peaking above 13 cents during the 1988 drought). A condition of this sweet deal for U.S. sugar producers is that it costs the Treasury no money. This production allocation, guaranteeing no less than 90 percent of the American sugar market to American producers, supports a rich price premium of 10 cents over 8 cents over 100 percent! Yet consumer complaints are nil.

America produces sugar (plus her new strains of corn sweeteners) in no fewer than 35 states. Consequently, no fewer than 70 members of the Senate's agricultural bastion have a proprietary interest in sugar matters. Acceptance of the Sugar Act has been thoroughly bipartisan. Jimmy Carter appointed Bob Berglund, universally respected as "a good man," deservedly so, to be his secretary of agriculture. Berglund had represented the Minnesota congressional district producing the highest-cost sugar in the United States; any part of the Sugar Act benefiting his constituency was bound to do even more for the other sugar growers around the country. Reagan accepted sugar protectionism as an exception to his rhetorical flourishes about free markets (though, in fairness, he may never have heard of its existence).

A system of oil quotas paralleling America's familiar sugar system is needed to support a two-price oil structure modernizing the forgotten pre-OPEC price differential enjoyed by domestic oil, which was standard practice for years. Of course, only Congress could set the terms of any quota and price system. But a guideline reserving 50 percent of the American market for American producers and allowing 50 percent to foreign producers would show more generosity than the paltry 10 percent Washington accords foreign free-world sugar producers. A 50 percent import quota would also allot foreign oil producers a larger share of the U.S. market than their high-and-mighty pricing bought for them while they were manufacturing the glut that swamped them, even though supertankers were burning in the Persian Gulf. When OPEC needed customers, its members and allies pursued a self-defeating price-fixing strategy. A U.S. quota system would subject their unrealistic priorities to a healthy correction.

At any rate, America is in a position to offer foreign oil producers more leeway than foreign sugar producers, because she is under no pressure to bar any of them, Russia included; her ban on Cuban sugar stands as a useful warning. Such evenhandedness would serve America's interest. It would achieve the double purpose of moving more American food products to oil-producing countries for value received on the barrelhead, while warning of retaliatory import cutbacks in response to skittish behavior, and do so at no risk to American levels of supply.

The weakness of the world's glutted sugar business is a given. Although the oil market was left even more weakened by OPEC's orgy and false recovery, in the 1980s the foreign producers as a group could be expected to maintain a more stable world price with 50 percent of the U.S. market assured to all of them (Russia included) than the sugar growers of the world do with only 10 percent. Why need accepted practice in the sugar trade be suspect when applied to oil, especially when oil enjoys an immeasurably broader demand base and a stronger growth trend than does sugar? Why considering the desperation level of what passes for life and politics in so many countries where oil and sugar are just additional burdens? When oil-import licensing can give foreign oil producers, as well as American oil consumers, a better deal than foreign sugar producers or American sugar consumers even ask for?

Diehard free-trade devotees will no doubt demur, maintaining that defensive mechanisms such as product quota allocations impose somewhat higher food and fuel costs on the consumer. They forget that consumers are income earners needing money to spend. Pennies more for food and fuel would be a bargain alongside the danger of unemployment and bankruptcy spreading everywhere from the farm and oil belts in the wake of the stock market crash, the paralysis of real estate, and the overborrowed condition of the consumer economy. Often the very voices raised in protest against an oil-import quota system are heard advocating a consumer tax on petroleum products. Many protesters have certainly remained callous to the plight of American agriculture. Such high-minded purists seldom stop to reckon the value of freedom from hunger and disease that a U.S. food-for-oil-trade quota system could bring to pestilence-ridden slums in the Third World.

At what price could realistic bargaining-for-barter begin? The domestic oil price that was deemed necessary to reverse the domestic oil depression and assure protection for the capital of the banks, as well as the debt structure and taxpaying capability of real estate in the oil patch, approximated $18 a barrel at 1987's cost levels; the price deflation did not prevent the stubborn inflation of the debt structure. Certainly the false recovery above $20, even with interest rates down and oil patch banks temporarily propped up, failed to lift the siege. Assuming as an example a $5.00 price for foreign oil, payable in food, and a 50 percent allocation of the American market to American producers, the net domestic price would come to $11.50 (averaging $18.00 and $5.00 in equal parts). America has been consuming close to 16 million barrels of crude oil a year. An annual import bill for half as many, or 8 million barrels, at $5.00 each would come to only $40 million, all of it payable in food products, adding export revenue to an annual cash export level falling short of $20 billion.

Would $5.00 be unrealistically low, considering that OPEC's disastrous Vienna meeting in December 1987 broke the world price to $15 before the ministers of the member countries left for home or acknowledged their loss of control over the market? And remembering the world price break under $9.00 in 1986? The producing governments made all pricing calculations irrelevant by raising production even faster than they lowered quotations. This frenzied spectacle of sellers' breaking their own market provoked their customers to suspend all bids. It duplicated the frustration America suffered when she negotiated against herself to sell grain to Russia in 1986. The experience demonstrated that stability at any price level is a bargain compared to the alternative of bidless pricing.

The consumer does not benefit automatically, if at all, from breaks in the price of crude oil, though jumps are passed along promptly. When the oil companies failed to stave off the 1986 break in crude prices, they succeeded in keeping price concessions on gasoline and heating oil moderate. Even a $15 average price for domestic crude would leave the American consumer better off than any other approach (short of the catharsis of world depression, which would necessarily subordinate all consumer interests to income needs). The first belated antitrust probe of "stickiness" in refined product prices would sensitize the structure of refined product prices to downtrends in the crude market, guaranteeing quick and full pass-throughs to consumers. Extension of the barter legislation to include the extensive inflow of refined products would have the same effect.

Admittedly, a favored alternative—an oil product sales tax—enjoys the advantage of familiarity. But this preference harks back to the days when curtailing oil consumption beckoned as a formula for survival, not disaster. The quota system recommended here offers instead the double advantage of protecting the income of the productive work force as well as the living costs of the consuming public. In addition, it anticipates a return to boom conditions, encouraging more people to buy more cars and trucks, and to use them more, especially for longer-distance commuting, without limitation by train and bus schedules.

America's outworn assumptions have run her into a cross fire on her embattled consumer front. The years of affluence encouraged the country to live on the road (and to do without public transportation). This life-style popularized the nostrum of a gasoline tax as the easy fix for austerity. But this same mind-set inspired the unrealistic free-trade vision of a tranquil world expanding its reciprocal dealings. Conveniently ignored was the conflict of purpose between targeting auto use as the culprit for high living and shielding the consumer from the ravages of Rustville.

Moreover, advocates of a tax on gasoline have allowed the related assumptions of oil shortage and farm shrinkage under conditions of stagnation to blind them to the promise of America as the land of plenty. Her enormous usable open spaces provide ideal conditions in a boom environment for at least a semirural solution to urban congestion. Meanwhile, the feature of the Tax Reform Act of 1986 bringing the oil companies into the U.S. taxpaying community provided a partial, if indirect, alternative to a consumer tax on prices at the pump.

Tragically, countries that profiteered with oil at America's expense—notably Nigeria and Mexico—happen to be starving. In all the annals of social and political negligence, none matched the opportunity America lost lavishing cash on Mexico for oil, yet blinded to Mexico's need for the modernized agricultural infrastructure only America could provide. Angry complaints over American machinery exports wanted and imports and "wetbacks" not wanted diverted attention from America's wanton waste of private sector cash encouraged by the malign neglect of public sector policy, while Mexico disintegrated into chaos.

What's more, the oil price collapse has depleted the excess cash of the sheikhdoms, which have been not only running deposits out of banks counting on limitless gushers of oil liquidity but also cutting back on allowances to thickly populated Moslem countries chronically short of food (let alone of sanitized handling and storage facilities). Food-for-oil barter is discounted on the grounds that the sheikhdoms are too thinly populated to become sizable customers for American agriculture. But the teeming Moslem population centers outside the sheikhdoms breed terrorists and inflate the cost of protection. Russia shares this felt vulnerability with each of the congested Moslem countries, is also short on food, and is equally terrified of terrorism.

Any power that is strong enough to make a deal with Russia and make it stick is assured of facing the rest of the world in a posture of strength. Russia is America's most promising prospect for food-oil bargaining—obviously, because the world's oil markets into which Russia sells are glutted and, ironically, even though the world's food markets from which it buys are glutted too. In 1987, Gorbachev's pre-Christmas public relations conquest of Washington committed Reagan to put up cash for crop buying in the midst of a U.S. budgetary crisis, and just at a time when the Soviets were using the Saudis as sales agents, in a deal to dump their joint accumulations of oil and refined products on the demoralized American market.

While Russia, the world's number one oil exporter, has been a thorn in the side of the OPEC producers, its decision to team up with the Saudis conceded the priority of maintaining production over maintaining prices in the national strategy of the Saudis, the least unsophisticated of the oil-producing governments. This decision also accepted a return to single-digit world prices as inescapable, a prospect as unpleasant for Russia as for any of the weaker oil-producing countries caught in a free cash market for oil. But if war is the business Russia is always prepared for, barter is the business the Soviets are always in. In terms of public relations, beating the Kremlin at its own game would be hailed by American muscle flexers as the coup of the century. Accepting payment in Russian oil, and using it to get a handle on the politics of world oil marketing, would be better business for America than simply putting up cash to ship occasional cargoes of crops to Russia. Although such an ongoing barter would benefit both sides and improve U.S.-Soviet relations to America's profit, it would not, however, make a dent in Soviet need to buy or American need to sell.

Reaganite ideologues talked a good game against Russia. But they put their feet in their mouths spouting Reaganomics, which committed them to wait for "free markets" to impose their discipline on all players. They reckoned without the techniques Russia uses, as the player without money, to change the market rules. Russia is always ready to use its muscle to bend markets: routinely sealing a sale with a political arrangement, selling where its buying has built credit balances, and, above all, profiting from skillful manipulation of commodity and currency markets, as the Kremlin's arbitrageurs have been doing in America for years. A long-term food-for-oil barter contract, measured by volume, would belatedly put up the meter on Russia's rides to market in America.

Dwayne Andreas, the knowledgeable head of the Archer Daniels Midland food-processing giant, told *The Wall Street Journal* in 1987 of Gorbachev's determination "to put more food on Russian tables." When Gorbachev asked him how to avoid familiar pitfalls, Andreas advised him that "high-protein soybean meal for livestock feed is the key and explained the technology we use to produce it." Andreas was instrumental in arranging the official reception given Gorbachev by American industrialists interested in accepting Gorbachev's invitation to explore a basis for joint-venture investment in Russia. America has nothing to fear, and everything to gain, from transferring this technology to Russia; climate bars Russia from achieving soybean self-sufficiency.

The food-for-oil double play would not only work as a steady and effective policy; it would take hold quickly. Nowadays, markets look to government for their leads. The barter solution is tailor-made to turn the trick. It would revive realistic strength in the stock market, behind the leadership of the oil stocks, which are always primed for leadership, and it would provide a trustworthy economic lift in regionally sensitive commodity futures markets, which are always primed to jump—especially against swollen short positions.

The positive effect of self-sustaining advances in the national speculative markets and in the regional economies, based on reliquifying regional banks, would open the door to a solid recovery from a chain reaction of deflation and bankruptcy. Such a reversal would certainly activate a business buying move, which is the indispensable recipe for a broad-based recovery in the country's industries and markets. Such a domestic recovery solution for America is the way to defuse the foreign depression danger. It would combine two depression-ridden regional negatives—oil and agriculture—into one vigorous new macroeconomic positive.

I am indebted to my wise old friend, former Senator Russell Long, longtime chairman of the Senate Finance Committee, for a characteristically practical suggestion, thanks to his kindness in reading this chapter in an earlier draft. Long called my attention to my oversight in failing to take account of the peacetime military need for oil. As he pointed out, the armed services are never willing to lavish funds for such basic insurance, even in the absence of austerity pressures, because of their assumption that any emergency will automatically give the brass the power to preempt whatever supplies it wants. His suggestion was to link the oil-import side of the food-for-oil swap to a nominal cash tax on importing governments, and to earmark the money for payment into a military reserve petroleum fund. His flair for innovative legislative ideas showed up in the "financial engineering" he proposed for the earmarked fund: to be spread as deposits among sick oil patch banks and paid out to independent producers as a way to finance their drilling for new reserves and to buy standby supplies for the military stockpile.

America is the only world power exporting food also able to industrialize, sanitize, and package food exports. Access to vast tonnages of sanitized food end products (mainly soy-based) on a non-cash basis—that is, for value received in oil—holds the key to economic stability for Third World countries. Their stability, in turn, is essential to the export industries of Europe and Asia, which have drifted into dependence on dumping more volume at lower prices into their expanded American market. America would be better off to divert price-cutting by her industrial competitors into the Third World, which has traditionally been victimized by the overpricing of industrial imports.

America oil reciprocity offers America a golden opportunity to increase and upgrade American exports of food products to the

322 DISTINCTIVE SOLUTIONS FOR AMERICA

Third World, beginning with its key oil-producing countries, for less
money and more oil; these countries cannot afford to pay cash for
crops, but oil barter would make their demand for the products of
America's food industries effective, as it has not been for years (including the years of oil inflation). Other American manufacturing
industries could also benefit from oil barter. A two-tier American oil
price structure, lowering the cost of importing oil to America, would
end by spreading cheap oil to the oil have-not, export-dependent
industrial economies of Europe and Asia and enable them to offer
more generous terms to their insolvent customers for industrial exports.

In the relatively stable world of the 1970s, the balance of power
between the White House and Capitol Hill lay in the skillful hands
of Chairman Wilbur Mills, of the House Ways and Means Committee. At the time, I urged on him the advisability of extending farm-
export assistance operations to include sanitized and packaged food
products, which would fetch more oil in payment than raw crops.
His native Arkansas had taken the lead in the mass production of
packaged poultry. I pointed out that chickens are raised on soybeans.
His rejoinder was that an administration under fire could not concentrate on an innovative lead. He added that the old-line farm organiza-
tions, clinging to antediluvian vested interests and dominating the
Department of Agriculture, would not hear of any such approach.

But the farm depression of the 1980s forced a hearing for this broadened approach to bring America's distinctive food technology into
play. No member of Congress from the farm belt spoke with more
authority than Senator J. James Exon of Nebraska. Fortified by
reelection as a Democrat despite Reagan's 1984 landslide, and
buoyed by *The New York Times*'s identification of his vote as critical
to the Senate's delicate 1986–87 balance, he uttered the three simple
magic words—*food-for-oil.* Until then, they had been taboo, or suspect, in the farm belt. By 1988, with agriculture desperate and new
opportunities for superior American food engineering ready to supplement the superior quantity of American farm production, political support for the idea had developed. Qualifying food products as
well as raw crops for food-export assistance programs will forge a
formidable trade weapon with which to activate Washington's economic arsenal.

America has every incentive—ranging from humanitarian to plain business sense to the esoterics of power politics—to pursue the food-for-oil solution. Economic collaboration with a Soviet empire seeking to restructure its society and open its economy to new concepts offers an unprecedented opportunity to move away from the dangerous politics of nuclear stalemate and the doctrinaire economics of free trade toward a hopeful future bolstered by a sound, because reciprocal, economic base.

HULL'S
HALO
Modernizing Reciprocal Trade for America

The Swiss bankers, normally thought of as stern, have an indulgent expression addressed to governments whose financial policies come under fire: "At least do something." In September 1985, the government in Washington got this message, and it did something big. It eyed its two deficits, targeted the trade deficit as easier and quicker "to do something" about than the budget deficit, and adopted the standard cure: devaluation. Bankers—not just in Switzerland, but everywhere—approved the action. Markets went wild with enthusiasm over the assumed accomplishment.

By October 1987, the markets, never quick, registered their recognition that Washington, after drawing cheers from its critics for selecting the right target, had come under suspicion for picking the wrong weapon to fire with. First, the dollar collapsed, then the stock market followed. For the second time in two years, Washington was on notice to "do something" about its deficits.

Its first shots having backfired, Washington quite logically changed both its target and its weapons. Whereas, in 1985, the government had picked the right target in the trade deficit but taken the wrong shot with devaluation, in 1987, it spent a blank cartridge on a decoy. This time, it selected the budget deficit to cut, but offered just a token solution implied in the agreement to cut nuclear missiles and, inferentially, spending on them.

No meaningful, quantifiable solution for a direct attack on the budget deficit can be mounted unless and until the trade deficit is brought under control. Once the trade deficit is, however, the improvement in America's foreign dealings will spill over into the budget deficit, as neither direct cutbacks on spending nor direct step-ups in tax rates can. The trade deficit offers an opportunity to mount a flanking attack on the budget deficit, avoiding political controversy by opening valves permitting cash flows into the legal economy and into the Treasury that guarantee continuing improvement. Moreover, improvement in the trade deficit will yield two immediate and continuous magical by-products: cuts in interest rates and in tax refunds. Each drain is large enough to offer abundant elbow room for improvement.

The professional policy debate in the 1988 presidential campaign centered on the critical importance of savings to investment and, therefore, to growth. Perversely, however, the consensus favoring a priority for budget cuts over cuts in the trade deficit invariably favored free trade as well. Their conventional view ignored the chain reaction that cuts in the budget deficit were bound to trigger in an economic environment dominated by high debt levels. To begin with, cuts in domestic consumption, resulting from lowered federal contributions to the income stream, would have had the immediate effect of intensifying import competition; and aggressive import price cutting would have had the follow-on effect of cutting corporate earnings and personal incomes: a counterproductive consequence of a well-intentioned effort at fiscal responsibility.

Nevertheless, the White House, luxuriating in a world of its own, assumed it was taking no chances in calling a budget summit. In preparation for its pageantry over peace with Gorbachev, it put defense spending cuts on the table along with social spending cuts. The president's calendar of visiting engagements guaranteed a swift conclusion for the budget summit. Gorbachev's arrival to celebrate the peace summit was scheduled to black the budget summit out of the headlines in exactly two weeks, providing swift media ratification of the massive arms spending cuts the Reagan-Gorbachev summitry was all about.

Not until the day after the glamour of the gala had worn off did Vice President Bush and Secretary of State Shultz renew the same call—the candidate to voters from the campaign trail in Iowa, the statesman to NATO in Copenhagen—for increases in U.S. defense spending. After Gorbachev returned to Moscow and warned that "nothing had changed," Reagan confirmed not only the change in course, but the size of the tab: as much as was needed to end Soviet superiority in conventional arms across Europe. He paid no lip service to cost as a restriction. Nor did he acknowledge that Europe's price for accepting more arms to maintain would be a quantum leap in American cash payments for arms manufactured in Europe, with staggering consequences for the trade deficit. Moreover, he showed no sensitivity to the key considerations in bargaining with allies over military support: assuring them of financial stability and economic strength.

Europe had felt the need to do well in its dealings with the Soviet bloc even before the dollar collapsed and the American economy wobbled. It had lived for years with the reality of overwhelming Soviet superiority in conventional military capability. The initiative for tactical nuclear disarmament had come not from Europe or Russia, but from Reagan himself. His discovery that his invitation to Russia to join America in nuclear disarmament had left America dependent on Europe to join her in a conventional rearmament race completely destroyed his economic and financial bargaining power— and not only with Europe. His new toughened stance on conventional arms also ignored his ongoing efforts to satisfy the markets of his ability to perform on his promise to cure America's two stubbornly pernicious dollar deficits.

The sad irony, however, lay in the fact that the more menacing America's trade deficit seemed, and the heavier it made her military burden, the more helpful the food-for-oil barter became to her overall trade operations in the national interest, not antagonistically toward America's trading partners but reciprocally. Moreover, the simplicity and speed of food-for-oil barter mechanics improved the chances of reversing foreign penetration of the American market before it broke America and, with her, the entire monied world. The same policy approach, applied across the spectrum of dealings responsible for the rest of America's trade deficit, would reduce it to manageable proportions and, therefore, the uncomfortable bulk of her interest payment drains abroad as well.

The Reagan administration repudiated traditional Republicanism when it rewrote history and put the entire blame for the 1929 Depression on protectionism. This about-face of Reagan's revived memories of a name widely known and venerated during the Roosevelt era: Cordell Hull. The approach I recommend resurrects and modernizes the philosophy of reciprocal trade implemented by Hull in 1933, when he took office as Roosevelt's secretary of state. His tireless battling against punitive tariff protectionism and his dedication to free trade had earned him a reputation for statesmanship while he was rising to national prominence, first on the House Ways and Means Committee, later representing Tennessee in the Senate, operating on the side as chairman of the Democratic National Committee and as its spokesman against Smoot-Hawley. His appointment dramatized FDR's decision to make the most of public resentment against the "closed door" Republican tariffs of the 1920s.

By the time Hull took office, however, trade movements were paralyzed. Hull had to settle for what he could get: a halfway house in the form of necessarily small-scale reciprocal trade agreements with countries close to home. He signed up twenty-seven of them in short order, arranging to revive two-way traffic on a "most favored nation" basis. While Hitler was shutting down Europe and Japan was building up its Greater East Asia Co-Prosperity Sphere, reciprocity worked in bilateral form for America as well as anything did during those unsatisfactory years. It put Hull on a pedestal among Roosevelt's appointees, beyond the reach of partisan attack. Reagan's political success demonstrated the salability of FDR's slogans in the face of the studied judgment of Democratic advisers that the country had moved into a conservative mood. Reviving FDR's call for reciprocal trade would be winning politics as well as workable economics. America needs reciprocity again. I recommend its renewal on a large scale with the entire world.

A note is in order on the findings and fears formulated by America's most eminent historian of the tariff during the years when Hull was mounting his crusade against it. His name was Frank Taussig, and he held a distinguished chair of economics at Harvard.

Hull venerated him, and free-trade opinion accepted his counsels of moderation in judging the controversies that dominated America's tariff history. To begin with, Taussig, a dedicated free trader, knew too much about that history to denounce tariffs outright as a blight in America's history. Instead, he acknowledged the legitimacy of the tariff principle, handed down from Hamilton through Clay to Lincoln, on the grounds that the infant industries a young country needs to grow cannot flourish without protection. Moreover, with a careful eye to the gestation period that America's burgeoning industries had required, Taussig postulated a fifty-year limit on tariffs as reasonable.

Accordingly, the very moderation of his approach strengthened the case for free trade by his time. So Taussig's opposition to the tariff was limited to protectionism wasted, as he argued, on industries that had matured and developed resources sufficient to enable them to bargain for reciprocity on their own. He took the reasonable enough position that strongly entrenched and richly capitalized industries that failed to recognize the prudence of timely investment in reciprocity would first harden and then rust, as indeed they did. But he always conceived of reciprocity as a lever for increasing America's trade flows in both directions. This is exactly the purpose recommended here; in the tradition of Taussig and Hull, not to exclude imports but to open the way for them to expand with the economy in step with exports. What is new and different about this recommendation of mine is my assumption that the negotiating offices of government to bank and broker—not just to legalize, as Hull did, the terms of reciprocity—are indispensable to program the results that markets can no longer be trusted to achieve in their random way.

Taussig noted too that the tariff battles of the 1920s were marked by a vigorous party competition to win tariff protection for agriculture as an equalizer against the benefits that had been legislated for industry. Under cover of America's overall trade surplus in the 1920s, American agriculture had suffered a trade deficit on which it blamed most of its troubles. Hoover, an outspoken free trader during his prepolitical years as a Democrat, had promised to deliver an "equalizing" protective tariff for American agriculture, but failed. A prime purpose of Smoot-Hawley had been to make amends for his failure. In my own initial interchange with the New Dealers during the 1930s, especially with the talented economics team Henry Wallace had assembled at the Department of Agriculture, enthusiasm ran high over the claim that the Roosevelt administration had delivered protectionism more far-reaching than a mere tariff could offer, in the form of the Agricultural Administration Act, which paid farmers to limit their production.

A simple procedure could reactivate reciprocity in the 1980s into a counteroffensive against sources of the trade deficit not attributable to oil and food. The initiative would be for Congress to take. Only Congress could draw the dividing line between present access and statutory reciprocity. In 1987, America's manufactured imports skyrocketed out of all reasonable proportion to expected, let alone historical, ratios. They accounted for a hair less than 50 percent of the manufactured products her domestic plants turned out. True, industrial raw materials were included in the import total, but oil was far and away the major import item; the food-oil swap proposed in Chapter 14 would inject a dynamic new balance into the food-oil relationship; oil imports would swing forward as a viable pivot for food-export expansion. But the saturation of America's domestic market for industrial products—not only end products, but components as well—calls for expanding the principle of reciprocity across the board, instead of limiting it to food and oil.

Foreign penetration of any number of American product lines accounts for much more than 25 percent by volume, as well as by value. (In many product lines, imports account for substantially more than 50 percent, and in some have preempted the entire market.) Moreover, the one conspicuous difference devaluation has made shows up in the sacrifice of foreign manufacturing profit margins. The dollar value of imports has expanded in stagnant domestic markets; that is, imports have increased their share of the American market, and foreign competitors for the American market have revealed their dependence on it by selling more but collecting less. Now suppose America were to activate the reciprocity trip-hammer automatically on any product after imports from all sources had penetrated 25 percent of domestic market share. Suppose also that the measuring rod past which mandatory reciprocity began were volume by unit, not value by dollars.

No suppositions are needed to see that this expedient would make the American economy dumping-proof in every domestic product market whenever the 25 percent import safety valve is hit. But, then, suppose America invoked reciprocity fortified by product quotas instead of simplistic protectionism. Suppose it gave importers of products penetrating over 25 percent of any domestic product market free choice of how to earn credits qualifying for further penetration of that product market: either by investments in productive assets anywhere in America, or by purchases of American products, or both, not necessarily in the same product line. Notwithstanding the political outcry against any effort to require reciprocity on imports, the open-ended opportunity to bring America's trade deficit under control lies on the export side of her trade problem. But any realistic hope of expanding America's exports hinges on her ability to engineer expansion of the size of her potential export markets. Her high hopes for a sustained 1988 export recovery assumed an expanding trend abroad; in fact, shrinkage was fostering protectionism in country after country.

Foreign businesses routinely operate as multinational trading companies—sometimes they are just that—accepting access into any product line market indicated by their marketing needs; foreign sellers will have no problem finding American products or properties attractive for collecting in payment. Suppose, further, that all applicants for import licenses above 25 percent of the U.S. market share for any product were obliged to reveal point-of-production terms of financing for items under application for import licensing. All of them have been using the American banks to carry their receivables from America. This technique of borrowing dollars to "buy import sales" added insult to injury.

American policy-making has been obsessed with a fancied need to defer to and copy Japanese business methods, seen as superior to American business practices. Washington accepts this notion as an article of faith because managements are stampeding to pay for advice on how to do things the Japanese way. American managements have not stopped to ask what good Japanese corporate methods would do them without Japanese government protection as underwriting, or whether the copying effort would be needed if Washington moved to even the balance in foreign dealings with overdue support. Nor have American managements contemplated the cost of "taking the long view" and "sacrificing short-term gains," as advocated by American academics saturated in Japanese lore, without counting on Washington to back and bank them with no thought of profits earned or taxes paid. Yet that's the Japanese way, and only it can support the fabled corporate luxury of "the long view" touted as Tokyo's copyright.

Back in the real world, the practicalities suggest that America need only set equitable terms for buying access to sell goods in her markets and the world will willingly meet them. But not until importers overrunning U.S. markets are obliged to pay for the privilege will the trade balance tilt back toward America's favor—not necessarily providing a surplus, but at least reducing America's severe imbalance at a steady rate to relatively nominal deficits that are readily financeable. "Paying for the privilege" involves no punitive measures to keep foreign goods out of the country. It calls for incentives to persuade them to use their dollar "take" to take American goods for the privilege of bringing in their goods. Yet precisely because the rest of the world has been cashing in beyond reasonable expectations on the import expansion of the American market, cutting the trade deficit to zero overnight in a world downtrend would have the effect of slamming on a car's brakes at one hundred miles per hour.

This suggested formula for reciprocity would not expose the world economy to any shocks. Certainly it would impose no ban on imports, or even on their growth. On the contrary, it would leave America's competitors, all dollar-rich, free to decide how many of their dollars to use to buy American market share of what products. It would offer all of them two attractive commercial alternatives: investing the proceeds of their American sales in American assets or spending them on the purchase of American goods. Either course would invite countries bringing goods into America to acquire tangible earning assets instead of wasting money on import licenses. Each would mandate reciprocity from any importer intent on exceeding the 25 percent limit on unrestricted imports for any product. The poorest of these importing countries have been permitting their own local *nomenklatura* to take dollars out in mind-boggling lumps.

In addition, this procedure would serve as a sharp reminder that America is the only megamarket offering cash in the one currency usable anywhere. For another, import license application forms, designed to emphasize the reciprocity feature, would serve as a fine stimulant for American exports—certainly more effective than dollar devaluation, because they would condition foreign governments, which habitually use their subsidy weapons to buy into American markets for their products at any cost, to begin reciprocal buying of American products in America.

All the export-dependent economies doing well in the American market are import-dependent themselves. All of them operate on notice that they are expected to do the buying they need in the countries where they do the selling they need; that is, in all their markets outside America. Outside America, they buy all that is expected of them. They would in America too, if America were not ignorant about how business is done abroad, and if she were not hangdog about asking customers dependent on her for their exports to buy her wares.

The moral for U.S. export policy is clear: let America stop relying on cheaper dollars to move more of her exports in a world that is increasingly nationalistic, and, therefore, unmoved by mere price considerations. Unless America develops policies calculated to exert leverage on potential customers in constant or even rising dollar terms, she will continue to collect less for shipping more.

A requirement to buy American, or to invest in America in amounts above the trigger set for free imports, would activate a simultaneous rise in exports and in domestic capital investment by importers. Plowing back import income into U.S. exports would be effective. So would a step-up in foreign capital investment in America and in joint foreign and American capital investment in America. The economy does not know or care which passport any investment dollar carries. But a switch—from U.S. cash outflow for all imports to U.S. cash retention from reciprocal investment or exports—would cut the trade deficit.

Contrary to popular prejudice, foreign investment in America is beneficial, but dumping is not. The simple expedient of sanctioning foreign investment as an offset to shutdowns and layoffs caused by dumping would quiet the political outcry against "foreigners buying up the country." Moreover, a follow-through effect within the American economy of a requirement on importers to buy American and/or invest in America could not fail to raise the level of Treasury tax collections without any need for tax increases.

By way of an extra dividend, U.S. export prices would tend to rise with U.S. export traffic as the rest of the world moved to protect its stake in selling to America. This procedure would shift the burden of market protection onto America's competitors by obliging them to buy shares above 25 percent of every product market they regard as important. It would free America to exploit trading advantages, always the best side of the table to be on. This procedure would promise the longest life expectancy for a recovery and the strongest investment follow-through. Investment—no matter whether by foreign or domestic capital—is the source of all significant leaps in incomes, and therefore, in tax revenues, which increase when earnings do.

16

ROSENWALD'S RESOURCEFULNESS
Paybacks for Taxpayers

The Treasury has jammed the revolving door that is supposed to move younger people into the taxpaying work force faster than older people move out of it. The U.S. Tax Code it administers bears a good deal of responsibility for the resultant squeeze on the Social Security system. The IRS imposes a punitive tax on legal employment, split between contributions from employers and withholdings from employees, and invites avoidance and evasion. This malfunction escaped the scrutiny of the authors of the Tax Reform Act of 1986.

The incentive the Tax Code gives to evasion establishes the Treasury as a principal source of its own revenue shortfall problem. The solution calls for creating incentives among the work force to seek, accept, and/or increase legal employment, subject to withholding. Withholding has been imposed as a cost of working. Yet the justification for it is as an investment. A cost is gone forever—in the case of income withheld, without ever having been enjoyed. But safety is the first attribute of an investment in a governmental instrument; a minimum calculable cash return is the second.

Holmes erred on the side of casualness when he exclaimed that he bought civilization with his taxes: he took too much for granted and not enough. He would have improved his credibility over time, as befits a sage, if he had explained that he had invested in civilization with his taxes: a chancier but more attractive prospect. But his own orientation was too amateurishly uneconomic for him to have grasped the difference. Half a century later, neither the government nor its wards and masters in the economy did either. Entries into the work force resented withholding for taking too much; retirees resented benefits for giving too little; and the country was taught to envy Japan, whose working people "participated" with no return to themselves. Americans will never participate in a social contract offering them no shot at a windfall. But they will always roll the dice for a payoff with a guaranteed minimum, subject to no maximum.

When Social Security began, continuous pay increases for newcomers to an expanding work force were expected to give the revolving door its first powerful self-accelerating push. A half century later, as increasing payouts became a problem for the funds, the coincidence of equal opportunity for women and for members of minorities was counted on to provide a continuing improvement factor. Accordingly, white male beneficiaries of Social Security, cherishing the prerogatives they grew up with, grew old increasingly dependent on the equal opportunity for women and minorities they resented to generate the withholding flows they needed to collect in peace.

Though Gompers had grunted early in the twentieth century that *more* was the one word summing up what American labor wanted, fifty years later, it remained the one word describing what Social Security needed to provide. But the statute limiting government investment of Social Security funds to Treasury securities has failed to earn monthly stipends sufficient to cover the escalating needs of the exploding population of beneficiaries. Retirees could not live on the allowances that the Treasury could not afford to dole out to them, but that the consumer economy could not afford for them to lose: geriatric spending has been supplementing yuppie spending to give the consumer economy what flickering semblance of stability it has retained. But oldsters, defying the stodgy projections in the mortality tables, have been retiring earlier, dying older, and suffering incapacitation longer and very much more expensively. Not only have more retirees been collecting more Social Security entitlements by the year over longer spans of years; all of them owe much of their increased longevity to the health bills they have grown accustomed to split with the government.

Nevertheless, social insecurity has grown with the exponential expansion in the size of the Social Security funds and in the increase of benefits paid out. No actuarial projections for Social Security ever took into account the possibility of one population explosion into the underground economy and another out of the work force altogether, let alone of premature retirement due to severances resulting from shrinkage. So the Social Security Administration has conducted business as usual while members of the work force have detoured around the taxpaying economy into the underground cash economy on a continually rising scale; withholding costs create irresistible incentives for "temping" and moonlighting. On top of all this, minority unemployment and underemployment, plus continuing segregation in low-paid jobs, has aggravated fiscal as well as social distress.

The problem confronting the U.S. Treasury in the fiscal crisis of its maturity has crystallized as a throwback to the problem that confronted it in the fiscal crisis that complicated its birth. The annals classify each as a crisis of money, but both grew out of a crisis of strategy: how to attract money from outside the system. No brave talk about big spending cuts or tax increases will substitute for massive injections from a new source as the only effective third alternative.

Nor will rushes of optimism inspired by recalculations for projecting jumbo surpluses for the next generation. As fast as future surpluses are sighted, pressures to raise present payouts will gobble them up. Projected surpluses are no substitute for the country's need to work the money in the Social Security trust funds harder—and, for that matter, more safely—than they can work in nothing but Treasury securities.

From its inception, the noble dream of freedom from want and fear that Roosevelt's New Deal proposed has suffered frustration. FDR's flair for political arithmetic was at least a generation ahead of his feel for economic arithmetic. Political animal that he was, he was hypersensitive in anticipating money problems brought on by the insensitivity of others to people problems. He persuaded Congress to buy a package of idealistic aims financed by conservative means. Their most effective selling point was that the ambitious Social Security program would cost the Treasury no money, yet spread future trillions among its beneficiaries by the simple expedient of transferring dollars, through trust funds, from a work force always larger than its predecessors in retirement. The Treasury was designated as the custodian of the money the work force accumulated.

The timing was uncanny. Voltaire, who had a shrewd eye for money as well as for satire, might have applied his devastating irony to the script. Investment had won a bad name in 1929. People insisted on safety after they had lost their money. But the government, asserting its ability to care for everyone from "the cradle to the grave," was demonstrating its inability to manage its own books. Its opening whirl with Social Security threw the budget into an unanticipated surplus, and the abrupt withdrawal of cash threw the economy into an equally unexpected deficit. Before and after this shaky start, however, Roosevelt needed an abundance of new Treasury paper to accommodate the prodigious investment appetite projected for the Social Security trust funds. Thus, the deficits accumulated, first by the prewar New Deal, then by the war administration, served a dual purpose: providing the new trust funds as a ready, steady captive investor, satisfied with rock-bottom rates, in the Treasury's continual borrowings, and providing an equally ready, steady spillover of payroll cash, withheld for the trust funds, in need of eligible investment.

If ever there was a case of an economic myth imprisoning the political mind, the limitations on the Social Security system illustrate it. The decision to freeze assets paid into the Social Security system entirely in Treasury securities was made when interest rates were at their lowest point of the half century to come and fated to remain frozen in that nominal range through two war booms, both good for stocks.

The stock market was crawling at a low in 1938. Dividend yields, accordingly, were high and rising, thanks to the combined stimulus of renewed spending and preparedness programs on that smaller economy. Moreover, the key spread between cash returns on stocks and bonds hit an all-time high, favoring stocks. As late as 1942, when the stock market took its first realistic peek past the bloody business at hand and saw the great postwar bull market, AT&T was yielding more than double bonds in a 2.6 percent bond market. The Dow Industrial Average was yielding somewhat above 5 percent.

Suffice it to say that during the first long leg of the postwar bull market, income earned by the Social Security trust funds on their bulging portfolios stagnated on the wrong side of the yield spread. The trust funds missed the incremental benefit offered them by the yield premiums that dividends offered over interest from 1938 to 1958, when bonds started to yield more than stocks; though the yield spread between the two remained narrow enough to ensure the investment attractiveness of stocks over bonds all through the 1960s.

During that long twenty-year stretch and its ten-year aftermath, the young and hearty bull remained free to roam the investment range without risk of more than very temporary loss and without dependence on throws of the speculative dice. Nevertheless the trust funds operated subject to the officially imposed handicap in the race to keep up with the needs of the people to whom the money belonged. Under cover of the sense of security the country won from going back to work, winning the war, and enjoying the subsequent era of affluence, it lost the opportunity to put its money to work as well as it could have. The stubborn grip of the 1929 bust inhibited the money earned during the midcentury boom from maximizing the opportunity to work for the people who paid it in.

Social Security would be better off if some small fraction of its trust funds had been allowed flexibility for ongoing participation in the benefits that investment-grade stocks showered on their owners. Instead, a sterile marriage of convenience started by relying on the growing bulge in the Treasury's deficits to protect the trust funds from running into a shortage of Treasury securities to hold. But by the time the marriage was cracking up in the 1980s, the trust funds had failed to carry out their part of the bargain. The Treasury, in spite of its control over them, had run out of independent domestic buyers with ready and willing cash bids for its offerings.

No computers can calculate how much higher Social Security benefits could be on a funded basis if 5 to 20 percent of the money put into them since 1938 had been put to work in good stocks, and free to earn the dividends reinvested from stock splits. Nor is there any need to calculate where the overriding commitment to "safety first" has landed Social Security. Professors Richard E. Neustadt and

Ernest R. May, of Harvard, in their valuable commentary on presidential decision making, *Thinking in Time: The Uses of History for Decision Makers,* focused on the Reagan administration's management of the Social Security crisis that greeted it in 1981. As a portent of its dedication to fiscal responsibility, it improvised an emergency expedient: "tapping general revenues." Neustadt and May explained that "the conservatives opposing this in principle were forced into connivance with the liberals who liked it to find means of doing it in practice without saying so." This bipartisan consensus missed a simple solution: to increase trust fund earnings with dividends and gains. Neustadt and May, being clinicians and not strategists, did not consider this possibility either.

Even before the successive collapses of bonds, the dollar, and stocks in the late 1980s, the rate at which the Treasury debauched the credit markets with its fixed-interest offerings made an urgent case for the use of stocks by its trust funds. The case is reinforced by the government's loss of budgetary flexibility. If the merits are not sufficient reason for considering this recommendation of mine to end the ban on common stocks in the Treasury trust funds, subject of course to sensible guidelines, the example of Japan may be. Tokyo can do no wrong by accepted American analytical standards. Its official trust funds are permitted to use common stocks, even though Japanese dividend payouts are negligible: prima facie evidence, according to the new wisdom, that imitation would represent progress.

The speculative blow-off and bust of 1987 confronted the government with a make-or-break challenge to combine its two functions: managerial responsibility for the economy through the budget, and fiduciary responsibility for maximizing the performance of its trust funds on behalf of its beneficiaries. Standard Treasury practice, from the inception of the funds, has abused them by using them to hold its securities presumed to be distributed to independent third parties in the open market. But this administrative expedient has defeated the Treasury's larger purpose. It has developed a desperate dependence on its managerial ability to inspire high stock prices and, even more, to inspire expectations of still higher stock prices that will give a lead to the economy.

Yet it has never tried to engineer the same result in its fiduciary capacity. Instead, it has gone along with the fiction that the Federal Reserve Board manages the economy from its supposedly independent bastion, even though the Fed is the Treasury's agent for the management of debt, which has itself inflated into the pivotal factor in the economy. As Chapter 17 will show, negotiation of the terms of foreign debt placement with sovereign creditors of the United States, which only the Treasury can handle, is a vital ingredient in any realistic strategy for reduction of the budget deficit.

Successful Treasury debt management in lowering the deficit would also lower interest rates and improve the economy. But lower rates would cut trust fund earnings to beneficiaries, while improvement in the economy would increase cash inflows from withholdings into the trust funds. Given the traditional rules of the road governing Treasury management of the Social Security trust funds, high marks for discharge of its responsibility for the economy would earn it low marks for its responsibility to the beneficiaries of the trust funds. A no-win game for payers of withholding and collectors of benefits. Their frustration reflects the practicalities as well as the equities of the "double-dip" game the Treasury has played. Too many innings lost for the beneficiaries would foredoom future innings as lost causes for the economy.

Successive Treasury secretariats have had eyes only for ensuring orderly markets for the float of outstanding securities and the sale of new issues. In time-honored tribal fashion, one Treasury team after another, divided by political ideology but tied together by common superstitions of ancestor worship, looked on in awe during the good decades at the spectacle of common stocks funding benefits for corporate pension funds and showering liquidity on corporate sponsors as well. But the tug of tradition has dulled Treasury incentives to try what has worked in the responsible management of corporate trust funds.

The taboos on stocks have prevailed; the Treasury's sorry performance with bills and bonds has persisted. So has the sharp bite of the problem: Social Security benefits insufficient when received, but provoking a storm if threatened, and failing to attract holdouts from the system into the work force. Hence the need for a new social contract: to buy the participation, on a mass scale, of people put off by withholding. The New Deal and the Fair Deal were battle cries promising protection to the underdog. Fear was their target. But greed is the problem posed by the underground economy. It was also the driving force blessed by Reaganomics. In fairness to this premise, a rich deal would be needed to provide, not only meaningful inflows for cash from the work force into the trust funds, but also meaningful outflows to satisfy the greed of the new participants the system needs yet fails to attract, as well as the escalating needs of beneficiaries.

The legendary wolves of Wall Street still loom larger than life in the chronicles, but they were small potatoes alongside the institutions that took over anonymous sponsorship of the great bull market that started in midcentury. The institutional market made the coups of the old-time Wall Street pirates look like a series of penny ante forays. But it too lost its luster once the institutions allowed their portfolio managers to debase the "cult of performance" into a high-speed crap game too erratic for even professional speculators to win, and too erratic for professional investors to play.

The government has influenced the workings of the stock market with direct cash contributions before. Its emergency $1.4 billion in loan guarantees to Chrysler in 1979–80 picked the entire market up by its bootstraps when it was falling apart. But forty years before, with great fanfare, the government negotiated the buy-out of key electric utility operating companies. The price TVA paid set a basis for valuing all electric utility operating companies owned by the controversial holding companies with which the New Deal had been at war. These yardsticks, in turn, enabled prudent investors to comparison-shop for hidden values lurking behind the depressed price quotes for the preferred stocks of the utility holding companies.

Roosevelt had pressing political reasons for putting a high price tag on one of them: Commonwealth and Southern. Wendell Willkie was using his grievances against "TVA socialism" as a soapbox for his presidential campaign. FDR could not have cared less about how much he had the government overpay for Willkie's properties; nor could he have known less about how his move to put Willkie out of business as a candidate was about to bring the stock market back into business.

At the generous prices Roosevelt had the TVA pay for Commonwealth and Southern's rundown properties, every utility preferred stock was quickly seen to be worth accumulating for the break-up value of its properties; therefore, the accumulation started. The utility preferreds gave the nascent bull market the indispensable lead any bull market needs. Every equity-based portfolio for at least the next fifteen years started with a 25 percent nucleus of prime utility operating stocks distributed as the SEC forced dissolution of the holding companies. Social Security was already in operation by the time the New Deal underwrote this bonanza for utility investors. Its ultraconservative limitation on "Treasuries only" for its trust funds penalized the beneficiaries because one official hand did not know what the other official hand was doing.

An instructive experience of mine in the 1950s that helped whet the institutional appetite for stocks comes to mind as a model of how the Treasury can make up for lost time in the 1990s. I had been assuring the more venturesome institutional managers in my orbit that Boeing and Armco had become as safe as Treasury bills, but immeasurably better paying and more profitable to hold. My unrestrained bullishness for stocks led me to be dubbed "the astronomical statistician" by the foot-dragging consensus of institutional managements.

In 1953, an institutional investment firm invited me to a gathering at the mecca of the investment fraternity, the John Hancock Mutual Life Insurance Company. I made an impassioned exposition in favor of unprecedented highs for stocks that I foresaw in a continuing bull market. A speaker can generally tell when an audience is receptive to an unfamiliar message that, though supported by unorthodox reasoning and evidence, produces a welcome recommendation.

Confirmation came promptly and incisively the following morning with a gratifying endorsement over the phone from the executive vice president of another major Boston insurance company, the New England Mutual, crowing that on the basis of his summary of my presentation his investment committee had given him an unqualified go-ahead to add 10,000 shares of Telephone and 10,000 of Standard Oil of New Jersey to his quarterly bond buying list—a whopping zero-plus of its huge investment portfolio.

Even so, baptism in the new institutional faith in stocks started with this very gradual immersion. My convert at New England Mutual did not try to persuade his colleagues to switch their entire portfolio from bonds to stocks. He settled for their willingness to spread minuscule sprinklings of just two bluest of the blue chips into their company's massive bond portfolio. One good reason why this token move produced such rich dividends is that the sums that were insignificant in the institutional bond portfolios added up to big money in the stock market—not merely for the two stocks involved in the initial experiment, but as a prelude to institutional buying for all stocks.

A remedy for the country and its flawed system in the 1990s will be for the Treasury to grasp the same cautious insurance policy that this cautiously progressive insurance company adopted for its investment portfolio in the mid-1950s: to mix just a small fraction of top-rated common stocks with its bonds in order to raise its overall net return by giving its trust funds some chance of appreciation.

The procedure for the trust funds to follow was pioneered in the early phase of institutional stock accumulation. One of the fathers of modern American capitalism embodied it in his will. Julius Rosenwald built Sears, Roebuck and made retail selling on the installment plan respectable. He wanted control of his company to pass from his heirs to his employees. As a young man, I was privileged to observe the workings of the mechanism for doing so devised by the manager of the family funds.

Goldman, Sachs was employed, not to make judgments about likely market price trends, but simply to compute the median price of transactions in Sears stock each month. In an arm's-length transaction, once a month, the family manager sold a bloc of Sears to the company's funds at the median price, independently computed. This systematic transfer of control never disturbed the market, though the transfer was public knowledge. As it happens, a successor to the Rosenwald family fund manager subsequently took over the management of Ronald and Nancy Reagan's personal investment funds. The fund managing operation trusted to carry out the will of Julius Rosenwald by starting the experiment that transferred control of a prime blue chip into employee trust funds without roiling the stock market was selected to protect the interests of the Reagans: a politically convenient irony that gives my procedural recommendation ecumenical credibility.

A degree of quasi-official fortification, or at least comfort, for the thinking behind my proposal came from the new Federal Retirement Thrift Board late in 1986. Its purpose was to offer federal employees a broader choice of investment alternatives for their pension plans than just government securities, although all its start-up contributions, amounting to over $1 billion, were still in Treasury

344   DISTINCTIVE SOLUTIONS FOR AMERICA

bonds. *The Wall Street Journal* suggested that this new retirement fund was on its way to becoming the largest in the country, offering participation to the 2.2 million employees participating in the Civil Service Retirement Plan. It invited competitive bids from pension fund managers for strategy proposals and awarded the entire contract to Wells Fargo on the basis of its performance.

The *Journal,* in reporting the award, described the investment arm of the San Francisco bank as "the leading provider of U.S. index funds." On the day of the 1987 crash, Wells Fargo sold no less than 10 percent of all stock index futures traded: reason enough to earn top performance standing. An operation such as Wells Fargo, gaited to speculate in index funds, runs on a faster track than one content to hold Treasury bonds through thick and thin. Without questioning the merits of the award, or the qualification of Wells Fargo, a happy medium would call for a simple mix of investment grade stocks with Treasuries, in line with my proposal, and this does indeed appear to be the strategy proposed for this new government employees retirement fund.

Of course investment, not control, would be the guiding purpose of any Treasury trust fund accumulation of stocks; certainly, any diversion of trust funds for operating projects, especially in the inescapable risks of real estate, would be outlawed. Moreover, the Treasury could not launch the same experiment on its necessarily larger scale by employing any single investment from a group of firms for guidance, nor could it execute market orders on the stock exchange floor for the trust funds. The same ban would apply to any procedure smacking of "chosen instruments" in the selection of stocks the funds might own. Stock tips could not be tolerated. Direct dealings with most major pension funds certified by ERISA would set a net median price for monthly acquisitions of predetermined dollar amounts of stocks.

Happily, authentic investment application of the modern technique of indexing a portfolio and direct buyer-seller transfer of entire market baskets of stocks can eliminate concerns over manipulation and favoritism. Market baskets simulate the popular averages in precisely the same proportion as the weighting assigned each stock in each average. Hard cash buying to hold for income would sanitize the stock market from the institutional speculation in financial futures that contributed to the crash of 1987.

Bears never kill a bull market. Only mass suicide by speculative overdose kills off the bulls. Speculative exploitation of market baskets of stocks exposed the market to its 1987 crash. During the doldrums of the 1930s, the Wall Street operators of 1929, with a shrewd and powerful lead from the New Deal, figured out how to convert market baskets of stocks from suicide swords into plowshares needed to plant profitable salvage values in the burnt-over soil of market wreckage. These instruments were none other than the utility holding companies, and they played the same speculative role that their computerized descendants did in the 1980s.

Luckily, the utility operating stocks that filled the holding company market baskets survived to give the next bull market its healthy start when investors accumulated holding company preferreds as cheap options on the operating stocks. In the late 1980s, when trading the averages killed the appetite for direct investment in the stocks represented in the computations, the market-basket technique was needed to fill a constructive need again.

Injection of common stocks into the Social Security trust funds on a conservative schedule, and subject to prudent safeguards, would free stock selection from the rumor mongering and self-dealing that corrupted the takeover game; market-basket accumulation of the quality averages would be automatic and continuous, with no latitude for playing the market. Selling would be automatic too, mandated on the rare occasions when the average-compiling concerns (Dow Jones or Standard & Poors) eliminated a stock or when the dividend yield on an accepted average fell below a predetermined norm—say 3.5 percent (3 percent being a historic bust point). Schedules of the sums of money to be invested and the numbers of shares to be accumulated in each quarter, as well as the dividends receivable, would be publicized beforehand and afterward. So would Treasury announcements, subject to the appropriate congressional committees, of the tiny fractional percentage of trust fund resources earmarked in that period for investment in stocks (preferred as well as common).

Investing a minuscule fraction of trust fund resources in stocks would expose them to less risk than they routinely run from exposure to interest rate fluctuation but would make an enormous difference in their performance. The market would quickly make a bullish adjustment to anticipate regular trust fund accumulations of stocks in direct transactions with institutions off the market—just as it did when the pension funds first adopted their stock-buying schedules. Step-ups in government trust fund accumulations would have the same bullish effect. Brokers would willingly forgo commissions on government trust funds as a trade-off for the business a revived bull market would bring them from commission-paying customers.

Actually, institutional business is seldom profitable for brokerage firms. On the contrary, competition for it lures them into the speculative operations that again and again cost more than they could earn. Dealing with the public is where the profit is. No purely institutional investment in stocks can restore trustworthy vigor to the stock market, not even with the Social Security trust funds adding a new dimension to previous yardsticks of institutional leadership. Only a renewal of preference for stocks among the income-minded investing public can start the next bull market.

No doubt the public's first round of long-term reaccumulation would favor the blue chips, once they pay acceptable yields again. The brokerage business went broke chasing institutional business. It will come back to life when the stock market does; the stock market will revive when the investing public returns; and the investing public will do so when it sees a Hamiltonian solution for Rooseveltian priorities. No doubt, however, the first commitments would be to the big stocks the trust funds necessarily favored. Nevertheless, the yield-minded public would resume the stock habit by subordinating yield to quality.

Then too, a decision to permit the Treasury's trust funds to own common stocks would provide a buyer of last resort for institutional funds under pressure to dump blocs of stocks onto a bidless market. Historical precedent can be found in the Investment Company Act of 1940, when Congress provided the SEC with an escape hatch in the event of market collapses, although this provision was for mutual funds. While it did not put the mutual funds into business, it legitimized them and provided a firm and conservative launching pad for their subsequent takeoff. Within two years, the stock market got the message and took off on its long cycle-spanning climb.

Frank was chairman of the SEC; Douglas was on the Court, though on continuous tap for avuncular consultation; and the Corcoran-Cohen team shuttled between the two. Senator Maloney remained the SEC's legislative sponsor and, along with Douglas and Corcoran, retained the closest of ties with Joe Kennedy, then convinced that Britain and the market were both about to collapse. This cast of characters persuaded Congress to authorize fund managements beset by redemptions in a market collapse to make direct distributions of their portfolios in lieu of cash to shareholders. Such a fallback procedure envisioned the pro rata allocation of fractional shares of all stocks to fund shares held. The precaution was taken; the need for the mutual funds to use it never arose.

But the crash of 1987 did provoke fears of the forced liquidity for pension funds on a vastly larger scale and put the market in real jeopardy of the false alarm anticipated for the mutual funds in 1940. The funding problems looming before the corporate pension funds were beginning to mirror those of the Social Security funds; hirings and pay levels were falling and, therefore, so were corporate contributions to their funds, while severances and retirements were accelerating disbursements. Some major corporations—notably United Airlines—were liquidating their pension funds entirely in order to supplement their cash availability. The entire Fortune 500, confronted with huge write-downs in portfolio values and sharp accelerations in severance payments, were obliged to charge supplemental contributions against operating earnings. Corporations with funding problems began to substitute their own stocks for cash in making contributions to their pension funds. Others, fearing takeover raids, used their credit to finance contributions to their pension funds, reducing their supposedly independent fiduciaries to the status of captive proxy signers while inflating their own interest charges.

The preliminary "Boesky bust" of November 1986 provides official confirmation that the speculative practices of the institutionalized market had exposed it to a panic. Boesky was just one operator, with less than $2 billion, and yet the SEC braved criticism by allowing him to liquidate in secrecy "to avoid a panic." It then accentuated this market sensitivity by giving him an additional eighteen months to withdraw from all his institutional entanglements. Imagine what would happen if the entire institutional fraternity, with its portfolios of the same stocks and its portfolio managers conditioned to stampede, tried to dump a trillion dollars worth of stocks on a market they dominated! As it was, only incisive intervention by the

New York Stock Exchange to stop program trading at noon on the second day of the Blue Monday rout of October 1987 reversed the market into a rally.

As early as 1913, Congress had empowered the Federal Reserve Board to intervene as lender of last resort in a banking emergency, though at that point its role, while indispensable, was merely defensive. As a result, the bond market plays with an official backstop, while the stock market does not. But the corruption of the stock market, under official auspices, into a high-powered gambling casino has created its need for a buyer of last resort. Congress can give the Treasury the power to do more than hold the line in a financial emergency. Freeing it to mix a modest fraction of stocks with its bills and bonds would turn the Treasury into a pivot for reversing a market bust into a business boom.

So far so good. But saving the stock market from its folly is not enough of a solution to pump the energies of the economy through the mechanism of government, and to accomplish the country's overseas financial purposes. The larger goal outlined here calls for the trust funds to use the stock market to provide regular annual step-ups in payouts to beneficiaries and to sell income earners in the underground economy on the incentives of going legal: paying taxes as a trade-off for invaluable perks, beginning with old-age benefits, and including health insurance as well as child and geriatric care. In the summer of 1988, health-insurance premiums jumped some 27 percent across the board.

Hamilton had discerned, as is detailed in Chapter 2, that the states ratifying the strange new federal sovereignty neither could nor would put the money up themselves to fuel the new system. Furthermore, the new sovereign lacked the military muscle and will to overrun opposite numbers abroad with classic rapacity. As the Louisiana Purchase showed, takeovers, far from netting money, need financing. So Hamilton pursued the obvious in desperation when he appealed to the greed and insecurity of the loyalists to the Crown. They had the money, and he gave them the motivation, to buy into the system and to invite their principals overseas to follow. When Hamilton made the loyalists an offer they could not refuse to buy a piece of the system, the fiscal problem was solved and, therefore, the problem of sovereignty was too. Neither was—nor is—manageable without the other.

Stocks that perform as cash machines create their own free advertising. Public relations penetrated the underground economy in Wall Street's 1986–87 heyday. Then, when the tape was late and attracting new customers in droves, Wall Streeters regaled one another with tales of immigrant customers from around the world who were unable to fill out forms in English or unwilling to supply information on bank accounts, but who plunked down $50,000 in cash at a time. The trick, today and always, is to persuade newcomers to the system, as well as exiles from it, to route their earnings through it in response to an offer that is more than merely better: that is distinctive, as only the government of the United States could make it, because of the dimensions and dynamism of the New York stock market.

My proposal would activate a four-step chain reaction. The first phase of marginal stock investment by the trust funds would revitalize the stock market and equip it with a new dimension of stability. The second would transmit the revived momentum of the stock market to the economy—immediately, by relieving corporations of their considerable concerns for the liquidity of their pension funds and gradually, by satisfying them that a stock market pointed on an upward course without speculative excess signaled the all-clear for higher employment, pay, and dividend levels. The third—resulting from bottlenecks, especially (but not only) of skills—would press employers to pay more to attract people otherwise tempted underground part- or full-time. The fourth would send Treasury collections surging with the renewed cycle of expansion. The trust funds would get the stock market moving; the stock market would get the economy moving; bottlenecks would get employers to pay more and business to buy more as well as invest in more capacity; and the economy would get the Treasury to collect more and refund less: a self-accelerating cycle of prosperity. Reaganomics at its height stopped short of the third and fourth payoffs to the system.

Once the trust funds were seen to enjoy continual leaps in independent earning power, a free-for-all to share in the increased cash benefits would begin. All the pressure groups in the system would be scrambling for more—for the aging, for health, for agriculture, for the unemployed (whose numbers would, for a time, increase with the economy, as an honest realistic count would acknowledge the "nonpersons" earlier dropped from the record). Only exiles in the underground would miss out—unless and until they opted to come back, or into, the system. None of the increased benefits would load the budget or impede the scheduled participation of the funds in the stock market.

Admittedly, however, another managerial dilemma would confront the Treasury at the outset: finding the middle road. Too cautious a start-up commitment to stocks would miss the opportunity to combine the market turnaround with the buy-in and lose precious time in converting the vigor of the country's underground economy. Too aggressive a chase, such as forays into options trading, could jeopardize the effort and compromise the method for years to come.

A middle course would aim, for example, at achieving a 1 percent participation in stocks in the first year. No more than $1.3 billion would be involved in the first year, a token exercise in diversification from trust fund holdings in Treasuries, and a token entry into the buy side of the stock market (representing, to take just one conspicuous yardstick, somewhat less than what the market, on its upward surge, absorbed from Boesky when he came under fire). In 1953, New England Mutual's nominal investment in just two blue chips demonstrated the powerful ripple effect that any token start at institutional buying is likely to have on a disturbed market, even when a business slump has not materialized. Such a start would support improvements in benefits where the need is most urgent, where fear is most crippling, where any bonus makes the biggest difference, where the demographic burden is least costly, and where the distribution is entirely uncontroversial: among the aging. Extra dividends in benefits from stock market performance in an expanding economy, distributed in the first year to the oldest group of Americans and spread in subsequent years as each lower age group was covered, would achieve maximum sales impact with minimal dollar output.

Once the stock market won credit for cooking and passing out really nourishing free lunches through the compounded dividend earnings and capital appreciation of the Social Security funds, refugees from the system would beat the doors down in their eagerness to get in. Fringe benefits, not just for health, could ice the cake. Once national economic policy tackled the twin problems of the deficiency in the Social Security trust funds and the dynamism in the underground economy, momentum would take over, as it has again and again so distinctively for America in the past.

Americans like beating the game, and they also like watching other Americans beat the game. But the better they do at beating the game, the more they need and use their government. The stock market started out as the central arena in which individual operators pitted their judgments against events. It wound up as the central rumor mill on the big plans the government had for gunning the market (which it sometimes shoots instead). The customers fed the rumors, some bragging that they caught a hot stock and some complaining that they either missed the last move in the market or were trapped in it. But all are always on the lookout for a new start-up IBM, cheap but topless. Anytime the government parades a bandwagon, they will jump on it. Wall Street is their favorite parade ground for cheering messages from Washington, especially if Washington knows how to turn them into economic trends.

17

NUNN'S NUANCE
Inventiveness in Defense

Technology is mediating the ongoing battle over the defense dollar. Changes in methods of weapons design (and in the weapons themselves) enable much less to buy much more—and along the many bastions America is committed to protect and patrol. From the perspective of a mere generation of advance in military technology, the Eisenhower administration can be rated as timid in its call for more bang for less buck. More bang for fewer bucks has become easier to plan than living with the high cost of obsolescence. More money is now buying less military muscle. Contrary to popular misconception, newer can be cheaper and simpler. Nor need it be a dud; it can work. Veblen anticipated this evolution early in the century, for the engineers continually modernize weaponry along with all technology. But vested interests defend their stake in obsolete weaponry behind fortresses of paperwork.

In Reagan's last year in office, his televised pageants in summitry inspired the Gibbonses of the screen to speculate about his place in history. But his marketing of Reaganomics established it at the outset of his term. He created a vocal majority clamoring for more from government, paid for with less from its beneficiaries: a tour de force of radical conservatism. Three simple expedients of his did the trick. He declared defense out of bounds for the discipline of spending; he established tax increases as out of bounds for political winners; and he ruled Social Security cuts out of bounds for political discussion.

Altogether, the Rooseveltian twist Reagan's rhetoric gave to standard Republicanism and standard conservatism offered hard-core Democrats an unmatchable promise of protection in exchange for their loyalties; and they succumbed to his appeal while ignoring its substance. Meanwhile, ironically, developmental military technology was catching up with televised political technology. The sleight of hand that Reagan's political platform could not deliver for expectant taxpayers, patriots, and pensioners was more than matched by the miracle that emergent technology proved able to perform for the country as a whole. By the time the 1988 election campaign swung into high gear, the most stridently conservative constituencies in the country—most conspicuously Orange County in California—were clamoring for more from government (beginning with freeways, financed with borrowings) but vowing extermination for candidates countenancing advice to raise taxes. Yet for the obsolescence-ridden, cost-swollen Pentagon, a switch to the purchase of strength had turned into an opportunity to seize a bargain.

Thanks to the common industrial base shared by weaponry and machinery, the speedup in the arms race has accelerated de facto technological détente. Russia has become too dependent on the economies of Western Europe and Japan to risk the precarious redirection of its own primitive economy by threatening them in any way. In addition, as we have seen, the economies of Western Europe have erected their own peaceful deterrents to invasion in the form of grids of commercial nuclear power plants. Any shoot-out over Western Europe would scorch Russia from the skies with a giant-size Chernobyl reaction more deadly and more difficult to defend against than any launched by Genghis Khan, Napoleon, or Hitler on land. For Russia, the lesson of Chernobyl needs no reminders. West Germany made a particularly shrewd nonmilitary investment in deterrence by locating one potential Chernobyl fifteen kilometers from the East German border, on the unavoidable Berlin-to-Hamburg artery.

This technological standoff extends to conventional weaponry.

The classic rule that the defensive side always catches up remains in force. The overwhelming bulk of weaponry in use by potential belligerents is metallic. So are the plant and equipment for reproducing and improving it. All these assets are militarily obsolete. Infrared detection nullifies their use and makes their reproduction counterproductive, except for the military equivalent of public works to politically entrenched arms producers.

If the trained ignorance that Veblen identified at the root of all managerial evil had not plagued the Reagan White House during its wild-goose chase after SDI, it could have identified the technology of infrared as a practical basis for rapid disarmament. Like any number of modern technologies, infrared is broad-based and indispensable to military and commercial use. Instead, Reagan took the bait of nuclear disarmament that Gorbachev dangled and trapped America into the astronomic cost of replacing not only her metallic arsenal but her metalworking arms plants as well in order to design and deploy weapons made of infrared-proof compound fibers. The Soviets could hardly be expected to sit on their conventional arms lead during the interim. Throughout their checkered career on the defense front, the Reagan ideologues were obsessed with the idea that complexity is the copyright of formidability, but simplicity is.

Three policy aims will command America's attention when she pursues modernization of her conventional weaponry: strength, reciprocity, and collaboration against terrorists. Engineering is the common denominator of all three aims. Although start-up investment in the military engineering of conventional weaponry is unavoidable, it pays rich dividends in commercial by-products. Not that commercial progress is necessarily the justification for military research; rather, America's commercial vigor withered once her military research turned sterile. Whether this sequence was coincidental or causal is arguable. The result is not.

Russia's national emblem, the bear, offers a forceful metaphor delineating the distinction between competition and confrontation. Anyone wanting to get close to a bear is on notice to use a larger approach than that contained in the jocular advice on how to make love to a porcupine: cautiously. So strength, America's first policy aim, is a must in undertaking the risky business of accepting or inviting a bear hug. Moreover, any American effort to establish a continuing, working relationship with Russia will work only as well as the defense it erects against international suspicions and terrorist trickery, while packing the necessary military punch where needed. Politically, the Reagan precedent provided protective cover for his successor to continue negotiations without inviting conservative charges of mush-headedness. Yet stale promises to cut defense fat will not do—especially because cost-effectiveness is now a precondition of the weapons modernization needed to restore the negotiating position of American bear huggers. Lest we forget, the American emblem is the eagle. Where the bear prods and squeezes, the eagle swoops. Mobility aloft is its strategic advantage.

America's cumbersome, aging inventory of weaponry has bogged down her armed forces. The high cost of immobility is eating the Pentagon out of house and home. In translating American-Soviet rapprochement and détente into more familiar terms of bargaining, America would do well to remember that she was the world's dominant power in the area of conventional arms until Russia superseded her. Russia has looked down on America for losing her old lead, and the Kremlin is determined to hold on to its new lead. Reagan's decidedly uncertain call for regaining parity in conventional arms was delayed until it became clear that the Gorbachev deal meant sacrificing parity in nuclear arms, and then it was an afterthought, echoing a stern alert from Sam Nunn.

All this said, Russia nonetheless remains the political equivalent of a familiar kind of potential business partner, wary to the point of paranoia, who demands matching strength as a condition of confidence. Russia is not likely to trust America to help with its problems until America has demonstrated her ability to solve her own problems, beginning with the problem of reengineering the Pentagon, but including the problem of bringing strength to bear in her dealings with Soviet power, while incidentally reasserting her economic and financial strength. Except for the interlude of necessity forced on Moscow and Washington by Hitler, Washington has found arguing with Moscow a ready justification for dealing with the Kremlin's professional apparatchiks from amateurish ignorance, inertia, or downright weakness.

Reciprocity in dealings with all foreign military establishments is America's second policy aim. It parallels her trade deficit crisis. By definition, all other countries, especially Russia, rely primarily on conventional forces. America has invited not only competitors but her allies as well to make too much of a good thing out of her inertia, her unwariness, and her lack of coordinated bargaining stance. In 1987, military imports from Europe accounted for no less than 20 percent of America's shrinking military procurement, with Japan gearing up to accept Washington's offer to get into the game.

Europe has learned the hard way how to grade arms customers: by whether a customer's domestic arms product is competitive; whether new arms acquisitions—domestic and imported—improve the combat effectiveness of a customer's armed forces; and whether a customer's economy benefits or suffers from acquisitions of weaponry. By these interrelated European standards, America rates a demerit as a customer, recommended by the size of its checkbook rather than by the usability of its arsenal and the likelihood of repeat orders. Not that Europeans are given to snubbing eager, well-heeled customers: just that Europe is hardened enough not to trust arms customers who don't know what they're doing but continue paying arms bills.

Israel offers a provocative case history, at once inviting and forbidding, for American defense and fiscal policymakers to ponder. Under siege and on a war footing, Israel proclaims how poor it is. Yet its creditors advertise how limitless and gilt-edged its credit is. Israel, a socialist democracy, indulges its upper strata the conspicuous luxury, shared by all poor countries (including China), of keeping their accumulated cash "take"—guesstimated by Israeli insiders at a minimum of $20 billion—in international tax havens. Israel relies on exports to supplement its special relationship with America in putting its defense burden on a cash-and-carry basis. To be sure, opportunism born of success did lead Israel to jeopardize its security in its dealings with China. Its coup in selling the Silkworm to China boomeranged into its crisis when the Silkworms turned up in Syria. Nevertheless, the analogy remains relevant for America. She need not play the "merchant of death" game, or even expand her defense exports enough to make a significant contribution to her defense costs.

Admittedly, America faces resistance in pushing her allies into burden sharing. Reagan's agreement to fold up America's nuclear umbrella over Europe leaves them exposed to Soviet conventional arms superiority. Still, America has a solid inducement to offer, over and above the assertion of food-for-oil bargaining recommended in Chapter 14. The arms business is the best one in Europe, by any standard, and the fastest-growing. Moreover, it is not only depression-proof but guaranteed to grow when commercial markets go bad. The Reagan administration, by way of making amends to Japan for going along with restrictions on commercial imports, dangled back-door entry to the U.S. military market; however, Japan played coy by breaking into the Russian military market first: a major blunder. America and NATO between them offered the largest single market for conventional arms in the world even before America undertook to modernize her conventional capability.

America's third policy aim is to implement an effective technology for dealing with terrorists. This is an immeasurably more elusive problem than matching muscle with Soviet power or ensuring reciprocity from allied and friendly governments. Terrorism has put every other regime in the world, no matter how relatively progressive or reactionary, in urgent need of the antiterrorist, policing, and patrolling technology being developed, if not adopted, in America.

It is axiomatic that terrorism grows from deep roots in social confrontation and violence. It is also clear that terrorism enjoys a long head start over effective defenses. Terrorists dramatize issues (and pseudo-issues). They appeal to true believers and rivet universal attention. Though the apparatus of terrorism is small, its effects are widespread. Because terrorism can appear almost anywhere, terrorists can conduct one hit-and-run play, then scurry back underground.

Whereas defense is capital-intensive and requires enormous commitments to overhead, terrorism enjoys the highest ratio of results to investments of any business in the world; the worse the oil business gets, the more irresistible the lures of subsidized terrorism are bound to grow. Dealing with terrorists in a planned and coherent fashion is enormously difficult for any organized system of defense. No military blitz, however devastating, can begin to offer the cost-benefit ratio of terrorist strikes, as Iran and Libya can testify. Even Honduras, America's Central American base, erupted into a wave of terror against the American Embassy in April 1988. By midyear, in the five years since 1983, terrorism had registered no fewer than 88 unresisted attacks on American embassies and consulates.

Realism calls for recognition of terrorism as a long-term constant in any society in transition. The good news is that Russia shares America's need for technological defenses against terrorists, not least, as we have seen, because some terrorists are in charge of governments and fortified with nuclear capabilities. The bad news is that trustworthy Soviet-American détente, as pressing a need as any in the world, might incite record outbursts of terrorism against both Russia and America. In my *Prescriptions for Prosperity,* I summarized the proposal Senator Nunn put to President Reagan in 1982 to activate a joint American-Russian command/communications center in space to mobilize emergent technology to defuse large-scale nuclear threats from centers of terrorism. Reagan ignored this bipartisan initiative during his various exercises in dramatics with the Russians. Coping with smaller-scale terrorist activities requires earthbound weaponry at close quarters, beginning with the overdue job of defending America's elaborate worldwide network of embassies and bases and including every major airport and utility operation in America.

Behind these three policy aims, an urgent shift in strategic purpose is overdue. For decades, America has been overprepared for the all-out nuclear war neither she nor Russia dares to start. America has lavished inordinate waste on "high-intensity warfare": military gobbledygook for intercontinental nuclear missilry. America's stance as the fanged dinosaur has distracted her from her need to simulate the capabilities of the wolf, the fox, and the eagle, while displaying the potential for counterterror of the dragon. But her instinct to be the target, instead of to aim at it, has led her to underprepare for the "low-intensity warfare" that she has not managed to avoid. Granted that sweeping changes in Pentagon procurement practices are the order of the day; nevertheless, no changes limited to the procedural level will help unless and until America implements a workable reversal of her strategic priorities into line with the realities confronting her.

America's three military policy aims—strength vis-à-vis Russia, reciprocity from allies, toughness against terrorism—rest on the broad base provided by the American economy. The solutions proposed for the problems confronting America's armed forces will draw on her resources and strengthen her economy as well. Again, not that modernizing American armed forces is needed as a stimulus to the American economy; rather, indulging their unfundable obsolescence will continue to devitalize the American economy.

A word is in order on how America can anticipate Soviet military power, and prevent it belatedly, with disincentives, to play the aggression game. As Soviet economic and financial dealings with Europe grow, the odds will rise against a Soviet military strike to the west; the threat of muscle in the background, combined with the market lure in the foreground, are enough to fortify Soviet superpower status in Europe. But taking the best case of prospects for Washington-Moscow dealings, all the way from markets to space, the odds will also rise on Soviet aggressive operations aimed through Third World surrogates at or near "choke points," as Admiral Alfred Thayer Mahan identified them at the turn of the century, sensitive to American security; the Panama Canal is the most obvious, but the Caribbean, the Mediterranean, and the Bering Strait also qualify. Taken together with the tension between Israel and its neighbors, alternating between chronic and acute, the prospect is that the 1990s will unfold as a decade of wars, but not as a replay of past wars. Instead, realism calls for directing American preparedness against small wars fought with conventional weapons on congested Third World terrains adjacent to choke points, and against terrorism. The Pentagon closed out the 1980s entirely unprepared for wars of either character.

Procurement method. The present method compensates contractors for weight of metal fabricated, multiplied by time spent. But a system obligated to pay billions against costs run up for pounds hammered, times hours spent, is a relic of the passing era of metallic weaponry. It creates incentives to load weapons with heavy, complicated, and fragile electronics. In recent years, new versions of old fighter planes have been redesigned into disequilibrium in the air and, consequently, into dependence on fancy electronics to reequilibriate them, straining the ability of flight personnel to manage instrumentation. Politically, this method invites lobbyists for manufacturers of electronic systems to combine with lobbyists for military contractors in making weapons more complicated than they already are; cost, not use, is the driving incentive. Military contracts are an electronics manufacturer's delight because they weigh so much, take so long to tinker with, and require more time, labor, and money to repair and replace than to install.

Functionally, overuse of electronics inflates the weight and, therefore, the cost of weapons and foredooms them to insupportable requirements for repairs and parts replacements that are impossibly inefficient and time-consuming to support. In the Osprey helicopter under development since 1983, the imbalance between use and repair swelled the number of maintenance hours to flying hours to the ridiculous ratio of 65 or 70 to 1. Financially, the nonfunctional procurement method of paying for pounds times hours has elevated the accounting departments of corporate contractors to the status of supreme command and demoted their engineering departments to mere adjuncts of their accounting departments.

The Defense Department, in its official 1988 Almanac, numbered its total uniformed force at 1.85 million; its civilian work force of 1.2 million was running a respectable second and by no means losing ground. In the breakdown of skills and specialties, only 266,000 were itemized as cleared for combat. Of the other categories, personnel on electrical and mechanical repair duty were counted at 384,000. On top of this formidable shop-centered corps, however, another 172,000 specialists were detailed to electronic repairs: evidence of the extent to which electronification had come to dominate the design and performance of weaponry. Uniformed personnel relegated to administrative and clerical duties numbered just under 300,000. Between the functions of communications and intelligence, plus supply and service, another 300,000 warriors were girded for action. But this bare-bones classification of the insupportable ratio of noncombat to combat troops understated the inflation of military support and overhead costs, because it omitted any reference to the prodigious work force employed by civilian contractors handling the overflow of assignments (not to be confused with contractors producing weapons).

An alternative approach to procurement pricing, based on existing surveillance of standard Soviet military procurement, would reintroduce the old American value: competition. Without competition, the Russians have been building weapons in quantity for years that are superior to America's. Let the Pentagon award contracts on the basis of demonstrated, fulfilled claims to beat, or at least match, Soviet norms.

The pursuit of perfection is a phantom haunting the present system. Contractors bid and perform on the pretense that they can satisfy the standards of all the specialists on procurement committees: in terms of firepower, fuel, electronics, costs, and so on. Realism fortified by costly experience suggests that contractors be invited, instead, to offer performance trade-offs, in order to avoid time-consuming changes resulting from personnel turnover in committees.

In 1987, Nunn suggested a set-aside in the procurement system for the technological *gadfly,* to use his apt term. Any present-day Henry Ford or Donald Douglas, who ran the gauntlet of America's modern procurement system and managed to land a contract for a miracle weapon, would be overwhelmed by demand. In order to meet delivery schedules, while waiting for reimbursement of costs, he would be forced to team up with an established contractor—and established contractors have a vested interest in accumulated obsolescence. The result would have the proverbial consequence of entrusting a chicken coop to a fox. The critical problem of supporting new concepts and factoring them into the system can only be solved by giving creativity a chance and a reward, on the clear understanding that, by virtue of being entrepreneurial, creativity is sparked by corporate vehicles that are small and, therefore, lack cash to burn or even to operate. The way to do it is to guarantee the vesting of proprietary rights to Nunn's gadflies in their own developments and requiring established contractors to accept subcontractor status, with no claims to the transfer of technology for repeat orders.

Reciprocal incentives. The reciprocal trade system outlined in Chapter 15 suggests one method of compensating America for military assistance and protection. America's need for creative weapons engineering, described previously, suggests another. America has been consistent in her reversal of priorities: she has bought the products of her allies' military technology, not the technology itself, while she has allowed her defense production costs to inflate and her skilled production forces to scatter. As a result, she has had to cut back or stretch out delivery schedules on her own weapons; each stretch-out inflates overhead costs, leaving the Pentagon paying more per unit for fewer units. Financial incentives to allies and friends are simpler to count and would work better.

Financial engineering has produced results as revolutionary as product engineering in the decades since the end of the 1939–45 war. America needs to renew her flair for financial engineering. Past breakthroughs with the greatest continuous effects—notably the evolution of the markets in financial futures—suited the purposes of the public markets rather than that of the U.S. Treasury. Nevertheless, the Treasury did engineer one remarkable innovation and has had occasional recourse to another that originated in the markets. The first useful new practice has been the invention of nonmarketable long-term Treasury bonds, placed directly with friendly creditor central banks. The financial markets have adopted the zero-coupon bond, paying no current interest, placed at a deep discount, paid off in profit at par on maturity, in lieu of interest in dollars at par on maturity. Combining the two, zeroes and nonmarketables, would provide a strategy that could ease the Treasury's burden and help it to restabilize without wasting billions on market support operations.

Lip service to the cause of stabilizing the dollar will not succeed this side of never-never land unless and until America solves the problem of her two deficits. Neither will open market operations by central banks free, or pressed, to resell dollars or Treasuries. But massive private placements of nonmarketable Treasury zeroes with foreign central banks would restore steadiness to currency markets.

Such expedients as private placements of conventional bonds abroad by the Treasury provide temporary detours around the market. Standard operating practice, such as dollar support operations of central banks, are counterproductive because they accentuate the volatile susceptibility of the dollar to rumors and runs while increasing the interest rate burden. Neither placements nor supports can have the pragmatic effect of cutting the interest load on the Treasury as if it were benefiting from a simultaneous reduction in the deficit and in interest rates. But the massive and continual private placement of nonmarketable zeroes would create the pragmatic effect of accomplishing both purposes at once.

Late in 1987, the Treasury demonstrated its awareness of the potential of zeroes in intergovernmental financial dealings. But in the process, it affirmed its instinct for using the right instrument for the wrong purpose. It played the zero hand in order to guarantee the debt of the Mexican government to the New York banks. The spectacle of the world's largest debtor going bail for the debt of the second biggest debtor, and to the first debtor's most conspicuously busted wards, prompted *Barron's* to share a cynical belly laugh with its sophisticated readers. If any debtor needed the relief zeroes can provide, the U.S. Treasury itself did.

At a 9 percent interest rate for Treasury bonds—below the crisis level, but above the comfort level—placement of each $100 billion of nonmarketable Treasury zeroes would displace $9 billion of interest expense; it would avoid $45 billion a year routinely squandered in interest. A $500 billion goal spread around the world is not impractical. The Kremlin might even make a contribution; it could always borrow the necessary for an investment of any size from the Eurodollar market and recoup the cash by collateralizing its zeroes bought with the borrowings: a standard device for looking good on the cuff. Access to proprietary U.S. oil and gas technology, a longtime Kremlin frustration, might sweeten the ante. Gorbachev complained bitterly at the 1986 Iceland summit because Washington persisted in withholding Soviet access to fuel technology. Access to U.S. proprietary food technology as well would give the Kremlin a powerful incentive to buy such Treasury nonmarketable zeroes. Any Soviet participation would put a powerful lever to work in the market. Interest expense is traditionally treated as a fixed charge beyond managerial reach. But a matching contribution from the rest of the world on the massive $500 billion scale could restore instant flexibility to the Treasury and stability to the dollar.

Any practical expedient adopted as a novel solution at a difficult time will provoke troubled questions. Granted that the financial markets—particularly the Eurodollar market—have a large appetite for zeroes, that appetite is limited to borrowings in currencies enjoying good standing. Admittedly, the collapse of the dollar, though engineered by the U.S. Treasury, left the dollar suspect. Hence the obvious anxiety over whether this expedient would merely postpone the day of reckoning and inflate the cost of postponement—especially if deep discounts below 50 percent were offered.

Alexander Hamilton had the answer for President Washington. Before him, Robert Walpole did for the Duke of Wellington. The answer was not financial, but political. It still is. The day of reckoning need never arrive for a power whose economy and whose military command confidence. Even before the growth of massive insurance and pension claims, historically such a power could carry a large floating debt and also enjoy premium investment demand for access to its instruments. The problem centers on finding current earning power to carry the debt. If the equity float is expanding, a large stable debt load can be carried with comfort, especially if an increasing portion of it is guaranteed not to be a supply burden on the market, or a cash drain on the issuer—in this case, the U.S. Treasury.

America's standard inducement for defense cooperation from client governments was to offer industrial and commercial advantages, but financial incentives would have served both America and her allies much better. All of her allies have been dollar-rich but arms-poor. All of them have been ready and certainly able to buy protection under the American defense umbrella, protection that America in fact had been providing in the common cause for no consideration. Nothing could have come more naturally from Reagan than an offer to American allies (and Switzerland) to buy new issues of zero-coupon Treasury securities especially designed for each of them. Because the investor buys zeroes at a discount in lieu of collecting interest during the waiting period until maturity, zeroes are tailor-made for foreigners, who have abundantly demonstrated their preference for gain over interest. So long as American policy keeps America strong, and American strategy demonstrates its viability, negotiated refundings can always anticipate and defer maturities.

Creative weaponry. When Defense Secretary Weinberger was caught dispatching a convoy force to the Persian Gulf with no minesweepers, Nunn told me that the Senate Armed Services Committee, aware of the lack of these vessels, had been after the Pentagon for seven years to design a modern squadron of them. America's armada in the Gulf also lacked air cover. In the twenty years since Lyndon Johnson had cannibalized the Sixth Fleet (patrolling the Middle East) to compensate for the losses of the Seventh Fleet (ranging across the South Pacific) suffered in Vietnam, the Pentagon had had ample time to anticipate the crisis in the Gulf. Sheikhdoms had served abundant notice on Washington of their entitlement to protection as well as their refusal to open an inch of sacred Moslem soil to infidels, even if allied, by granting basing rights to American aircraft. The armada was left to patrol the Gulf supported by no land-based aircraft with the range needed to arrive on the scene, to engage in minimal combat operations, and to make the 2,400-mile trip safely from land to Gulf and back. The American aircraft carriers had to be stranded at the mouth of the Indian Ocean, beyond the reach of Iran's Silkworms.

But the Pentagon Weinberger left behind him managed to top this achievement of amnesia. It awarded a fifteen-year contract, with $45 billion before sweeteners, for a Navy attack bomber, with a down payment of $241,000, split between two major prime contractors. One reason given for the award was that two new carriers already authorized might have been launched without planes to carry. Such shoestring finance explains the standard Pentagon practice of digging itself deeper into debt by the billions every time it puts up a dollar. The contractor incentive that feeds on the practice helps the Pentagon to do so by making customers pay through the nose for costly, time-consuming changes that substitute procurement for production. The practice is a stern reminder to advocates of a budget-balancing constitutional amendment that the obstacle is not the money the Pentagon spends, but the money it obligates in the present and spends in the future—long after the contracting officers have departed.

America's inertia-bound, overhead-ridden procurement system has been designing complex weapons to perform specialized missions. But it needs simple weapons to do double duty or better in performing multiple missions. As an example, the glaring failure of the system in force is advertised by the absence of an American antitank weapon capable of outmaneuvering and outgunning a Soviet tank. We need not posit the engulfing disaster of a Soviet-American war to envision circumstances under which American expeditionary forces might collide in combat with Soviet tanks.

Libya, when danger was skirted in 1986, was loaded with Soviet tanks manned by well-paid, trained operatives of many nationalities. Nicaragua offered an even more urgent warning. The Reagan administration pressed its holy war against the Sandinistas with no sensitivity to their ability to activate fifty Soviet Hind gunships against the obsolete American equipment on the spot there. The Hind has held the world's helicopter speed record since 1975. It is literally a flying tank, not only able to carry up to eighty troops, depending on the model, but bristling with firepower as well. Against it, American weaponry could offer only one helicopter model able to carry troops but not guns, and another able to carry guns but not troops: a monument to America's longstanding commitment to buy enough models to keep her many contractors in business (though not against one another) and a safe-conduct pass for Soviet tanks.

Trade-offs being standard operating practice in weapons design, the strength of Soviet tanks is concentrated against land attack, while they also operate in liaison with close air support, as German troops were trained to do in the 1939–45 war, and as American troops learned to do under fire during the battle for Europe in 1945. The built-in weakness of tank attacks, however, is their vulnerability to dive-bombing. No specialized landborne antitank weapon will work as effectively as a double-purpose fighter-diver, capable at once of outrunning the formidable new class of deadly but slow (again, a trade-off) Soviet gunships and swooping down on the tanks they protect.

The Pentagon has not faced this dual-purpose challenge. The U.S. Air Force rejected the Pentagon's request that it assume responsibility for close air support. Fast, high-flying planes cannot find tanks, let alone fight them. Meanwhile, America's armed forces have no weapon ready to stand off Soviet power in any air-land battle. America's obsolete helicopter fleet is no match for Soviet gunships and has no defenses against Soviet land-based air power. The Pentagon's two ambitious efforts at helicopter modernization, budgeted at $90 billion, have been candidates for cancellation. Each was a model of procurement for the sake of procurement, and each is single-purpose, not conceived to dominate a multipurpose target.

Avoiding airports.   Another task for creative weaponry is aircraft designed to avoid airports, and also to hit a multiplicity of targets. Combat units equipped to operate without dependence on airports would enjoy mobility and flexibility. No less important, they would enjoy immunity from paralysis when their landing strips were bombed out. They would be free to use air power as striking power without being choked by it on the ground, as the German Luftwaffe was; its planes ran out of fuel and crashed looking for safe landings after Allied bombings, which had failed to wreck Germany's production, succeeded in blowing craters into its airfields.

A new type of plane designed to avoid airports has become as urgent a need for a modern, airborne economy as for the armed forces it needs to support. Every era of economic expansion is fired up by its own distinctive transportation breakthroughs: ships, canals, highways, railroads, automobiles, airplanes. The more high-powered the vehicle, the more dependent it becomes on its base, and consequently, the more high-powered its base grows as well.

America's megametropolitan airports have made a distinctive contribution of their own to the expansion of the American economy in the air age. Every period of expansion depends on the leadership of a growth industry, and airports provided it from the midcentury on. In one metropolitan region after another, the opening of a major airport staked out a two-phase explosion in growth within the region itself. First, the area between the airport and the center experienced explosive growth. Then the explosion radiated outward from the airport to new segments of outer suburbia, with the airport as a new center. During recent decades of growth in America, land use along new corridors, to and from megametropolitan airports, paced it. When stagnation set in, metropolitan airports were gradually identified as traffic choke points, breeding breakdowns of all kinds, from air and land traffic congestion to centers for dope trafficking. Worse still, airports, even more than airplanes, offer ripe targets for terrorists. Every other economy is in urgent need of alternatives to megametropolitan airports. Even countries lacking them, notably China, are eager to grasp shortcuts around this cumbersome, costly infrastructure.

Creative weaponry and practical traffic management call for a common vehicle as a common answer: a lightweight, nonmetallic, slow-moving, fast-lifting, vertical takeoff and landing plane—as a weapon, for flexibility and freedom from airstrip bombings; as a carrier, to decentralize traffic and freight away from megametropolitan airports; and to take over intercommunity transportation. In the past two decades, America's major aerospace contractors have designed not one new weapon. In commercial competition, one new growth area of demand has developed: the commuter plane. A sad and startling commentary on the failure of American aeronautical engineering is dramatized by the fact that America's 350-odd commuter airlines all fly foreign models.

The vertical takeoff and landing plane needed to provide multiple-purpose weaponry for America's armed services also would open a new channel of unsubsidized profitable growth in the commercial economy. The moment America launches a new common-purpose vehicle, capable of ensuring America's security and relieving her arteries from the congestion imposed by the megametropolitan airport, she will be on her way again—not least, with an irresistible export product, wanted everywhere, on America's terms, for cash.

Though America is distinctive in her characteristics and in her potential for leadership, the congestion centered in her megametropolitan airports is not. Every country in the world—from the most to the least advanced—suffers from the same complaint. A solution to this problem, in the Pentagon to begin with, then with her mega-airports, will give the world the new American lead it knows it wants and is waiting for. The lead will begin with a technological breakthrough. It will reconfirm America's distinctive role as the military superpower which is also the economic superpower.

Competitors who rely on price-cutting to market their wares are weak when they try and weaker still when they succeed. But market powers that muster their reciprocal assets to make and finance their sales on their terms are strong when they try and still stronger when they succeed. Proprietary export sales will silence the clamor from the managerial-academic complex for American cost competitiveness with old technology. Such niche sales aimed at felt worldwide wants will provide the best protection the American economy can hope for: market protection for the dollar and for the living standards of America's work force.

The Russians enjoy the twin advantages in aeronautics of playing to win and not being able to afford formulas for losing. Yet the U.S. government indulges its largest supplier, the aeronautical industry, the luxury of structural uncompetitiveness: designing complicated and costly planes for which engines are still bought from catalogues. Industry on its own will not break out of this comfortable but uncompetitive cocoon. Only the government can change the rules of games that American industry has been playing to lose. The Pentagon could make billions for its contractors and save itself billions more; the moment it started buying like a business, its suppliers would start selling like competitors.

CONCLUSION
Holmes's Homily on Winning the Race Against Time

Patience was the order of the day from Washington in response to the crash that staggered Wall Street in 1987 and the slump that started to squeeze all of America in 1988. Economists advising both parties fortified one another with the same clichés assuring the country that the devalued dollar was slowly and soundly doing its painful job of freeing America to recover her export competitiveness and, therefore, to recover, as all stricken economies are supposed to do, via the export route. Chairman Alan Greenspan of the Federal Reserve Board spoke for the conventional wisdom when he pronounced the economy "in equilibrium." His counsel confirmed the deep-rooted confidence in patience; although he himself pivoted overnight in exactly the wrong direction when the Fed raised the discount rate to stem the tide against inflation, which it was told was rising.

But in the face of urgent evidence that costs were continuing to rise out of equilibrium with incomes, and, simultaneously, that incomes were continuing to fall out of equilibrium with costs as well, the logic of patience was insane. Urgency, not patience, is the need when an abrupt change of direction is overdue. This is one rule of motion that applies with equal force to the intangibles of policy-making and to the arithmetic of logistics on land, at sea, and, above all, in the air, where supplies—beginning with fuel—are critical and miscalculations are fatal. Once direction is recognized as the problem, speed in changing course becomes the key to the solution; all the professions subject to rigid performance tests agree. The soldiers reduce this maxim to a grim rule of war: time takes over as the enemy once the enemy takes aim. Wall Street has a simple colloquialism for the same strategic principle: "Run from your first loss—it is the cheapest." The medics act with equal incisiveness in taking diabetics off sugar, and alcoholics off booze: never a little at a time, but entirely and at once.

In the classical world, the role of "prime mover" was reserved for the gods. But the omnipotence attributed to them was not envisioned as a one-time shot. On the contrary, the mythic logic of the classical world alternately expected and invited the gods to meddle continuously in every nook and cranny of human life, from cosmic visions to daily housekeeping. Reaganomics flawed this cultural heritage in the name of classical economics. It adopted a "go-and-stop" logic that alternately invokes the prime mover role of the gods and turns it off. Admittedly, the spectacle of Reaganomists studying case histories from Homer defies imagination. So does their naïveté in assuming that government could intervene in currency markets without disrupting them. Their amateurishness went on display when they compounded the disruption by withdrawing the intervention.

Government functions within our society as the modern parallel to the classical gods. While its operations also range from the cosmic to the mundane and routinely stir up scandal, as the behavior of the gods did, its impersonation of God running the exchange markets is also an invention of imaginations running riot. When the government believed that it was deregulating the most important of its market responsibilities—for the exchange value of the dollar—it soon discovered itself engaged in continual and frustrating efforts to reregulate the exchange markets—just as the gods, once conjured up to explain the meaning of life, were relied on to run it.

The logic of politics has made less progress across the centuries than meets the eye "in this best of all possible worlds," as Voltaire's prototypical optimist, Dr. Pangloss, says. If Dr. Pangloss could be with us today, Voltaire might have dedicated a modern edition of *Candide* to a demonstration that each source of marketplace dislocation was confirming Chairman Greenspan's complacent judgment by offsetting the other. Voltaire's quixotic hero would cheerfully have cited all the evidence of disequilibrium in force within each sector of the economy as proof of equilibrium prevailing across its macrocosm. Without the benefit of Voltaire's satirical bite, however, the official statisticians have learned to employ precisely the same mumbo jumbo to measure the growth of the overall economy as the sum total of the shrinkage of its working parts. This manifestation of progress armed Greenspan with one telling advantage over Dr. Pangloss: the ability to cite official statistics to prove that all victims of prevailing economic dislocations saw prosperity blooming through the glasses they wore, presumably rose-colored.

Unarguably, America needed market readjustments all through the late 1980s. But this conventional rationale of reliance on the markets to know best and, consequently, to do better ignored America's need to change her policies in order to change her luck. Conventional American impracticality has put the market cart before the policy horse. The theoretical rationale behind America's political impracticality has stretched illogic into lunacy. Officialdom, reinforced by bipartisan approval, has ignored America's need for a new policy horse. Instead, it has sighted her market carts moving again, all in the right direction, on their own, without leads from any practical new policies to pull and guide them.

True, the conventional rationale does assume the primacy of the need for bargaining between America and the outside world. Hence the pivotal role of the dollar in modern laissez-faire theory and the mischief its devotees made by presuming to lay hands on the dollar and then let it float. But the conventional price-oriented approach to foreign trade ignores the distinctive operating difference between America and all her competitors with smaller-than-continental, industrial economies. The latter depend on their nationals to accomplish more abroad than at home and, accordingly, are geared to help them try. America is not and does not.

Governments outside America habitually intervene in private commercial, industrial, and financial dealings abroad, and they certainly police foreign participation inside their own jurisdiction, as the

American government is not geared to do. Government in America limits its nosiness to goings-on among Americans at home but welcomes all dealings by accredited foreigners in America. Washington blinds itself to whatever American businesses are doing in their dealings abroad, both to the dealings in need of support and to the kind in need of scrutiny.

Consequently, Americans have expected the foreign bargaining role of their government to be limited to political and military matters. They welcomed the historic dollar devaluation of 1985, but as a one-time shot. The accompanying ballyhoo had promised that, once the planned readjustment was decreed, Americans bargaining with foreign principals would deal from strength and in private, trusting market incentives and competitive responses, without further U.S. government intervention. But to leave the national security, not to mention what the Constitution speaks of as "the general welfare," dependent on corporate management of America's economic affairs, is to run afoul of the plain meaning of the Logan Act.

This start-up statute had seen as far into the future as it did into the complexities of the economics of sovereignty. Its framers, though unarmed with any road map to the future, made provision for the global economy to come of age dominated by foreign governments underwriting their own corporations and banks; that's why it banned private American dealings in behalf of the United States with foreign governments. Even if they were up to competitive standards domestically, American corporate managements would be no match for the political gang-ups against them abroad. Behind the myth of the Japanese corporate mystique, the question is not one of market smarts, but of market power: foreign governments routinely put power behind their nationals in foreign dealings, and the American government never does. Success by IBM and the rest of the Fortune 100 in buying their way into cushy deals overseas is by no means a reversion to dollar diplomacy. Their strategy has worked for them because they have been sharing slices of the American market with foreign governments and/or partners for no consideration! Their corporate progress abroad has left America way behind and stuck in a rut.

Hence the sane logic of urgency in switching course at the negotiating table and for insisting that the American government learn how to play host there with chips that its official guests from abroad will be eager to accept for value given. How to negotiate around the world looms as America's top-priority problem, more urgent than even any of her specific emergencies. Her distinctive advantage as the world's only diversified superpower is offset by her confusion over the embarrassment of riches with which she can bargain: a burden every other government in the world would eagerly shoulder. Washington, relying on arrogance born of naïveté, has been taking offers of the wrong resources to the bargaining table: summits in some cases, weapons in others. Both approaches are synonymous with emergencies but are symbolic of America's failure to mobilize her economic and financial resources to normalize her political and military emergencies. The way for her to do so is to bring purposive, interlocking national policies into play to prevent the emergencies in the first place.

During the 1914–18 war, at the height of the Mata Hari scandal, the word in Paris counted her charms worth more to the German general staff than its artillery batteries of Big Berthas, then rated the ultimate knockout weapons. Without denigrating the valor of the United States Marines, the American economy is an even more formidable strategic asset, on tap to prevent the alarms that call for response by the Marines. Though Washington has missed the opportunity to use the American economy as America's bargaining trump, the debacle of dollar devaluation under Reagan, like the disaster of oil inflation under Carter, has demonstrated that her presidents cannot avoid involving the economy even in their most hapless efforts to deal internationally from self-imposed weakness.

Once the Carter people, and the Nixon people before them, accepted the OPEC fiction of oil as a free-market commodity, they forfeited America's ability to bargain the terms of trade with all parties to the Gulf oil war into the terms of peace. Once the Reagan people, in their turn, debased the dollar into a commodity, they forfeited the one proprietary asset America had the power to stabilize as her lever for bargaining over everything else.

Because the Carter people failed to recognize the urgency of the opportunity to bargain from market strength over the price of oil, they were dragged by the scruff of their necks to bargain from weakness over the cost of peace in the Middle East. Because the Reagan people failed to recognize the urgency of the opportunity to bargain from market strength with America's dollar resources, they were dragged by the scruff of their necks to bargain from weakness over the exchange value of the dollar. Because both administrations failed to recognize the abundance of economic assets America has ready to put into play in troubled areas of the world, each administration, in turn, was dragged by the scruff of its neck to put inadequate military assets into play. In each case, ignorance in high places mistook the bargaining table over which the White House presided as political, when it was economic. In each case, the president misconceived his role as that of the mediator and failed to perceive himself as the principal armed with the chips the economy behind him was dealing to him.

Yet, win or lose, prepared or unprepared, the president of the United States is endowed with the economic and financial chips that are decisive in dealing with governments eager to trade political and military assets for access to them. For America to make such trades is not to buy agreement. Quite the contrary, it is to use America's distinctive powers of persuasion as a bargaining lever to negotiate agreement. America has the power to do so. The world is mystified by the hangups that alternately prevent her from trying and detour her into relying on shows of force instead.

Every American president starts out armed with economic and financial powers of persuasion; whether he knows enough to use or is ignorant enough to waste them, he starts out with them; he plays out his hand with them; and his opposite numbers at and between television opportunities at summits bargain for access to them. So the acid test for any crisis presidency—and all fin de siècle presidencies have been—is whether it will lead the American government at long last to the bargaining table forearmed, knowing what it is prepared to give in order to get full value in return in coin of its designation, by no means necessarily financial, just as likely political, hopefully not military, until economic extensions of political warfare have given the process of reciprocity a chance to defuse inescapable conflicts of interest from flaring up into avoidable conflicts.

Alternatively, reliance on America's familiar stance of offering without demanding will continue to invite successful bluffing from countries with specialized strengths and weaknesses, costly to her in more than mere money terms, eroding her sense of national purpose or, more precisely, exposing its erosion. The challenge of preparedness to bargain calls for the American government to draw on the resources of the private sector in flexing its bargaining muscles abroad: neither to rely on the private sector to go it alone on a pure and simple business basis, as if markets abroad were free, as claimed, nor to go it alone on a government-to-government basis, as if America's private sector were not an enormous business and political asset, when backed by the American government, as foreign businesses are everywhere by their governments.

When America was coming of age, big-business bullies gave dollar diplomacy a bad name by battening on little brothers abroad. Colonialism was the name of that game. But as soon as America discovered that it was a no-win exercise for suckers, she stopped playing it. Only big brothers abroad are the targets of the new style of dollar diplomacy counteroffensive that America is on notice to formulate for the 1990s. Any qualms of guilt are out of order: America's competitors infiltrated her economy by aiming the tactics of dollar diplomacy against her years ago. American fear of foreign retaliation is a rhetorical exaggeration.

In Nixon's phase, America is "the helpless giant, swinging in the wind." America has waited for her embattled realists to drag her, wounded and whining, into a stance of self-defense. Belated efforts by America to salvage what is left of her splintered industrial base are not threats to initiate economic aggression against her triumphant invaders. They have fed so well at America's expense that they face leaner pickings themselves. Accordingly, the pressures on America to move from acceptance of passivity to insistence on reciprocity is a hope for her competitors. None of them cannot afford to score more breakthroughs against the American economy.

Yet America can count herself lucky, and so can the world, that technology has now carried Clausewitz's dictum—"War is an extension of politics by other means"—to "a higher stage of development," to adopt Leninist jargon. In capital cities everywhere outside Washington, markets are battlefields routinely fought over by "other means"—that is, by political methods. Dollar diplomacy is due for legitimized resurrection as America's indispensable dual-purpose key: simultaneously to reciprocal trade with nonmilitary powers and to profit sharing with military powers, instead of wasting money on military overkill.

Only the American government is endowed with this distinctive dual power to sanitize dollar diplomacy as an investment instrument for the export of peace and prosperity. But the know-how even to try to activate America's limitless arsenal of economic, financial, and technological assets for such an effort is not lodged in Washington. America's private sector owns them. America's businesses and banks are thoroughly at home in the outside world's one nonpolitical market: the Eurodollar market, where the dollar is "stateless money" and, therefore, where government backup is no consideration. In the past, the urgencies of war drove America to mobilize the resources of her private sector. The urgency of going to war against depression calls for America to repeat the performance to preserve the peace, which depressions always shatter. The crisis precipitated by the last depression culminated with the unveiling of America's nuclear monopoly. The crisis threatened by any recurrence would begin where the world war of 1939 left off in 1945: with a round-robin of nuclear anarchy. Moscow's defensive stance toward this source of nuclear aggression is motivating the Kremlin to seek cooperative access to America's technological head start in defensive nucleonics. Negotiation is the Kremlin's stated objective. But dependence is its reality.

To recall how America won her wars as wars of production is to be on notice that America wards off her threats and makes the most of her opportunities by harnessing her economic resources to her needs; although, in her smugness over her long-standing nuclear strategy, since discarded, she has allowed Russia to outproduce her in conventional weaponry. She dare not scratch herself as an entry from the growth race into the 1990s.

America's resources offer a potential that defies Malthusian limits, but her needs again and again grow desperate from malign neglect. The entire pack of competitive entries is running against her. The faster America puts on her economic running shoes, the less need she will have to fall back on military gear. Back when weapons were simpler, the Chinese spoke of "silver bullets"—money replacing missiles. (As with dollar diplomacy, but with better reasons, the old Chinese custom of paying off warlords gave silver bullets a bad name.) America is the one country in the world enjoying not only free choice of market weapons but, more important tactically, first choice of which to use. Her economy need not be stripped to stock an arsenal; her economy is her arsenal, if she will but use it, beginning with the gentle reminder that none of her contestants in the race for position in the 1990s, no longer even Russia in its race to go capitalist, can do without ready and continuous access to her resources under Washington's oversight and on reciprocal terms acceptable to Washington.

In the late 1700s, Adam Smith taught that slumps are built into the vicissitudes of economic life. Two hundred years later, blessed with central banks, monetary policies, budgetary strategies, and macrostatistics (badly needed, but badly applied), America has taught herself to plan them. Another mark of progress! The surefire conventional governmental recipe for slumps is to force cuts in consumption—the basic, not the conspicuous kind. A bipartisan clamor to do just this greeted America in 1988. Permitting cuts in consumption, let alone forcing them, would devitalize America's effectiveness in the world as surely as scrapping Russia's elite tanks corps, shutting Japan's harbors, or liquidating Switzerland's banks would strip all three of their respective limited powers.

Any Washington effort to cut America's consumption when it was already on the decline would advertise America's unawareness of the unique bargaining advantage inherent in the distinctive size of her import capacity. Any Washington success would cripple efforts by Americans to hold their own in the marketplace, let alone regain their losses. America's ability to increase her consumption will measure her allure for her trading partners, and it will broaden the range of reciprocal advantages to be wrung out of them by the urgent application of the return to reciprocity, government by government, in the spirit of Cordell Hull that I recommend.

Classic wisdom encourages savings. It assures a classic discipline in avoiding debt. The world of Adam Smith had no idea of private debt. Consequently, it made no provision to institutionalize debt for economic man. It is a far cry from Smith's legendary fable embodying his version of Robinson Crusoe landing in a semirustic community and starting to craft rudimentary items to trade for subsistence, to today's Herculean private debt load under which the economy staggers. Given the double bite of rising debts and falling incomes, America cannot afford the luxury of learning the hard way that forced cutbacks in consumption will not lead her across the magic threshold to spending speedups, let alone activate new savings into new investment.

Ironically, the advocates of consumption cuts as the conventional solution to America's debt problems ignore the chain reaction their advice would activate: from consumption cuts, to income cuts, to private debt defaults (especially on junk bonds, and, consequently, by S & Ls and even banks), to cuts in Treasury revenues and jumps in Treasury cash refunds, invariably hoarded until consumption is seen to rise above paying for bare-bones necessities again. A new government policy start is needed to free private borrowings for the big-ticket items on which the American economy runs and on which the living standards of Americans depend. America cannot hope to reduce or even carry her private debt load by cutting the consumer incomes needed to support it.

Increased incomes offer the only way to cut the relative burden of her debt load. Like it or not, incomes and consumption travel together. There's no way to cut consumption without cutting incomes. Savings are a luxury once the wolf is kept from the door: income deflation in a high-debt environment provokes debtors and creditors to cannibalize one another. Higher consumption offers the only support for higher incomes in the high-debt society on which America's structure of business debt has come to hinge. Moreover, business will not increase capital investment until consumption strains the capacity to deliver, so that consumption feeds investment as surely as it feeds on incomes.

The conventional rationale has compartmentalized the business of policy-making. Politics is politics; economics is money; the Pentagon is a world unto itself; people—beginning with the helpless old and the helpless young—are left to fend for themselves; and never the quatrain shall meet. In the real world, however, policy-making can succeed only if it is conceived and carried out with all four sides of the parallelogram making a simultaneous fit and offering policy-makers a manageable vehicle.

Wall Street's October 1987 crash sounded the shot heard round the world that the race against time was on, not only in markets and with weaponry but also as a challenge to statesmanship, forearmed with both and, therefore, forewarning partners at the bargaining table that pronouncements from America were no longer just mouthings. Presidents "standing tall" as they pose for photographic opportunities do not win races for national competitiveness. In today's world, the race to exert political leverage is run by racking up breakthroughs in the economy, including the defense economy.

The final score in the race is reckoned in the lives of people: satisfactions salvaged for the aging, against whom the clock is ticking, and opportunities opened for the young fated to forfeit them without the start in life that only community-organized health and education can provide. Every year lost raises the count of casualties from the failure of policy-making to maximize the limitless potential of the people who offer the American economy its richest resource. By the same token, every new reach added to America's assets will stop the clock on lives wasting and start the clock on lives waiting to be reclaimed. The race for time is the key to harnessing the power to create the wealth to put new dimensions on life. Americans will be the first beneficiaries but also will provide the conduits to others everywhere.

A basic paradox has inspired America's achievements from the beginning. The efficiency built into her momentum has consistently borne prodigious cost inefficiencies as a fixed charge claimed by her distinctive national commitment, enshrined in the Bill of Rights, to give greater priority to human values than to property values. America's recurrent seductions by the ethics of greed represent costly diversions. A striking contrast is apparent with the Soviet system, which subordinates human values to program goals.

Fighting slogans pep up declarations of policy, but neither suffices to hit targets. "More bang for the buck" was dynamite, coming from the Eisenhower administration, but it changed nothing and accomplished nothing. "The Great Society," proclaimed by LBJ, the self-adopted son of FDR, sounded better still, but fell flatter still, against the backdrop of Vietnam. Earlier, by contrast, lend-lease stirred up no passions and dealt in no ambiguous abstractions, but it did signal a clear objective, which it met: to send American war equipment to unequipped Allied troops—the more equipment sent, the less American manpower needed to follow. Ironically, FDR, though a dashing sloganeer, won a more effective success for this laconic, uninspired label of his, which bridged his transition from the

New Deal to his war presidency, than for his phrase-making coups—
beginning with the New Deal itself.

The progress and problems of aeronautics have dominated the second half of the twentieth century. The further aerodynamic breakthroughs have taken us into the space age, the more complications
they have proliferated on the ground—all the way from the airport
to the bank. To factor the problems and often the opportunities
looming ahead for the space age into the crisis at hand for the
aeronautical era is to limit the competition for leadership—as well
as the needed collaboration—to today's two military superpowers as
they stretch and fit themselves toward political and economic accommodation with each other.

A basic triangle within America's political economy connects
her airports and its banks with the government. Both the manufacturing and the service sectors of the American economy travel on
wings. Consequently, any realistic search for the terrain from which
to launch any new American policy calculated to revitalize the
American economy and, with it, America's purpose in the world and
at home is bound to start with the aeronautical segment of her
political economy. It links the vital components of her besetting
challenges of science, technology, finance, defense, skills, traffic, education, leisure, and crime; of congestion, sprawl, and frontier; of
management, political, economic, engineering, and negotiation; of
obsolescence and competitiveness, internationally and domestically.
Functionally, fear of flying is fatal to the economy, and so is failure
to organize aeronautics. To be grounded in the twentieth century, let
alone in the twenty-first, is to be stranded. The terrain of the challenge fits the substance of the challenge, which is aeronautical.

When the world roamed at sea and on horseback, it meandered.
When it learned to drive, it went where roads were paved to lead it.
When it began to fly, airports set its flight plans. As aeronautics
outgrows its supportive but confining network of landing fields, a
sense of direction on the ground and in space will be the first need
of flight planning and planners. The need to land, on earth and in
space, will outpace the growth of landing fields. No American policies that hit their targets will work unless and until they work for
America's aeronautical problems—all the way from crime and drugs
to employment and, therefore, Social Security. By the same token,

any new American policy that does work in her aeronautical era will achieve miraculous breakthroughs for her in her earthbound activities.

Yet, thanks to America's distinctive geography and sociology, no other country is equipped to assert the pattern-making leadership awaiting aerodynamics. Not that other entries in the international aeronautical derby lack for innovative aeronautical engineering or aerodynamic research behind it. On the contrary, America has fallen behind in the scoring of aeronautical competition, as in so many others, and the world has visibly suffered as, and because, she has. To score America's aeronautical performance at its worst, standard U.S. jet engines are the direct descendants of the original British imports of the 1939–45 war. Ironically, the Pentagon, symbol of America's lead in the air, bears the responsibility for losing it, notwithstanding the primacy of air power in American military doctrine. The reason the world is waiting for a dramatic and decisive new American launch into aeronautical and aerodynamic leadership is simplicity itself and goes back to the governing premise of this book: America is the world's only diversified superpower.

Traditionally, politics makes strange bedfellows. But the marriage between politics and aeronautics was made in the heavens within full sight of everyone on earth. All roads into the 1990s lead through the Pentagon—not to make war, but to manage peace, not by police methods, but by facilitating, rather than obstructing, the new engineering revolution that has started in America's private sector. Thanks to America's mismanagement, the Pentagon is the focal point for any new start at technological modernization, just as the armed services became during the 1939–45 war.

Any presidents who continue to look to the performance of the economy for their policy leads will preside over accelerated disintegration into depression. Instead, the economy will learn to look to America's presidents to finance its revitalization. Beneath the triple burden of debt, depreciation, and dumping, the economy is stifling. The Pentagon is the choke point for all three. Only the president can free the American economy from this exposure to strangulation—by management domestically and by negotiation internationally. Only the American economy offers America the vehicle for the leadership that the world is awaiting, but only the president can provide the piloting.

The same grim logic, unanticipated by Smith and his successors of all schools, holds the president responsible for dealing with his designated opposite number in the Kremlin. This logic resolves the argument over waiting for the markets to make history or urging policy initiatives on the government. Remember, Soviet power can deal in markets, but markets cannot deal with Soviet power. Only American presidential power can.

The tempo of life has accelerated immeasurably, along with its complications and its risks, since that grim March day in 1933 when Holmes told Roosevelt, "Young man, you have a chance, but hurry." Back then, the presumption was that the terrain of competition was land. In the 1990s, the terrain starts above the skyscrapers. "Don't just hurry, fly," would be the sage's admonition to Roosevelt's would-be successors in the space age, "with a flight plan."

Kennedy excelled in using the White House as a backdrop for the profile in leadership he offered. His impact for calling for legislation was immense, but Johnson's was greater still in passing the legislation Kennedy called for. When Reagan's turn came, he set the White House clock back from Johnson's legislative activism to Kennedy's emphasis on eloquence. Reagan reversed JFK and LBJ. He concentrated his eloquence on explaining that anything the government did would interfere with the country's ability to keep moving.

In the aeronautical age, a theatrical presidency will not manage an explosive takeoff and sustained momentum for the economy. Nor will a purely political presidency. Only a presidency with a world flight plan for negotiating happy landings with other governments can fulfill the potential of the American economy, as well as its obligations to society, while revitalizing and rechanneling the sources of growth around the world.

BIBLIOGRAPHY

AGANBEGYAN, ABEL. *The Economic Challenge of Perestroika.* Blooming-
ton, Ind.: Indiana University Press, 1988.

AYRES, LEONARD P. *The Economics of Recovery.* New York: The Macmillan Company, 1933.

BATRA, RAVI. *The Great Depression of 1990.* New York: Simon and
Schuster, 1987.

BEARD, CHARLES A. *An Economic Interpretation of the Constitution of the
United States.* New York: The Macmillan Company, 1941.

BECKER, CARL L. *The Declaration of Independence: A Study in the History
of Political Ideas.* New York: Alfred A. Knopf, 1942.

BERLE, ADOLF A., JR., and MEANS, GARDINER C. *The Modern Corporation
and Private Property.* New York: Commerce Clearing House, 1932.

BILLINGTON, JAMES H. *The Icon and the Axe: An Interpretive History of
Russian Culture.* New York: Alfred A. Knopf, 1966.

BUKHARIN, NIKOLAI. *Imperialism and World Economy.* London: Martin
Lawrence, 1927.

CALLEO, DAVID P. *Beyond American Hegemony: The Future of the Western
Alliance.* New York: Basic Books, 1987.

CLAUSEWITZ, KARL VON. *On War.* London: K. Paul, Trench, Trubner &
Co., 1911.

COHEN, STEPHEN S., and ZYSMAN, JOHN. *Manufacturing Matters: The Myth of the Post-Industrial Economy.* New York: Basic Books, 1987.

THE CUOMO COMMISSION ON TRADE AND COMPETITIVENESS. *The Cuomo Commission Report.* New York: Touchstone, 1988.

CURRIE, LAUCHLIN. *The Supply and Control of Money in the United States.* Cambridge, Mass.: Harvard University Press, 1934.

DORFMAN, JOSEPH. *The Economic Mind in American Civilization, 1606–1865.* New York: The Viking Press, 1946.

———. *Thorstein Veblen and His America.* New York: The Viking Press, 1934.

DOUGLAS, PAUL H. *Real Wages in the United States.* Boston: Houghton Mifflin, 1930.

———. *The Theory of Wages.* New York: A.M. Kelley, 1964.

ECCLES, MARRINER S. *Beckoning Frontiers: Public and Personal Recollections.* New York: Alfred A. Knopf, 1951.

EICHENGREEN, BARRY. "The Political Economy of the Smoot-Hawley Tariff." Working Paper No. 2001, NBER Working Paper Series. Cambridge, Mass.: National Bureau of Economic Research (August 1986).

FAUX, JEFFREY. "The Austerity Trap and the Growth Alternative." *World Policy Journal* (Summer 1988).

———. "Getting Rid of the Trade Deficit: A Cheaper Dollar Is Not Enough" (briefing paper). Washington, D.C.: Economic Policy Institute (March 1988).

FISHER, IRVING. *Booms and Depressions: Some First Principles.* New York: Adelphi, 1932.

FORRESTAL, JAMES. *The Forrestal Diaries.* Edited by Walter Millis. New York: The Viking Press, 1951.

FRANK, JEROME. *Fate and Freedom.* New York: Simon & Schuster, 1945.

———. *If Men Were Angels.* New York: Harper & Row, 1930.

———. *Save America First: How to Make Our Democracy Work.* New York: Harper & Row, 1938.

FREIDEL, FRANK B. *Franklin Delano Roosevelt.* 3 vols. Boston: Little, Brown, 1952–56.

GALBRAITH, JOHN KENNETH. *Economics in Perspective.* Boston: Houghton Mifflin, 1987.

GARRISON, FIELDING H. *History of Medicine.* Philadelphia: W. B. Saunders Co., 1929.

GILLISPIE, CHARLES COULSTON, ED. *Dictionary of Scientific Biography.* 14 vols. New York: Scribner's, 1970.

GOLDMAN, MARSHALL I. *Gorbachev's Challenge.* New York: Norton, 1987.

GOODWIN, DORIS KEARNS. *The Fitzgeralds and the Kennedys.* New York: Simon & Schuster, 1987.

GORHAM, LUCY. *No Longer Leading: A Scorecard on U.S. Economic Performance.* Washington, D.C.: Economic Policy Institute (December 1986).

HAMILTON, ALICE. *Exploring the Dangerous Trades: The Autobiography of Alice Hamilton, M.D.* Boston: Little, Brown, 1943.

HARDEMAN, D. B., and BACON, DONALD C. *Rayburn: A Biography.* Austin, Tex.: Texas Monthly Press, 1987.

HARRIS, WILLIAM H., and LEVEY, JUDITH S., EDS. *The New Columbia Encyclopedia.* New York: Columbia University Press, 1975.

HARROD, R. F. *The Life of John Maynard Keynes.* New York: Harcourt, Brace & Co., 1951.

HAWLEY, ELLIS W. *The New Deal and the Problem of Monopoly.* Princeton, N.J.: Princeton University Press, 1966.

HAYEK, FRIEDRICH A. *Prices and Production.* London: Routledge, 1935.

HOBSON, J. A. *Imperialism: A Study.* Ann Arbor, Mich.: University of Michigan Press, 1965.

———. *Physiology of Industry.* New York: Kelley & Millman, 1956.

———. *Veblen.* New York: A.M. Kelley, 1971.

ICKES, HAROLD L. *The Secret Diary of Harold L. Ickes.* 3 vols. New York: Simon & Schuster, 1953–54.

JAEGER, WERNER. *Aristotle.* Oxford: Oxford University Press, 1948.

JAMES, WILLIAM. *The Letters of William James.* Boston: Atlantic Monthly Press, 1920.

JANEWAY, ELIOT. *The Economics of Crisis.* New York: Weybright & Talley, 1968.

———. *Prescriptions for Prosperity.* New York: Times Books, 1983.

———. *The Struggle for Survival: A Chronicle of Economic Mobilization in World War II.* New Haven, Conn.: Yale University Press, 1951.

JANEWAY, ELIZABETH. *Improper Behavior.* New York: William Morrow, 1987.

JANEWAY, MICHAEL C. "Who Is He?" (a study of Michael Dukakis). *New England Monthly* (December 1987).

KENNEDY, PAUL. *The Rise and Fall of the Great Powers.* New York: Random House, 1987.

KEYNES, GEOFFREY. *A Bibliography of Sir William Petty F.R.S.* Oxford: Oxford University Press, 1971.

KEYNES, JOHN MAYNARD. "The Economic Consequences of Mr. Churchill." *Essays in Persuasion.* New York: Norton, 1963.

———. *The Economic Consequences of the Peace.* New York: Harcourt, Brace & Co., 1920.

———. *The General Theory of Employment, Interest and Money.* New York: Harcourt, Brace & Co., 1935.

———. *A Treatise on Money.* 2 vols. New York: Harcourt, Brace & Co., 1930.

KLEIN, LAWRENCE. *Economic Theory and Econometrics.* Philadelphia: University of Pennsylvania Press, 1985.

KONDRATIEFF, NIKOLAI. *The Long Wave Cycle.* New York: Richardson & Snyder, 1984.

LA FOLLETTE, ROBERT M., SR. *La Follette's Autobiography.* Madison, Wis.: La Follette's Magazine, 1913.

LARRABEE, ERIC. *Commander in Chief: Franklin Delano Roosevelt, His Lieutenants and Their War.* New York: Harper & Row, 1987.

LENIN, V. I. *Imperialism.* New York: International Publishers, 1939.

———. *Materialism and Empirio-Criticism.* New York: International Publishers, 1927.

LEONTIEF, WASSILY. "Why Econometrics Needs Input-Output Analysis." *Challenge* (March–April 1985).

LILIENTHAL, DAVID E. *The Atomic Energy Years, Journals,* vol. 2. New York: Harper & Row, 1964.

LOCKE, JOHN. *Works.* London: W. Otridge, 1812.

LOWE, CHARLES. *Prince Bismarck: An Historical Biography.* 2 vols. New York: Cassell & Co., 1886.

MACAULAY, THOMAS BABINGTON. *The History of England from the Accession of James II.* New York: Harper & Bros., 1849–61.

MARX, KARL. *Capital.* 3 vols. Chicago: Charles H. Kerr, 1919.

MCDOUGALL, WALTER A. *The Heavens and the Earth: A Political History of the Space Age.* New York: Basic Books, 1985.

MENCKEN, H. L. *The American Language.* New York: Alfred A. Knopf, 1963.

MILL, J. S. *Three Essays.* London: Oxford University Press, 1971.

———. *Principles of Political Economy: With Some of Their Applications to Social Philosophy.* London: John W. Parker, 1848.

MINSKY, HYMAN P. *Stabilizing an Unstable Economy.* New Haven, Conn.: Yale University Press, 1986.

MOLEY, RAYMOND. *The First New Deal.* New York: Harcourt, Brace & Co., 1966.

MONAGAN, JOHN S. *The Grand Panjandrum: The Mellow Years of Justice Holmes.* Lanham, Md.: University Press of America, 1988.

MOYNIHAN, DANIEL PATRICK. *Family and Nation.* The Godkin Lectures, Harvard University, 1985. (Reprint, Office of Sen. Daniel Patrick Moynihan)

NEUSTADT, RICHARD E., and MAY, ERNEST R. *Thinking in Time: The Uses of History for Decision-Makers.* New York: Free Press, 1986.

NOVICK, DAVID. *Beginning of Military Cost Analysis 1950–1961.* Santa Monica, Calif.: The Rand Corporation, 1988.

O'BRIAN, JOHN LORD, and FLEISCHMANN, MANLY. "The War Production Board: Administrative Policies and Procedures." *The George Washington Law Review* (December 1944).

PERKINS, FRANCES. *The Roosevelt I Knew.* New York: The Viking Press, 1946.

PETERSON, MERRILL D. *The Great Triumvirate.* New York: Oxford University Press, 1987.

POTTER, DAVID M. *The Impending Crisis 1848–1861.* New York: Harper & Row, 1976.

————. *People of Plenty: Economic Abundance and the American Character.* Chicago: University of Chicago Press, 1954.

REICH, ROBERT B. *Tales of a New America.* New York: Times Books, 1987.

Reports of the Senate Committee on Armed Services. Washington, D.C.: U.S. Government Printing Office, 1970–88.

RICARDO, DAVID. *The Works and Correspondence of David Ricardo.* Edited by Piero Sraffa. 9 vols. Cambridge, Eng.: Cambridge University Press, 1952.

RICHEBÄCHER, KURT. *Currencies and the Credit Markets,* no. 176. West Palm Beach, Fla.: Weiss Research (December 10, 1987).

ROBERTSON, JAMES OLIVER. *America's Business.* New York: Hill & Wang, 1985.

SCHUMPETER, JOSEPH A. *Business Cycles: A Theoretical, Historical, and Statistical Analysis of the Capitalist Process.* 2 vols. New York: McGraw-Hill, 1939.

————. *History of Economic Analysis.* Edited by Elizabeth Boody Schumpeter. Fair Lawn, N.J.: Oxford University Press, 1954.

SKIDELSKY, ROBERT. *John Maynard Keynes: Hopes Betrayed, 1883–1920.* New York: The Viking Press/Elisabeth Sifton Books, 1986.

SMITH, ADAM. *The Correspondence of Adam Smith.* Edited by Ernest Campbell Mossner and Ian Simpson Ross. Oxford, Eng.: Oxford University Press, 1977.

————. *An Inquiry into the Nature and Causes of the Wealth of Nations.* New York: E. P. Dutton, 1977.

SOROS, GEORGE. *The Alchemy of Finance: Reading the Mind of the Market.* New York: Simon & Schuster, 1987.

STALIN, JOSEPH. *Leninism.* New York: International Publishers, 1928.

STOWE, HARRIET BEECHER. *Uncle Tom's Cabin.* New York: Dodd, Mead, 1952.

STRAUSS, E. *Sir William Petty: Portrait of a Genius.* London: The Bodley Head, 1954.

TAUSSIG, FRANK W. *Principles of Economics.* New York: The Macmillan Company, 1927.

THUROW, LESTER C. *The Zero-Sum Solution: Building a World-Class American Economy.* New York: Simon & Schuster, 1985.

TURNER, FREDERICK JACKSON. *The Frontier in American History.* New York: Henry Holt, 1920.

U.S. CONGRESS. *Investigation of Concentration of Economic Power.* Hearings before the Temporary National Economic Committee, Congress of the United States, 76th Cong., 1st Sess. Washington, D.C.: U.S. Government Printing Office, 1939.

VEBLEN, THORSTEIN. *The Engineers and the Price System.* New York: B.W. Huebsch, 1921.

————. *Imperial Germany and the Industrial Revolution.* New York: The Macmillan Company, 1915.

————. *An Inquiry into the Nature of Peace and the Terms of Its Perpetuation.* New York: B.W. Huebsch, 1919.

————. *The Theory of the Leisure Class.* New York: B.W. Huebsch, 1919.

WAKEMAN, FREDERIC. *The Hucksters.* New York: Rinehart & Co., 1946.

WALKER, PATRICIA. "Do the Figures Make Cents?" *Platt's Week* (May 2, 1988).

ACKNOWLEDGMENTS

While this book has been more than three years in the writing, its critical diagnosis of America's impending crisis, like the pragmatic cures it recommends, have been drawn from experience, observation, and judgment. Accordingly, the personal and professional debts I have accumulated in preparing it for publication are largely tactical and factual. Revisions too extensive to enumerate have claimed one priority; updated explorations in quantification, another. For the first, my endless thanks go to three members of my staff: Ms. Cynthia Werthamer, a redeemed poet, my patient editorial conscience; Ms. Maureen Armstrong, my shield from and conduit to the round-the-clock pressures of the outside world, the personification of upwardly mobile, self-trained immigration responsible for America's roots and renewals; and Mr. Frank Buono, another self-trained virtuoso whose patience and intelligence in piecing together clean manuscript from my meandering hieroglyphics have filled the gap in the literary side of my life between faith and works.

For continuity of strategic guidance, wisdom, and selectivity, and patience in dialogue, my long-term indebtedness has compounded to my wife and what's more, my friend and mentor, Elizabeth Janeway; and to my two sons, Michael C. Janeway, the executive editor of Houghton Mifflin and Company, and Dr. William H. Janeway, a partner in the investment banking firm of Warburg, Pincus and Company. Mr. David Fromkin, the author of *A Peace to End All Peace,* and our lifelong family friend, Carolyn Kizer, an unredeemed poet, in fact the Pulitzer Prize–winner for poetry in 1985, have been generous enough with their time to read considerable portions of the manuscript in draft. So has the gray eminence of the financial cognoscenti around the world, Mr. Charles Brophy, formerly the financial columnist of the old *New York Herald Tribune,* subsequently the editor of *The Bond Buyer,* for some years the financial press's source of intelligence at Salomon Brothers. Mr. Paul Niculescu, of Salomon Brothers, has been helpful in unraveling the Treasury's jumbled accounts. Former Senate Finance Committee chairman Russell Long was generous with his time in reading an early draft of my recommendation that America embark on a systematic barter of food for oil and with his shrewd and experienced counsel on how to sharpen its thrust.

To my dear friends Nellie and Albert Sindlinger, the redoubtable team of fellow explorers, goes gratitude more endless than I can measure, for their steadfast and cheerful standby beyond the call of duty for functional probes in quantification. Mr. Martin Sorkin, America's Lone Ranger among realistic economic analysts of her troubled yet rich agricultural political economy, has been supportive in checking intricate and elusive market data. Mr. Richard Kaufman, the assistant director of the Joint Economic Committee, and Mr. Bill Buechner, one of its senior economists, have been stalwart and stubborn in their pursuit of obscure fiscal and monetary trails. Mr. Jeff Faux, the resourceful director of the Economic Policy Institute in Washington, D.C., alerted me to the painstaking research of Barry Eichengreen of Harvard and his exposé of the folklore of American tariff taboos. Mr. Greg Pallas, chief legislative assistant for Senator J. James Exon, helped me through the thorny underbrush of Pentagon finance. Ms. Melinda Rothel, for many years executive vice president of the Janeway Publishing and Research Corporation, now an investment broker with the venerable St. Louis firm of Newhard Cook and Company, has dug up statistical lore on market behavior.

Professor James H. Billington, the Librarian of Congress, to whom my Princeton son Bill is indebted for his sense of modern European history, has incurred my indebtedness by retrieving from the archives the invaluable gem of American macroeconomic exploration represented by Dr. Lauchlin Currie's pioneering testimony to the Temporary National Economic Committee in May 1939. Mr. Orlando Potter, of the staff of Chairman Claiborne Pell of the Senate Foreign Relations Committee, verified that Secretary of State Cordell Hull signed twenty-seven reciprocal trade agreements during the first phase of his memorable tour of duty. Ms. Patricia Walker, managing editor of Platt's Week, McGraw-Hill's oil-trade publication, provided invaluable assistance in foraging data from a jumble of oil-world crosscurrents.

This effort at interdisciplinary linkage has depended throughout on the immense resources of the scholarly arsenal around the corner, the New York Society Library, and the indefatigable resourcefulness of its librarian, Mr. Mark Piel. Ms. Ellen Futter, president of Barnard College, Columbia University's bastion of independent women's studies of which Elizabeth Janeway was a trustee for many years, and its helpful library staff, furnished emergency aid in the form of essential bibliographical assistance.

Last but by no means least, my warm appreciation goes to my publisher, Truman M. Talley, for his unfailing support recalling the memorable flair he and his late partner, Victor Weybright, gave to my Economics of Crisis in 1968, a best-seller that still influences the course of events. Mr. Talley's assistant, Mary Wagstaff, helped in many ways in getting this book into print.

INDEX

Academic establishment, 55, 238, 240, 266–67; in Reagan administration, 268–70; and technological revolution, 38, 39

Aerodynamics, 290, 380–81

Agriculture (U.S.), 306–10, 311–12, 322; deficit, 260; Department of, 194, 322, 328; exports, 313–314; in New Deal, 194–95; subsidies to, 260, 306, 308, 309, 313–14; slump in, 68–69; tariff protection for, 328–29; *see also* Farm economy

Agripower (U.S.), 120, 122, 135–39

Aircraft, 367; military, 249, 250, 364–66; vertical takeoff/landing, 367–68

Airports, 366–68, 380

American colonies: paper money issues, 24–30

American Economic Association, 62, 63

American Petroleum Institute (API), 219, 222–23, 225

American Revolution, 13, 20–36, 168

Arms business, 41, 101, 137, 356–57; Russia, 157, 277; *see also* Weapons

Arms race, 3, 353

Assets, distinctive (U.S.), 6, 97–119, 250–51, 272; developed, 120–41; changing liabilities into, 297–99; failure

to use, 373–75; hidden, 142–60; sociological, 109–17; use of, in international relations, 376–77

Atomic bomb, 4, 89, 256, 282–85

Australia, 102, 135, 314

Auto industry, 213–14, 237

Bacon, Francis, 47, 170, 172

Baker, James, 243, 263, 304

Bank credit, 20, 227, 228

Banking crises, xi–xii, 30, 39, 260, 261; federal government and, 33; Federal Reserve Board and, 348

Banking system, 24, 27, 37–38, 240; collapse of, 34–35; effect of food and oil crises on, 307; Eurodollars and, 151; regulation of, 30–31, 35–36, 128, 129, 229; shakiness of, 128–29; states' control over, 33–34

Bankruptcy, 10, 24, 29, 321

Banks, 23, 128, 380; central, 132, 362; multinational, 35, 36, 151; national, 33; need for government involvement, ix, 21, 214; regional, 321; state, 24, 30; statistics regarding, 233; Texas, and oil crisis, 223; and Third World debt, 224

396